STUDIES IN SOVIET SCIENCE

PHYSICAL SCIENCES

1973

Densification of Metal Powders during Sintering
V. A. Ivensen

The Transuranium Elements
V. I. Goldanskii and S. M. Polikanov

Gas-Chromatographic Analysis of Trace Impurities
V. G. Berezkin and V. S. Tatarinskii

A Configurational Model of Matter
G. V. Samsonov, I. F. Pryadko, and L. F. Pryadko

Complex Thermodynamic Systems
V. V. Sychev

Crystallization Processes under Hydrothermal Conditions
A. N. Lobachev

Migration of Macroscopic Inclusions in Solids
Ya. E. Geguzin and M. A. Krivoglaz

1974

Theory of Plasma Instabilities
Volume 1: Instabilities of a Homogeneous Plasma
A. B. Mikhailovskii

Theory of Plasma Instabilities
Volume 2: Instabilities of an Inhomogeneous Plasma
A. B. Mikhailovskii

LIFE SCIENCES

1973

Motile Muscle and Cell Models
N. I. Arronet

Pathological Effects of Radio Waves
M. S. Tolgskaya and Z. V. Gordon

Central Regulation of the Pituitary-Adrenal Complex
E. V. Naumenko

THEORY OF PLASMA INSTABILITIES

Volume 1: Instabilities of a Homogeneous Plasma

STUDIES IN SOVIET SCIENCE

THEORY OF PLASMA INSTABILITIES

Volume 1: Instabilities of a Homogeneous Plasma

A. B. Mikhailovskii

I. V. Kurchatov Institute of Atomic Energy
Moscow, USSR

Translated from Russian by

Julian B. Barbour

CONSULTANTS BUREAU • NEW YORK - LONDON

Library of Congress Cataloging in Publication Data

Mikhaĭlovskiĭ, Anatoliĭ Borisovich.
Theory of plasma instabilities.

Translation of Teoriîa plazmennykh neustoĭchivosteĭ.
Includes bibliographies.
CONTENTS: v. 1. Instabilities of a homogeneous plasma.—v. 2. Instabilities of an inhomogeneous plasma.
1. Plasma instabilities. I. Title. [DNLM: 1. Nuclear physics. QC718 M636t 1974]
QC718.5.S7M5413 530.4'4 73-83899
ISBN 0-306-17181-3 (v. 1)
ISBN 0-306-17182-1 (v. 2)

Dr. Anatolii Borisovich Mikhailovskii is one of the world's leading plasma theoreticians. He works at the renowned I. V. Kurchatov Institute of Atomic Energy in Moscow in the theoretical group led by Academician M. A. Leontovich. He has specialized in the field of plasma instabilities, and many of the more than 70 papers he has published in Soviet and Western journals are already regarded as classics.

The original Russian text, published by Atomizdat in Moscow in 1970, has been corrected by the author for the present edition. The translation is published under an agreement with Mezhdunarodnaya Kniga, the Soviet book export agency.

ТЕОРИЯ ПЛАЗМЕННЫХ НЕУСТОЙЧИВОСТЕЙ
1. Неустойчивости однородной плазмы
А. Б Михайловский

TEORIYA PLAZMENNYKH NEUSTOICHIVOSTEI
1. Neustoichivosti Odnorodnoi Plazmy
A. B. Mikhailovskii

A Division of Plenum Publishing Corporation
227 West 17th Street, New York, N.Y. 10011

United Kingdom edition published by Consultants Bureau, London
A Division of Plenum Publishing Company, Ltd.
Davis House (4th Floor), 8 Scrubs Lane, Harlesden, London, NW10 6SE, England

Printed in the United States of America

Preface

Plasmas are distinguished from other states of matter by the great variety of instabilities to which they are subject. Depending on the circumstances, instabilities may be desirable or undesirable. For example, instabilities make possible the turbulent heating of plasmas but they are also the main obstacle in the way of controlled thermonuclear fusion. However, the instabilities are not of the same nature in the two cases.

The theoretical studies of instabilities are mostly scattered throughout the scientific journals. A number of reviews have also been published on various aspects of the theory of instabilities. The time is now ripe for a systematic exposition of these results from a unified point of view. This is the aim of the two-volume monograph now presented. The first volume is devoted to the instabilities of a homogeneous plasma; the second, to those of an inhomogeneous plasma. The instabilities discussed in the first volume are those that are essentially of interest for the problem of turbulent heating of a plasma while the instabilities considered in the second volume affect the problem of plasma containment.

The theory of plasma stability deals with the spontaneous excitation of collective degrees of freedom of a plasma — plasma oscillations. Such effects are only possible if thermodynamically the plasma is in a nonequilibrium state. Oscillations can be excited either as a result of a nonequilibrium particle distribution velocity or a spatial inhomogeneity of the plasma or a combination of these factors.

The first volume of the monograph is devoted to oscillations that arise because of a nonequilibrium particle velocity distribution; the second, to oscillations due to spatial inhomogeneity.

Volume 1 consists of two parts. In the first it is assumed that a static magnetic field does not affect the plasma oscillations (the approximation of a vanishing magnetic field). Some of the results and most of the physical conclusions obtained with this approximation also remain in force when the magnetic field is not in fact weak. Effects peculiar to a plasma in a magnetic field are discussed in the second part. Each part is subdivided into several chapters.

Chapter 1 treats the simplest example of a plasma that is thermodynamically in a nonequilibrium state – a plasma consisting of two cold beams of charged particles.

The study of kinetic effects is begun in Chapter 2. The point of departure is the solution of the Vlasov equation by the Laplace transform method, and this is followed by a discussion of the collisionless dissipation of plasma oscillations (Landau damping). The general conditions for stability and instability of a plasma are then analyzed. The various instabilities associated with kinetic effects – kinetic instabilities – are discussed in Chapter 3.

In Chapter 4 an investigation is made of the behavior of spatially localized perturbations (wave packets) in an unstable plasma and the amplification of waves is discussed; Chapter 5 is devoted to the influence of plasma inhomogeneity on wave-packet dynamics.

In Chapter 6 the applicability of the approximation of a collisionless plasma in stability problems is analyzed and a study is made of the excitation of oscillations under conditions when collisions are important.

A plasma in a magnetic field – the subject of the second part – is a more complicated entity than a plasma in the absence of a magnetic field. The analysis of the instabilities of such plasmas is therefore preceded by the derivation of expressions for the permittivity (Chapter 7) and a discussion of the types of oscillation in the case of a Maxwellian particle velocity distribution (Chapter 8). A study is then made of plasma instabilities for various cases of thermodynamic nonequilibrium (Chapter 9-17).

Chapters 9-11 are devoted to purely electron instabilities due to the interaction with a plasma of a beam of electrons moving along the magnetic field (Chapter 9), a nonequilibrium velocity distribution of the electron component (Chapter 10), and the presence in the plasma of a group of non-Maxwellian electrons (Chapter 11).

The excitation of electron – ion oscillations when a longitudinal (Chapter 12) or transverse (Chapter 13) current flows in a plasma is discussed in the next two chapters.

In a plasma whose ion component possesses a non-Maxwellian velocity distribution an instability can arise with a growth rate that exceeds the ion cyclotron frequency. This high-frequency instability is discussed in Chapter 14.

The various different forms of ion-cyclotron instabilities are discussed in Chapters 15-17. Such instabilities can arise because of anisotropy of the ion velocities (Chapter 15), a nonequilibrium transverse distribution of the ions (Chapter 16), or the presence in the plasma of ion streams (Chapter 17).

The analysis of Volume 1 shows that beam effects and velocity anisotropy can lead to a large number of different types of instability. However, for any particular plasma the number of instabilities that are actually possible is greatly reduced – there may not even be any.

It should be borne in mind that the class of instabilities due to a non-Maxwellian distribution function is not unique. Instabilities may also arise as a result of inhomogeneity of the plasma. In general their growth rates are small compared with the growth rates of the instabilities of a homogeneous plasma. The instabilities of an inhomogeneous plasma must be taken into account if the approximation of a homogeneous plasma predicts stability.

The book can be read without special knowledge of plasma physics – all that is required is familiarity with the Boltzmann and Maxwell equations. The complete theory is a consequence of these equations. All the derivations are given in their entirely, and the book can therefore be read without reference to any of the numerous papers that have been published on the subject. In many cases the results are given in an approximate form in order to bring out more clearly the physical situation. The same also applies to the figures, few of which pretend to more than qualitative accuracy.

The monograph contains results obtained by a great number of authors. References to the pertinent papers are given at the end of each chapter together with a brief explanation of the bearing that these papers have on the subject under discussion. It is quite possible that papers worthy of mention have escaped my attention. I should like to take this opportunity of apologizing to the authors of these papers for my inadequate knowledge of the literature and also for any possible inaccuracies in citation.

The general plan of the book was discussed with Ya. B. Fainberg. The manuscript of the first volume was read by M. A. Leontovich and B. B. Kadomtsev, who made a number of helpful comments.

During my work on this book I have been greatly assisted by L. V. Mikhailovskaya and A. I. Chinnova, and also members of the Institute of Plasma Physics of the Czechoslovak Academy of Sciences.

I should like to express my gratitude to all these people.

Contents

Basic Equations

The behavior of the particles in a plasma is determined by their distribution function $f_\alpha(\mathbf{r}, \mathbf{v}, t)$, where α is the particle species (electron or ion). This function specifies the number of particles that are situated at time t in an element of unit volume near point **r** and have velocities in a unit interval near **v**. The distribution function satisfies the Boltzmann (kinetic) equation:

$$\frac{\partial f_\alpha}{\partial t} + \mathbf{v}\nabla f_\alpha + \frac{e_\alpha}{m_\alpha}\left(\mathbf{E} + \frac{1}{c}[\mathbf{v}\,\mathbf{B}]\right)\frac{\partial f_\alpha}{\partial \mathbf{v}} = C_\alpha\,. \tag{I}$$

The change in f_α due to collisions is represented by C_α, for which explicit expressions will be given later.

The remaining notation in Eq. (I) is as follows: e_α and m_α are the charge and mass of the particles; c is the velocity of light; and **E** and **B** are the electric and magnetic fields, respectively.

The fields **E** and **B** satisfy Maxwell's equations:

$$\left.\begin{aligned} \operatorname{div}\mathbf{E} &= 4\pi(\rho+\rho_0),\\ \operatorname{rot}\mathbf{B} &= \frac{4\pi}{c}(\mathbf{j}+\mathbf{j}_0) + \frac{1}{c}\cdot\frac{\partial \mathbf{E}}{\partial t},\\ \operatorname{rot}\mathbf{E} &= -\frac{1}{c}\cdot\frac{\partial \mathbf{B}}{\partial t},\quad \operatorname{div}\mathbf{B} = 0. \end{aligned}\right\} \tag{II}$$

Here, ρ and j are the charge and current densities summed over the electrons and ions,

$$\left.\begin{aligned} \rho &= \sum_\alpha e_\alpha \int f_\alpha d\mathbf{v},\\ \mathbf{j} &= \sum_\alpha e_\alpha \int \mathbf{v} f_\alpha d\mathbf{v}. \end{aligned}\right\} \tag{III}$$

The densities ρ_0 and j_0 correspond to additional (external) charges and currents produced in the plasma either artificially or as a result of fluctuations that are not taken into account by the Boltzmann equation.

Part I

THE APPROXIMATION OF A VANISHING MAGNETIC FIELD

Chapter 1

Instabilities of a Cold Homogeneous Plasma

§ 1.1. Hydrodynamic Approximation

This chapter can be regarded as an elementary introduction into the theory of plasma stability. The simplest example of a plasma in a state of thermodynamic nonequilibrium – a plasma consisting of two cold beams – serves to demonstrate the basic types of instability of a homogeneous plasma and to introduce many of the important concepts of stability theory. The mathematical apparatus employed is also very simple. Kinetic effects are ignored and the treatment is based on a hydrodynamic (or fluid) description of the plasma.

Nevertheless, the results of this chapter are of great importance for applications, since the instabilities are manifested most clearly if the beams have a small thermal spread. Their experimental study is therefore facilitated, and they can readily be used to excite and amplify oscillations and for turbulent heating of a plasma.

In this chapter we ignore not only the thermal motion but also plasma inhomogeneity, collisions between particles, and a static magnetic field. In addition, only the simplest kinds of disturbances – plane waves – are considered. The simple picture obtained in this manner is progressively ramified in the subsequent chapters.

The plasma will be assumed to consist of several groups of particles moving with respect to each other. The motion of the particles within each group will be neglected. This means that the electron (or ion) distribution function has the form

$$f=\sum_{\alpha} n^{(\alpha)}\,\delta(\mathbf{v}-\mathbf{V}^{(\alpha)}), \tag{1.1}$$

where α designates the group and $n^{(\alpha)}$ and $\mathbf{V}^{(\alpha)}$ are, respectively, the density and velocity of the particles of group α. A plasma with a δ-function distribution of type (1.1) is said to be "cold," in contrast to a "hot" plasma, in which the particle velocities have a thermal spread.

In the case of a cold plasma, the Boltzmann equation (I) reduces to a system of hydrodynamic equations for $n^{(\alpha)}$ and $\mathbf{V}^{(\alpha)}$. To see this, we substitute (1.1) into (I), neglect collisions, and integrate over the velocity with weight 1 and $\mathbf{v}$. We obtain

$$\frac{\partial n^{(\alpha)}}{\partial t}+\operatorname{div}\,(n^{(\alpha)}\,\mathbf{V}^{(\alpha)})=0, \tag{1.2}$$

$$\frac{\partial \mathbf{V}^{(\alpha)}}{\partial t}+(\mathbf{V}^{(\alpha)}\,\nabla)\,\mathbf{V}^{(\alpha)}=\frac{e_\alpha}{m_\alpha}\left(\mathbf{E}+\frac{1}{c}\,[\mathbf{V}^{(\alpha)},\mathbf{B}]\right). \tag{1.3}$$

These equations relate the density and velocity of the particles of group α to the electric and magnetic fields. In their turn, **E** and **B** are, in accordance with Eqs. (II), (III), and (1.1), functions of the density and velocity of the charges:

$$\operatorname{div}\mathbf{E}=4\pi\sum_{\alpha} e_\alpha n^{(\alpha)}, \tag{1.4}$$

$$\operatorname{rot}\mathbf{B}=\frac{1}{c}\cdot\frac{\partial \mathbf{E}}{\partial t}+\frac{4\pi}{c}\sum_{\alpha} e_\alpha n^{(\alpha)}\,\mathbf{V}^{(\alpha)}. \tag{1.5}$$

The summation over α in these equations is over the electrons and ions. In conjunction with the last two equations of the system (II), Eqs. (1.2)-(1.5) form the closed system of equations of a cold plasma.

§ 1.2. Characteristic Plasma Oscillations

Suppose that up to the time $t = 0$ the plasma remains in a stationary state, i.e., for $t < 0$

$$\left.\begin{aligned} n=n_0,\ \ \mathbf{V}=\mathbf{V}_0,\\ \mathbf{B}=\mathbf{B}_0,\ \ \mathbf{E}=\mathbf{E}_0 \end{aligned}\right\} \tag{1.6}$$

the quantities with subscript zero being independent of time. (The superscript α of n and **V** has been omitted for simplicity.)

At t= 0 the plasma is perturbed in some manner, so that for t > 0

$$\left.\begin{array}{l} n = n_0 + n', \ \mathbf{V} = \mathbf{V}_0 + \mathbf{V}', \\ \mathbf{B} = \mathbf{B}_0 + \mathbf{B}', \ \mathbf{E} = \mathbf{E}_0 + \mathbf{E}'. \end{array}\right\} \tag{1.7}$$

The primed quantities characterize the deviations, or perturbations, of the density, velocity, and fields from the corresponding equilibrium values.

We are interested in the time dependence of the perturbations, which we shall assume are small. Neglecting products of small quantities, we then obtain two subsystems of equations – one for the stationary and another for the perturbed quantities – from the system of equations (II), (1.2)-(1.5). We represent the time dependence of the linearized perturbations in the form

$$X'(t, \mathbf{r}) = X'(\mathbf{r})\, e^{-i\omega t}. \tag{1.8}$$

This greatly simplifies the investigation, since the operators ∂/∂t are transformed into numbers:

$$\frac{\partial}{\partial t} \to -i\omega. \tag{1.9}$$

This simplified approach to the problem of oscillations of a cold homogeneous plasma can be justified by the more rigorous Laplace method (see the problem in § 1.3). The linearized equations now become

$$\left.\begin{array}{c} -i\omega n' + \operatorname{div}(n'\mathbf{V}_0 + n_0\mathbf{V}') = 0, \\ -i\omega\mathbf{V}' + (\mathbf{V}_0\nabla)\mathbf{V}' + (\mathbf{V}'\nabla)\mathbf{V}_0 = \\ = \frac{e}{m}\left(\mathbf{E}' + \frac{1}{c}[\mathbf{V}'\mathbf{B}_0] + \frac{1}{c}[\mathbf{V}_0\mathbf{B}']\right), \\ \operatorname{rot}\mathbf{B}' = \frac{4\pi}{c}\Sigma e(n'\mathbf{V}_0 + n_0\mathbf{V}') - \frac{i\omega}{c}\mathbf{E}', \\ \operatorname{rot}\mathbf{E}' = \frac{i\omega}{c}\mathbf{B}', \\ \operatorname{div}\mathbf{E}' = 4\pi\Sigma e n', \\ \operatorname{div}\mathbf{B}' = 0. \end{array}\right\} \tag{1.10}$$

Note that not all the equations of this system are independent — the last two are a consequence of the other equations. The solution of the system of equations (1.10) with given boundary conditions yields the spectrum of frequencies ω for which these equations have nontrivial solutions. These frequencies are the (characteristic) plasma oscillation frequencies.

In this chapter we shall consider oscillations of a plasma in the absence of a static magnetic field, $\mathbf{B}_0 = 0$. In this approximation, the term $[\mathbf{V}', \mathbf{B}_0]$ vanishes in the second equation of the system (1.10). This will be understood in all that follows.

§ 1.3. Oscillations of a Homogeneous Plasma at Rest

Suppose a plasma is spatially homogeneous and the stationary velocities of all its particles vanish:

$$\nabla n_0 = 0, \tag{1.11}$$

$$\mathbf{V}_0 = 0. \tag{1.12}$$

In the case of a spatially homogeneous plasma, the coordinate dependence of the perturbations can, as follows from (1.10), be chosen in the form of a plane wave,

$$X'(r) \propto e^{i\mathbf{kr}}. \tag{1.13}$$

This transforms the system (1.10) into a set of algebraic equations. The relationship between the frequency of the plasma oscillations and the wave vector is found by equating the determinant of this system to zero. We shall call the equation that is obtained the dispersion equation.

Substituting (1.11) and (1.12) into (1.10) and forming the determinant, we find that the frequency of the plasma oscillations must satisfy one of the two equations

$$\omega^2 = \omega_{p_e}^2, \tag{1.14}$$

$$\omega^2 = \omega_{p_e}^2 + c^2k^2, \tag{1.15}$$

where $\omega_{p_e}^2 = 4\pi e^2 n_0/m_e$. In these equations, the small terms of order m_e/m_i due to the perturbed motion of the ions have been neglected. In this approximation the ions serve only to compensate the stationary charge of the electron component of the plasma, $n_0^i = n_0^e \equiv n_0$; the ions are assumed to be singly charged, $e_i = -e_e$. The frequency ω_{p_e} is known as the electron plasma frequency or electron Langmuir frequency.

Oscillations of the type (1.14) are known as electron plasma oscillations (in Russian literature also called electron Langmuir oscillations). These are purely electrostatic oscillations, so that

$$\mathbf{B}' = 0. \tag{1.16}$$

The electric field of the plasma oscillations can be represented as the gradient of the scalar potential:

$$E' = -\nabla \psi. \tag{1.17}$$

Equation (1.14) can be obtained without the complete system of equations (1.10). It is sufficient to assume that Eq. (1.17) holds, and the system (1.10) then reduces to

$$\left.\begin{aligned} -\Delta\psi &= 4\pi e_e n'_e, \\ -i\omega n'_e + n_0 \operatorname{div} \mathbf{V}'_e &= 0, \\ -i\omega \mathbf{V}'_e &= -\frac{e_e}{m_e}\nabla\psi. \end{aligned}\right\} \tag{1.18}$$

It follows from (1.18) that (1.14) is valid not only for plane waves but for an arbitrary coordinate dependence of the potential of the perturbation. In this sense, Eq. (1.14) for the electron plasma oscillations is a degenerate dispersion relation.

If $c^2k^2 \gg \omega_{p_e}^2$, perturbations of the type (1.15) are electromagnetic waves that are independent of the motion of the charges. In this limiting case, Eq. (1.15) yields the well-known law of propagation of light in vacuum:

$$\omega^2 = c^2k^2. \tag{1.19}$$

This simple dispersion relation is radically modified in the case of long-wavelength oscillations,

$$k^2 < \frac{\omega_{p_e}^2}{c^2}. \tag{1.20}$$

In particular, the phase velocity ω/k of the waves appreciably exceeds the velocity of light,

$$\frac{\omega}{k}=c\left(1+\frac{\omega_{p_e}^2}{c^2k^2}\right)^{1/2}. \tag{1.21}$$

It follows readily from the system (1.10) that, as in the absence of a plasma, the electric field of waves of the type (1.15) is solenoidal,

$$\operatorname{div}\mathbf{E}'=0. \tag{1.22}$$

The subsystem of the system (1.10) that corresponds to electromagnetic waves can be represented in the form

$$\left.\begin{aligned}
&\Delta\mathbf{B}'+\frac{\omega^2}{c^2}\mathbf{B}'+\frac{4\pi e^2n_0}{c}\operatorname{rot}\mathbf{V}'_e=0,\\
&-i\omega\mathbf{V}'_e=\frac{e_e}{m_e}\mathbf{E}',\\
&\operatorname{div}\mathbf{V}'_e=0,\\
&\operatorname{rot}\mathbf{E}'=\frac{i\omega}{c}\mathbf{B}',\\
&\operatorname{div}\mathbf{E}'=0,\\
&\operatorname{div}\mathbf{B}'=0.
\end{aligned}\right\} \tag{1.23}$$

These equations reduce to a single vector equation for $\mathbf{B}'$:

$$\Delta\mathbf{B}'+\frac{1}{c^2}(\omega^2-\omega_{p_e}^2)\mathbf{B}'=0. \tag{1.24}$$

If (1.13) is substituted into this equation, Eq. (1.15) follows as an obvious consequence.

Problem. Obtain Eq. (1.14) by the Laplace method.

Solution. In the linearized equations (1.2)-(1.4) set $\mathbf{V}_0=0$, $\mathbf{B}=0$, $\mathbf{E}'=-\nabla\psi$. Take the spatial dependence of the perturbations in the form exp(ikx). Then

$$\left.\begin{aligned}
&\frac{\partial n'}{\partial t}+ikV'n_0=0,\\
&\frac{\partial V'}{\partial t}=-\frac{iek}{m}\psi,\\
&k^2\psi=4\pi en'.
\end{aligned}\right\} \tag{1.25}$$

Replace n'(t), V'(t), and ψ(t) by their Laplace transforms n'_p, V'_p, and ψ_p, defined as follows:

$$X'_p = \int_0^\infty X'(t)\, e^{-pt} dt. \tag{1.26}$$

Integrate both sides of Eqs. (1.25) over t from 0 to ∞ with weight exp(−pt) to obtain the equations for the Laplace transforms:

$$\left.\begin{aligned} pn'_p + i\,kn_0V'_p &= n'(0) \\ pV'_p &= -\frac{i\,ek}{m}\psi_p + V'(0), \\ k^2\psi_p &= 4\pi e n'_p, \end{aligned}\right\} \tag{1.27}$$

where n'(0) and V'(0) are the perturbations of the density and velocity at t = 0. This system of algebraic equations yields, in particular,

$$\psi_p = -\frac{4\pi e}{k^2}\cdot\frac{i\,kn_0V'(0) - pn'(0)}{p^2 + \omega_{p_e}^2}, \tag{1.28}$$

where ω_{p_e} is the electron plasma frequency. Application of the formula for the inverse Laplace transformation,

$$X'(t) = \frac{1}{2\pi i}\int_{-i\infty+\sigma}^{+i\infty+\sigma} e^{pt} X'_p dp, \tag{1.29}$$

where σ is a positive number lying to the right of all the poles of X'_p, to Eq. (1.28) yields

$$\psi(t) = -\frac{4\pi e}{k^2}\cdot\frac{1}{2\pi i}\int_{-i\infty+\sigma}^{+i\infty+\sigma} [i\,kn_0V'(0) - pn'(0)]\frac{e^{pt}dp}{p^2 + \omega_{p_e}^2}. \tag{1.30}$$

For t < 0 this integral can be calculated by displacing the contour of integration into the region Re p → +∞. The integrand then vanishes, so that

$$\psi(t) = 0, \text{ for } t < 0. \tag{1.31}$$

For t > 0, displace the contour of integration into the region Re p → −∞. In this case the integral does not vanish because of the contribution of the poles of the integrand. These poles correspond to the zeros of the denominator of (1.28) and are situated at the points p = p_0, where

$$p_0 = \pm i\,\omega_{p_e}. \tag{1.32}$$

It then follows from (1.30) that

$$\underset{t>0}{\psi(t)} = \frac{2\pi e n_0}{k^2}\left\{\left[\frac{n'(0)}{n_0}+\frac{kV'(0)}{\omega_{p_e}}\right]e^{-i\omega_{p_e}t}+\left[\frac{n'(0)}{n_0}-\frac{kV'(0)}{\omega_{p_e}}\right]e^{i\omega_{p_e}t}\right\}. \quad (1.33)$$

Thus, the potential oscillates with the time with the frequencies ω_{p_e} and $-\omega_{p_e}$, in agreement with (1.14).

§ 1.4. Dispersion Equation for a Plasma Consisting of Several Directed Beams

We now proceed to investigate the oscillations of a plasma that consists of several directed beams moving with respect to each other. The relative motion has a particularly strong influence on the branch of electron plasma oscillations, i.e., the electrostatic oscillations. In what follows we shall therefore restrict ourselves to electrostatic perturbations.

We shall obtain the dispersion equation for the electrostatic perturbations ($\mathbf{E} = -i\mathbf{k}\psi$) as follows. From the continuity equation and the equations of motion (1.10), we find the perturbation of the charge density $\rho^{(\alpha)} \equiv e_\alpha n^{(\alpha)}$ of the particles of the species α as a function of the potential ψ, $\rho^{(\alpha)} = \chi^{(\alpha)}\psi$:

$$\chi^{(\alpha)} = \frac{e_\alpha^2 n_0^{(\alpha)} k^2}{m_\alpha(\omega - \mathbf{kV}^{(\alpha)})^2}. \quad (1.34)$$

We then substitute $\rho^{(\alpha)}$ into Poisson's equation, $k^2\psi = 4\pi\sum_\alpha \rho^{(\alpha)}$. Canceling $k^2\psi$, we obtain the desired dispersion equation

$$\varepsilon_0 \equiv 1 + \sum_\alpha \varepsilon_0^{(\alpha)} = 0, \quad (1.35)$$

where

$$\varepsilon_0^{(\alpha)} \equiv -\frac{4\pi}{k^2}\chi^{(\alpha)}. \quad (1.36)$$

In the given case

$$\varepsilon_0^{(\alpha)} = -\frac{(\omega_p^{(\alpha)})^2}{(\omega - \mathbf{kV}^{(\alpha)})^2}. \quad (1.36')$$

It can be shown (see Appendix to Chapter 1) that ε_0 is related to the components of the permittivity tensor ε_{ij} by the equation

$$\varepsilon_0 = \frac{1}{k^2} k_i \varepsilon_{ij} k_j. \tag{1.37}$$

It can therefore be called the scalar permittivity or simply permittivity.

Equation (1.35) shows that the dispersion equation for the electrostatic oscillations is obtained by equating the scalar permittivity to zero. The dispersion equation for the general case is given in the Appendix to Chapter 1.

§ 1.5. The Two-Stream Instability

Suppose that the electron component of the plasma consists of two beams moving relative to each other. The electron oscillations of such a plasma are described by Eqs. (1.35) and (1.36') with two values of the superscript α. Some of the solutions of the dispersion equation (1.35) have a positive imaginary part, $\operatorname{Im}\omega > 0$. This means that, in contrast to the strictly periodic oscillations of a plasma at rest, perturbations in a system consisting of two beams can grow spontaneously in time. A plasma in which the spontaneous growth of oscillations is possible is said to be unstable.

1. Instability of Two Beams of Equal Density. The existence of solutions of Eq. (1.35) with $\operatorname{Im}\omega > 0$ can be demonstrated most readily for two electron beams of equal density moving in opposite directions with the same speed:

$$n_{01} = n_{02} \equiv n_0, \quad V_{10} = -V_{20} \equiv V. \tag{1.38}$$

In this case Eq. (1.35) becomes

$$1 - \frac{\omega_p^2}{(\omega - k_\parallel V)^2} - \frac{\omega_p^2}{(\omega + k_\parallel V)^2} = 0, \quad k_\parallel \equiv \mathbf{kV}/|\mathbf{V}|. \tag{1.39}$$

This equation reduces to a biquadratic equation in ω. Its roots, $\omega = \omega(\mathbf{k})$, are

$$\omega = \pm\sqrt{(k_\parallel V)^2 + \omega_p^2 \pm \omega_p (\omega_p^2 + 4k_\parallel^2 V^2)^{1/2}}. \tag{1.40}$$

This shows that if the wavelength is not too short,

$$k_{\parallel} < \sqrt{2}\omega_p/V, \tag{1.41}$$

one of the four roots is purely imaginary with $\operatorname{Im}\omega > 0$. In particular, if $k_{\parallel} \ll \omega_p/V$, then

$$\operatorname{Im}\omega = |k_{\parallel} V| . \tag{1.42}$$

The imaginary part of the oscillation frequency is, as a rule, called the growth rate (sometimes decay rate, if negative) and denoted by the letter γ:

$$\operatorname{Im}\omega \equiv \gamma. \tag{1.43}$$

The maximal growth rate of oscillations of the type (1.40) is attained when

$$k_{\parallel} = k_{\text{opt}} \equiv \frac{\sqrt{3}}{2} \cdot \frac{\omega_p}{V} \tag{1.44}$$

and is

$$\gamma_{\max} = \omega_p/2. \tag{1.45}$$

The real part of the expression for the frequency of growing waves in the example of two beams that satisfy the conditions (1.38) vanishes. In this sense, the process corresponding to the perturbed motion of the plasma is purely aperiodic.

Bohm and Gross explain the mechanism of the two-stream instability as follows: "... a very small perturbation away from zero field at a given point causes a velocity modulation of each beam. In time, this produces a bunching of space charge in the direction of motion of each beam, which creates a much larger potential than that due to the original perturbation. The fields due to any one beam modulate the other beam, which then feeds the disturbance back to the source in a highly amplified form. Thus, the perturbation builds up accumulatively, and instability results."

2. Instability of a Low-Density Beam Passing Through a Plasma. Let us now assume that the density of one of the beams is low compared with the density of the

other, $n_1 \ll n_0$. We shall show that a system of this kind is also unstable.

We take a frame of reference at rest in the more dense component of the plasma. The dispersion equation (1.35) in this case is

$$1 - \frac{\omega_{p_e}^2}{\omega^2} - \frac{\alpha\omega_{p_e}^2}{(\omega - k_{\|} V)^2} = 0. \tag{1.46}$$

Here, $\omega_{p_e}^2 = 4\pi e^2 n_0/m_e$, $\alpha = n_1/n_0 \ll 1$; and V is the velocity of the less dense component of the plasma.

Assuming that $k_{\|} V$ in (1.46) is not too near ω_{p_e}, and remembering that α is small, we find that two of the four roots correspond to the plasma oscillations (1.14) and that the other two are equal to

$$\omega = k_{\|} V \pm \sqrt{\alpha} \frac{\omega_{p_e}}{\sqrt{1 - \frac{\omega_{p_e}^2}{(k_{\|} V)^2}}}. \tag{1.47}$$

If $k_{\|} V < \omega_{p_e}$, the roots (1.47) are complex [cf. (1.41)], one of them corresponding to an instability with growth rate

$$\gamma = \sqrt{\alpha} \frac{\omega_{p_e}}{\sqrt{\left(\frac{\omega_{p_e}}{k_{\|} V}\right)^2 - 1}}. \tag{1.48}$$

Oscillations of the type (1.47) have a group velocity $V_{gr} = \partial\omega/\partial k$ that is near the beam velocity, $V_{gr} \approx V$. We shall refer to these as beam oscillations.

If $k_{\|} \approx \omega_{p_e}/V$, the frequency of the plasma oscillations is approximately equal to the frequency of the beam oscillations and Eqs. (1.47) and (1.48) cease to be valid. The relations that characterize the growth of oscillations for $\gamma > |k_{\|} V - \omega_{p_e}|$, can be obtained as follows. We shall assume that in the zeroth approximation in α the frequency of the oscillations is

$$\omega^{(0)} \approx \omega_{p_e} \tag{1.49}$$

and $|\omega^{(1)}| \gg |\omega_{p_e} - k_{\|} V|$, where $\omega^{(1)}$ is the correction to the frequency. Equation (1.46) then yields

$$2\frac{\omega^{(1)}}{\omega_{p_e}} - \frac{\alpha\omega^2_{p_e}}{(\omega^{(1)})^2} = 0. \tag{1.50}$$

For the corrections to the frequency of the growing perturbations we then obtain

$$\operatorname{Re}\omega^{(1)} = -\omega_{p_e}\frac{\alpha^{1/3}}{2^{4/3}}, \tag{1.51}$$

$$\operatorname{Im}\omega^{(1)} \equiv \gamma = \omega_{p_e}\frac{\sqrt{3}}{2^{4/3}}\alpha^{1/3}. \tag{1.52}$$

At the limit of applicability of these equations, $(\omega^2_{p_e}/k^2_{\parallel}V^2) - 1 \simeq \alpha^{1/3}$, the expressions (1.48) and (1.52) for γ are of the same order. The expression (1.52) determines the maximal growth rate of oscillations excited in a plasma by a low-density beam. It can be seen that even for a low ratio of the density of the beam to that of the plasma the growth of oscillations may be appreciable, since $\gamma \sim \alpha^{1/3}$. This enhancement of the growth rate is due to a characteristic resonance between the oscillations of the dense plasma at rest and the beam oscillations. In this sense, the instability with a growth rate of the type (1.52) is a "resonant" instability, in contrast to the "nonresonant" instability of the type (1.48).

The general form of the oscillation branches $\operatorname{Re}\omega = \operatorname{Re}\omega(k)$ for $\alpha \ll 1$ is shown in Fig. 1.1. The growth rates are also plotted in this figure.

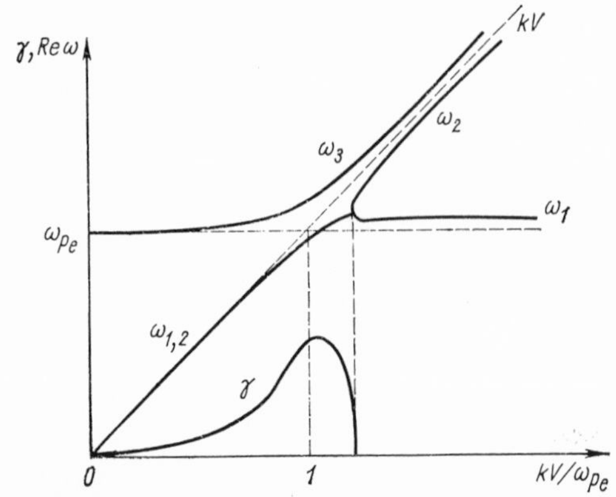

Fig. 1.1. Plasma–beam oscillation branches.

3. Instability of an Electron Beam of Arbitrary Density. Let us now consider perturbations in a plasma–beam system without assuming that the ratio $\alpha \equiv n_1/n_0$ is small. We shall assume that the component with subscript 0 is at rest ($V_0 = 0$) and that the component with subscript 1 moves with velocity V. The dispersion equation then has the previous form (1.46). Without loss of generality, the parameter α in this equation can be assumed to be not greater than unity,

$$0<\alpha\leqslant 1, \tag{1.53}$$

since the case $\alpha > 1$ is merely obtained by reversing the role of the components. We shall denote the plasma frequency of the more dense component by ω_p, omitting the indices 0 and e.

Note that for $k_{\parallel} \gg \omega_p/V$ all the solutions of (1.46) are real. Instability is possible only if $k_{\parallel}$ is less than some critical value:

$$k_{\parallel} < k_{cr}. \tag{1.54}$$

If $k_{\parallel}$ is slightly less than k_{cr}, two of the four roots of Eq. (1.46) are complex conjugates with $\gamma \neq 0$. For such $k_{\parallel}$, the left-hand side of Eq. (1.46), which we shall denote by ε_0, is

$$\varepsilon_0 = \frac{[(\omega - \mathrm{Re}\,\omega_{1,2}(k))^2 + \gamma^2](\omega - \omega_3(k))(\omega - \omega_4(k))}{\omega^2(\omega - k_{\parallel}V)^2}. \tag{1.55}$$

It follows from (1.55) that the stability boundary $[\gamma(k_{cr}) = 0]$ corresponds to a double real root of Eq. (1.46). This means that on the stability boundary not only Eq. (1.46),

$$\varepsilon_0(\omega_{cr}, k_{cr}) = 0, \tag{1.56}$$

but also the equation

$$\left(\frac{\partial \varepsilon_0}{\partial \omega}\right)_{\omega=\omega_{cr},\, k_{\parallel}=k_{cr}} = 0 \tag{1.57}$$

must be satisfied. Replacing ε_0 by the expression (1.46), we find

$$k_{cr} = \frac{\omega_p}{V}(1+\alpha^{1/3})^{3/2}, \tag{1.58}$$

$$\omega_{cr} = \omega_p(1+\alpha^{1/3})^{1/2}. \tag{1.59}$$

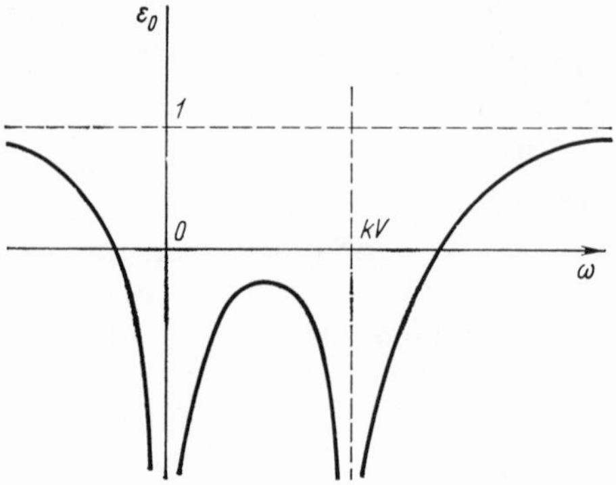

Fig. 1.2. Permittivity of the plasma–beam system for real ω.

The meaning of the result (1.57) is made clear by Fig. 1.2. The real roots of the dispersion equation correspond to the points at which the curve $\varepsilon_0(\omega)$ intersects the abscissa. If the wave number $k_{\parallel}$ is such that ε_0 vanishes only twice – as in Fig. 1.2 – then there are only two real roots. But Eq. (1.46) has four roots in all, and two of them must therefore be complex. The imaginary parts of the complex roots vanish at values of k for which the parabola depicted in the figure osculates the abscissa. This is precisely the condition (1.57).

§ 1.6. Instability Due to a Relative Motion of Electrons and Ions

Suppose that the electron component of the plasma moves relative to the ion component with speed V. In accordance with (1.35) and (1.36'), the plasma oscillations are then described by the dispersion equation

$$1-\frac{\omega_{p_e}^2}{(\omega-k_{\parallel}V)^2}-\frac{\omega_{p_i}^2}{\omega^2}=0. \tag{1.60}$$

Since $\omega_{p_i}^2/\omega_{p_e}^2=m_e/m_i\ll 1$, this equation becomes identical with the dispersion equation (1.46) for a plasma and a low-density beam if we write

$$\alpha=\frac{m_e}{m_i}. \tag{1.61}$$

In this sense, the "effective density" of the ions is $(m_e/m_i)n_0$.

The relations that characterize the electron–ion instability can be obtained by using the results of § 1.5.2. To this end we must make the substitution (1.61) in the latter and also the substitutions

$$\omega \to \omega - k_{\parallel}V; \quad V \to -V. \tag{1.62}$$

This is because the "more dense" component is the one that is moving in the given case. Making the substitutions, we find that the frequency of the beam oscillations of the type (1.47) is

$$\omega = \pm\, \omega_{p_i}\left[1 - \left(\frac{\omega_{p_e}}{k_{\parallel}V}\right)^2\right]^{-1/2}. \tag{1.63}$$

For $|k_{\parallel} V| < \omega_{p_e}$, this frequency is purely imaginary, having the value

$$\gamma = \omega_{p_i}\left[\left(\frac{\omega_{p_e}}{k_{\parallel}V}\right)^2 - 1\right]^{-1/2}. \tag{1.64}$$

The condition for resonance between the ion beam oscillations ($\omega \approx 0$) and the electron plasma oscillations ($\omega = k_{\parallel}V + \omega_{p_e}$), is [see (1.49) and (1.62)]

$$k_{\parallel}V = -\,\omega_{p_e}. \tag{1.65}$$

In the zeroth approximation in α, the frequency of the oscillations vanishes and we therefore obtain the following expression for the complete real part of the frequency of the resonant oscillations [cf. (1.51)]:

$$\mathrm{Re}\,\omega = \mathrm{Re}\,\omega^{(1)} = -\frac{1}{2^{4/3}}\,\omega_{p_e}\left(\frac{m_e}{m_i}\right)^{1/3}. \tag{1.66}$$

The growth rate of these oscillations is

$$\gamma = \omega_{p_e}\frac{\sqrt{3}}{2^{4/3}}\left(\frac{m_e}{m_i}\right)^{1/3}. \tag{1.67}$$

This instability is frequently called a Buneman type instability.

§ 1.7. Instability of a Stream with an Inhomogeneous Velocity Profile

Allowance for spatial inhomogeneity of a plasma leads to the appearance of new types of instability. This is illustrated below by the example of the instability of an electron stream with an inhomogeneous velocity profile. Such a stream is stable if the inhomogeneity is neglected.

Suppose that a stream moves along the z axis with a velocity $\mathbf{V}_0 = (0, 0, V_0)$ that depends on the transverse coordinate x:

$$\frac{\partial V_0}{\partial x} \neq 0. \tag{1.68}$$

In accordance with (1.10) the density and velocity perturbations of such a stream are governed by the equations

$$\left.\begin{aligned} -i\omega' n' + i k_z V'_z n_0 + i k_y V'_y n_0 + \frac{\partial}{\partial x}(n_0 V'_x) = 0, \\ -i\omega' V'_x = -\frac{e}{m}\cdot\frac{\partial\psi}{\partial x}, \\ -i\omega' V'_y = -i\frac{e}{m} k_y \psi, \\ -i\omega' V'_z + V'_x \frac{\partial V_0}{\partial x} = -i\frac{e}{m} k_z \psi. \end{aligned}\right\} \tag{1.69}$$

Here, $\omega' = \omega - k_z V_0$, and the perturbations are assumed to be electrostatic [the condition (1.17)]. It is possible that $\partial n_0/\partial x \neq 0$, and for this reason the equilibrium density has not been taken in front of the differentiation sign in the first of these equations. Using these equations to find the density perturbation and substituting it into Poisson's equation, we obtain

$$\frac{\partial}{\partial x}\left(\varepsilon_0 \frac{\partial\psi}{\partial x}\right) - (k_z^2 + k_y^2)\,\varepsilon_0\psi = 0, \tag{1.70}$$

where

$$\varepsilon_0 = 1 - \frac{\omega_p^2}{(\omega - k_z V_0)^2}. \tag{1.71}$$

Let us use Eq. (1.70) to study the stability of a stream whose velocity profile has the form

$$V_0(x) = \begin{cases} V_1, & x < 0, \\ V_2, & x > 0. \end{cases} \tag{1.72}$$

We shall assume that the stream density depends on x in the same manner:

$$n_0(x) = \begin{cases} n_1, & x < 0, \\ n_2, & x > 0. \end{cases} \tag{1.73}$$

The relations (1.72) and (1.73) mean that the velocity and density change abruptly over a layer of thickness a that is small compared with the transverse dimension of the stream.

We shall assume that far from the transition layer the potential of the perturbation decreases exponentially:

$$\psi(x) = \psi_0 e^{-\varkappa |x|}, \quad \varkappa > 0 \tag{1.74}$$

and that ψ is continuous on $x = 0$. This presupposes that the transition layer is thin compared with the characteristic length of decrease of the potential,

$$\varkappa a \ll 1. \tag{1.75}$$

Substituting (1.74) into (1.70), we find

$$\varkappa = \sqrt{k_y^2 + k_z^2}. \tag{1.76}$$

Recalling that ψ is continuous within the transition layer, we integrate both sides of Eq. (1.70) over the thickness of the layer. Neglecting terms of order $\varkappa a$, we obtain

$$\left(\varepsilon_0 \frac{\partial \psi}{\partial x}\right)\Big|_{-\delta}^{\delta} = 0, \tag{1.77}$$

where $a \ll \delta \ll 1/\varkappa$. Substituting (1.74) into (1.77), we obtain the dispersion equation

$$2 - \frac{\omega_{p1}^2}{(\omega - k_z V_1)^2} - \frac{\omega_{p2}^2}{(\omega - k_z V_2)^2} = 0, \tag{1.78}$$

where

$$\omega_{p1,2}^2 = 4\pi e^2 n_{1,2}/m_e.$$

Except for the notation, this equation is identical with the dispersion equation for the two-stream instability [cf. (1.39)]. However, the physical situation is different – Eq. (1.39) describes perturbations in two interstreaming beams, whereas Eq. (1.78) corresponds to two spatially separated beams.

Equation (1.78) yields an instability condition that is analogous to (1.58):

$$k_z V < \left[\left(\frac{\omega_{p1}^2}{2}\right)^{1/3} + \left(\frac{\omega_{p2}^2}{2}\right)^{1/3}\right]^{3/2}, \quad V = V_2 - V_1. \tag{1.79}$$

In particular, if the density is the same on both sides of the boundary ($n_1 = n_2 \equiv n_0$), an instability arises if [cf. (1.41)]

$$k_z < 2\frac{\omega_{po}}{V}. \tag{1.80}$$

In this case, the maximal growth rate corresponds to perturbations with

$$k_z = \sqrt{\frac{3}{2}} \cdot \frac{\omega_{po}}{V}. \tag{1.81}$$

It is

$$\gamma_{\max} = \frac{\omega_p}{2\sqrt{2}}. \tag{1.82}$$

This instability is sometimes called the slipping-stream instability.

§ 1.8. Instability of an Electron Stream between Cathode and Anode

Suppose that an electron stream with given density n_0 and velocity V_0 is formed at a certain plane that is perpendicular to its direction of motion (cathode) and is absorbed at another plane (anode) that is separated from the first by a distance L. The potentials of the cathode and anode are fixed. We shall assume that the space between the cathode and the anode contains only the particles of the stream and infinitely heavy ions that compensate the equilibrium charge of the electrons. In the limit $L \to \infty$ the stream is stable. We shall now show that if L is finite and the density of the stream is not too low its equilibrium state is unstable.

Let us consider perturbations with $k_\perp = 0$. The system (1.10) yields the following equations for such perturbations:

$$\left.\begin{aligned} -i\omega n' + V_0 \frac{\partial n'}{\partial z} + n_0 \frac{\partial V'_z}{\partial z} &= 0, \\ -i\omega V'_z + V_0 \frac{\partial V'_z}{\partial z} &= -\frac{e}{m} \cdot \frac{\partial \psi}{\partial z}, \\ -\frac{\partial^2 \psi}{\partial z^2} &= 4\pi e n'. \end{aligned}\right\} \quad (1.83)$$

If the cathode and anode potentials are fixed and the velocity and density of the stream at the cathode (at the point of injection) are specified, the boundary conditions are

$$\psi(0) = \psi(L) = V'_z(0) = n'(0) = 0. \quad (1.84)$$

The general solution of the system (1.83) is

$$\psi = Ae^{iz\frac{\omega + \omega_p}{V}} + Be^{iz\frac{\omega - \omega_p}{V}} + Cz + D,$$

$$V'_z = -\frac{e}{m\omega_p V}\left[A(\omega + \omega_p)\, e^{iz\frac{\omega + \omega_p}{V}} - B(\omega - \omega_p)\, e^{iz\frac{\omega - \omega_p}{V}}\right] - \frac{ie}{m\omega} C,$$

$$n' = -\frac{1}{4\pi e^2 V^2}\left[A(\omega + \omega_p)^2 e^{iz\frac{\omega + \omega_p}{V}} + B(\omega - \omega_p)^2\, e^{iz\frac{\omega - \omega_p}{V}}\right]. \quad (1.85)$$

Using (1.84) and (1.85), we obtain a system of four algebraic equations for A, B, C, and D. The dispersion equation obtained by equating the determinant of this system to zero is

$$2\xi\alpha\,(1 - e^{i\xi}\cos\alpha) + i\,(\xi^2 + \alpha^2)\sin\alpha\, e^{i\xi} + i\,\frac{\xi^2}{\alpha}\,(\xi^2 - \alpha^2) = 0, \quad (1.86)$$

where

$$\xi = L\omega/V, \qquad \alpha = L\omega_p/V.$$

It should be borne in mind that these oscillations are not purely electrostatic. The condition of approximate electrostatic behavior reduces to the requirement that the perturbed density does not vanish everywhere identically. It is a consequence of this requirement and the boundary condition $n'(0) = 0$ that only roots $\xi^2 \neq \alpha^2$, i.e., $\omega^2 \neq \omega_p^2$, satisfy the original assumptions.

We shall now show that an oscillation frequency that satisfies Eq. (1.86) and the condition $n' \not\equiv 0$ can have a positive imaginary part, i.e., the perturbations described by this dispersion equation can grow in time. Note that (1.86) has the solution $\xi = 0$ if

$$\alpha \equiv \frac{\omega_p L}{V} = \alpha^{(n)} \equiv \pi n, \qquad n = 1, 3, 5, \ldots, (2k+1) \ldots \tag{1.87}$$

Now suppose that α differs slightly from $\alpha^{(n)}$:

$$\alpha = \alpha^{(n)} + \alpha', \ \alpha' \ll 1. \tag{1.88}$$

In this case $\xi \ll 1$. The oscillation frequency is purely imaginary:

$$\left.\begin{aligned} \operatorname{Re}\omega &= 0; \\ \gamma &= \frac{\omega_p^{(n)}}{4}\alpha', \end{aligned}\right\} \tag{1.89}$$

where $\omega_p^{(n)}$ is the plasma frequency under the resonance conditions (1.87). We see that $\gamma > 0$ if $\alpha' > 0$; thus, there is an instability if α has values slightly greater than the resonance values. If $\alpha \ll 1$, solutions (1.86) corresponding to $n' \not\equiv 0$ do not exist. It follows that an instability is possible only for values of α that are not small, $\alpha \gtrsim 1$, i.e., for

$$\omega_p > \pi V/L. \tag{1.90}$$

An estimate of the maximal growth rate can be obtained by setting $\alpha' \approx 1$ in (1.89).

In the limit $\alpha \gg 1$, Eq. (1.86) does not have solutions with large positive $\operatorname{Im}\xi$. This means that for $\alpha \gg 1$ the growth rate is of the same order as for $\alpha \approx \pi$, i.e.,

$$\gamma \simeq V/L. \tag{1.91}$$

APPENDIX

The Permittivity Tensor

We obtain a representation for this tensor in the following manner. We first introduce the electric displacement vector,

$$\mathbf{D} = \mathbf{E} + \frac{4\pi i}{\omega}\mathbf{j}. \tag{A1.1}$$

We shall assume that the current density $\mathbf{j}$ is calculated and related to the field by the equation

$$j_\alpha = \sigma_{\alpha\beta} E_\beta. \tag{A1.2}$$

Here, $\sigma_{\alpha\beta}$ is the conductivity tensor. Substituting (A1.2) into (A1.1), we obtain the relation between D and E:

$$D_\alpha = \varepsilon_{\alpha\beta} E_\beta. \tag{A1.3}$$

The tensor

$$\varepsilon_{\alpha\beta} = \delta_{\alpha\beta} + \frac{4\pi i}{\omega}\sigma_{\alpha\beta} \tag{A1.4}$$

is the permittivity tensor.

Suppose the form of $\varepsilon_{\alpha\beta}$ is known. Then Maxwell's equations (II) can be reduced to the following three equations for the components of the perturbed electric field:

$$(\operatorname{rot}\operatorname{rot}\mathbf{E})_\alpha = \frac{\omega^2}{c^2}\varepsilon_{\alpha\beta} E_\beta. \tag{A1.5}$$

The subsequent procedure for deriving the dispersion equation reduces to the solution of just these three equations.

Thus, the introduction of the permittivity tensor splits the complete problem into two parts: the determination of the tensor $\varepsilon_{\alpha\beta}$ itself and the solution of Maxwell's equations with known $\varepsilon_{\alpha\beta}$.

In the case of a homogeneous plasma the dispersion equation can be obtained very simply by means of (A1.5). Taking the electric field in the form of a plane wave, $\mathbf{E} \sim \exp(i\mathbf{kr})$, we arrive at a system of three algebraic equations for the components of $\mathbf{E}$. The dispersion equation is obtained by equating to zero the determinant of this system:

$$\left| N^2\left(\frac{k_\alpha k_\beta}{k^2} - \delta_{\alpha\beta}\right) + \varepsilon_{\alpha\beta}\right| = 0. \tag{A1.6}$$

Here, $N^2 = c^2k^2/\omega^2$ and is called the square of the refractive index.

Let us consider how (A1.6) can be used to obtain the equations for the plasma oscillations and electromagnetic waves in a cold plasma at rest, i.e., Eqs. (1.14) and (1.15). To do this it is necessary to know the form of $\varepsilon_{\alpha\beta}$. Now it follows from (1.10) in the case of a cold plasma at rest that

$$\mathbf{j} \equiv e_e n_0 \mathbf{V}'_e = \frac{i e_e^2}{m_e \omega} \mathbf{E}, \tag{A1.7}$$

therefore, as follows from (A1.2) and (A1.4), the expression for $\varepsilon_{\alpha\beta}$ has the form

$$\varepsilon_{\alpha\beta} = \delta_{\alpha\beta}\varepsilon, \tag{A1.8}$$

where

$$\varepsilon = 1 - \frac{\omega_{p_e}^2}{\omega^2}. \tag{A1.9}$$

We take the **x** and **y** axes perpendicular to **k** and **z** along **k**. Then, using (A1.8), we can write Eq. (A1.6) in the form

$$\begin{vmatrix} \varepsilon - \dfrac{c^2k^2}{\omega^2}, & 0, & 0, \\ 0, & \varepsilon - \dfrac{c^2k^2}{\omega^2}, & 0 \\ 0, & 0, & \varepsilon \end{vmatrix} = 0. \tag{A1.10}$$

Equation (A1.10) evidently reduces to the system (1.14)-(1.15). It should also be noted that Eq. (1.15) is contained twice in (A1.10) – the electromagnetic waves described by Eq. (1.15) can be polarized in two different ways (for example, $\mathbf{E}\parallel\mathrm{x}$ and $\mathbf{E}\parallel\mathrm{y}$).

We shall now show how the simpler dispersion equation of the electrostatic approximation can be obtained from the general dispersion equation (A1.6). In the electrostatic perturbations the magnetic field is appreciably smaller than the electric field. Since $B' \approx ckE'/\omega$, this means that the parameter ck/ω is fairly large in electrostatic perturbations, i.e., $N^2 \to \infty$. It follows that the procedure for obtaining the electrostatic approximation in (A1.6) is as follows: first expand the left-hand side of (A1.6) in a series in $1/N^2$ and then let N^2 tend to infinity. Then

$$k_\alpha \varepsilon_{\alpha\beta} k_\beta = 0. \tag{A1.11}$$

Equation (A1.5) shows that this equation corresponds to electrostatic perturbations. Indeed, taking the divergence of (A1.5) and then using the condition $\mathbf{E} = -i\mathbf{k}\psi$ (approximate or exact), we arrive at (A1.11). If the oscillations are not strictly electrostatic, the substitution $\mathbf{E} = -i\mathbf{k}\psi$ in the left-hand side of Eq. (A1.5) could lead to incorrect results.

The quantity

$$\varepsilon_0 = \frac{1}{k^2} k_\alpha \varepsilon_{\alpha\beta} k_\beta \tag{A1.12}$$

is the scalar permittivity. This quantity has already been used (see, for example, § 1.4).

It is shown in § 1.4 that ε_0 can be represented in the form $\varepsilon_0 = 1 - 4\pi\chi/k^2$, where χ is given by $\rho = \chi\psi$. One can see that these two expressions are identical by taking into account the continuity equation of the electric charge, which follows from Eq. (1.2) (or Maxwell's equations):

$$\frac{\partial \rho}{\partial t} + \operatorname{div} \mathbf{j} = 0. \tag{A1.13}$$

Substituting **j** in the form (A1.12) into this equation, assuming $\mathbf{E} = -i\mathbf{k}\psi$, and comparing the expression for ρ obtained in this manner with the expression for $\rho = \chi\psi$, we obtain

$$\chi = -\frac{i}{\omega} k_\alpha \sigma_{\alpha\beta} k_\beta. \tag{A1.14}$$

Using (A1.4) and (A1.14), we find that (A1.12) does indeed reduce to (1.35).

The dispersion equation for electrostatic perturbations is much simpler than the general equation (A1.6). However, to justify the electrostatic approximation, one must, in general, first consider the general dispersion equation.

Bibliography

1. L. Tonks and I. Langmuir, Phys. Rev., 33:195 (1929). The dispersion equations (1.14) and (1.15) are derived in this paper.
2. A. V. Haeff, Phys. Rev., 74:1532 (1948).
3. J. R. Pierce, J. Appl. Phys., 19:231 (1948).
4. D. Bohm and E. P. Gross, Phys. Rev., 75:1864 (1949).
5. A. I. Akhiezer and Ya. B. Fainberg, Dokl. Akad. Nauk SSSR, 69:555 (1949). In [2-5] the dispersion equation of a system of cold streams is derived independently and the possibility of spontaneous excitation of oscillations is pointed out. A quantitative investigation of the growth of the perturbations is made in [4] and [5]. The instability of streams of equal density (§ 1.5.1) is investigated in [4]; the instability of a low-density beam (§ 1.5.2), in [4] and [5]. The maximal growth rate of the instability of a low-density beam [Eq. (1.52)] is derived in [5]. In [2] the treatment is restricted to streams of equal density (§ 1.5.1) and in [3] to a stream of electrons that move relative to the ions (§ 1.6).
6. V. A. Bailey, Phys. Rev., 83:439 (1951). The instability condition for streams of arbitrary density [Eq. (1.58)] is obtained.
7. F. D. Kahn, J. Fluid Mech., 2:601 (1957). The instability of streams of equal density is discussed (§ 1.5.1).

8. O. Buneman, Phys. Rev. Lett., 1:8 (1958).
9. O. Buneman, Phys. Rev., 115:503 (1959). In [8, 9] a study is made of the instability due to the relative motion of electrons and ions (§ 1.6).
10. V. S. Imshennik and Yu. I. Morozov, Zh. Tekh. Fiz., 31:640 (1961) [Soviet Phys. – Tech. Phys., 6:464 (1961)]. Detailed quantitative analysis of the two-stream instability. The form of the branches of the plasma–beam oscillations is obtained (shown schematically in Fig. 1.1).
11. J. R. Pierce, J. Appl. Phys., 15:721 (1944). The instability of a stream moving between a cathode and anode (§ 1.8) is established.
12. G. G. MacFarlane and H. G. Hay, Proc. Phys. Soc., B63:409 (1950). Investigation of the instability of a stream with an inhomogeneous velocity profile (§ 1.7).

Chapter 2

Kinetic Description of Plasma Oscillations and General Conditions of Stability and Instability

§ 2.1. Plasma Oscillations in the Kinetic Formulation

The thermal motion of the particles weakens the hydrodynamic mechanisms of instability growth but it by no means leads to complete stabilization of the plasma in all cases. The reason for this is the resonant interaction between the particles and oscillations that is manifested when there is a large thermal spread. This resonant interaction can lead to kinetic instabilities. (In a plasma with a Maxwellian distribution of the particle velocities, the resonant interaction is due to the collisionless damping discovered by Landau.)

The present chapter is an introduction to the theory of kinetic instabilities. We shall discuss the methods used to investigate the oscillations of a plasma with a continuous particle velocity distribution and the effects peculiar to such a plasma. We shall also formulate some general conditions that make it possible in a number of cases to state in advance whether a plasma with a given velocity distribution will be stable or not.

If allowance is made for the thermal motion of the particles, the perturbed behavior of the plasma can be described by means of the linearized equation (I):

$$\frac{\partial f'}{\partial t} + i\mathbf{k}\mathbf{v}\, f' = -\frac{e}{m}\left(\mathbf{E}' + \left[\frac{\mathbf{v}}{c}\,\mathbf{B}'\right]\right)\frac{\partial f_0}{\partial \mathbf{v}}. \tag{2.1}$$

Here, as in Chapter 1, collisions and a static magnetic field B_0 are ignored. The equilibrium state of the plasma characterized by the distribution function f_0 is assumed to be spatially homogeneous. This enables us to take the spatial dependence of perturbations in the form of a plane wave [see (1.13)] and this we shall do. We shall, as a rule, omit the primes of the perturbed quantities f', E', and B'.

If the system is disturbed at t = 0, i.e.,

$$f(t, \mathbf{v}) = \begin{cases} 0, & t < 0, \\ g(\mathbf{v}), & t = 0, \end{cases} \tag{2.2}$$

its behavior for t > 0 can be established by the Laplace method in the same way as in the problem of § 1.3 on the oscillations of a cold plasma.

Restricting ourselves to the case of electrostatic perturbations and adopting the Laplace transform procedure outlined in the problem of § 1.3, we obtain the following equation from (2.1) and Poisson's equation (II):

$$f_p(\mathbf{v}) = \frac{1}{p + i\mathbf{k}\mathbf{v}} \left[g(\mathbf{V}) + \frac{i\mathbf{k}e}{m} \psi_p \frac{\partial f_0}{\partial \mathbf{v}} \right], \tag{2.3}$$

$$\psi_p = \frac{i\, 4\pi e}{k^2 \varepsilon_0 (ip)} \int \frac{g(\mathbf{v})\, d\mathbf{v}}{ip - \mathbf{k}\mathbf{v}}. \tag{2.4}$$

Here, ε_0 (ip) is the scalar permittivity for

$$\omega = ip. \tag{2.5}$$

It is given by the expression

$$\varepsilon_0 (ip) = 1 + \frac{4\pi e^2}{mk^2} \int \frac{\mathbf{k} \frac{\partial f_0}{\partial \mathbf{v}} d\mathbf{v}}{ip - \mathbf{k}\mathbf{v}}. \tag{2.6}$$

Using (2.3), (2.4), and (1.29), we find the desired form of the functions $f(t)$ and $\psi(t)$ for t > 0:

$$f(\mathbf{v}, t) = \frac{1}{2\pi i} \int_{-i\infty+\sigma}^{+i\infty+\sigma} \frac{e^{pt}}{p + i\mathbf{k}\mathbf{v}} \left[g(\mathbf{v}) + \frac{ie\mathbf{k}}{m} \psi_p \frac{\partial f_0}{\partial \mathbf{v}} \right] dp, \tag{2.7}$$

$$\psi(t) = \frac{1}{2\pi i} \cdot \frac{4\pi i e}{k^2} \int_{-i\infty+\sigma}^{+i\infty+\sigma} \frac{e^{pt}}{\varepsilon_0 (ip)} \int \frac{g(\mathbf{v})\, d\mathbf{v}}{ip - \mathbf{k}\mathbf{v}} dp. \tag{2.8}$$

As in the case of a cold plasma, the integrals (2.7) and (2.8) reduce to the residues around the singularities of the integrand, and the frequencies ω of the oscillations with $\psi \neq 0$ are determined by the zeros of ε_0 (ip), i.e., $\omega = ip_0$, where p_0 is the solution of the equation

$$\varepsilon_0(i\,p_0) = 0. \tag{2.9}$$

If we wish to use Eq. (2.9) in practice, we must extend the definition of ε_0 (ip) to Re p ≤ 0. The structure of the integral (2.8) shows that to do this we must continue ε_0 (ip), defined for Re p ≥ 0, analytically into the region Re p < 0. This leads to a well-defined method for calculating the improper integral in (2.6) (Landau's rule).

The above treatment shows that the determination of the frequencies of the plasma oscillations in the case of an arbitrary velocity distribution reduces, as in the case of a cold plasma, to the determination of the roots of the dispersion equation. In this sense, the analysis of the initial-value problem is needed solely to establish the rule for calculating the improper integral with respect to the velocities in the expression for the permittivity. Having established this rule for the special case of the oscillations considered here, we shall assume that it is universal. Thus, when investigating other types of oscillation we shall adopt the method of Fourier harmonics used in Chapter 1 and not have recourse to the initial-value problem.

In what follows we shall regard the expression for ε_0 as a function of the oscillation frequency ω and not the variable p, i.e., we take ε_0 in the form

$$\varepsilon_0(\omega) = 1 + \frac{4\pi e^2}{mk^2}\int \frac{\mathbf{k}\frac{\partial f_0}{\partial \mathbf{v}}\,d\mathbf{v}}{\omega - \mathbf{k}\mathbf{v}}. \tag{2.10}$$

In the integral with respect to the velocities in this expression for ε_0, the contour is to be chosen as if Im $\omega > 0$.

In a number of cases it is convenient to use a different form of the expression (2.10), obtained by integrating by parts:

$$\varepsilon_0(\omega) = 1 - \frac{4\pi e^2}{m}\int \frac{f_0 d\mathbf{v}}{(\omega - \mathbf{k}\,\mathbf{v})^2}. \tag{2.11}$$

Problem. Obtain an expression for the permittivity of a hot plasma by using the hydrodynamic equations and assuming that the number of streams is infinite.

Solution. We start from Eqs. (1.35) and (1.36'). Since one stream differs from another by the magnitude of its velocity, the summation in (1.35) over all the streams (with given m and e) can be replaced by summation over the velocities. Assuming that each stream has its own density, we can represent the result in the form

$$\varepsilon_0 = 1 - \frac{4\pi e^2}{m} \sum_{V^{(\alpha)}} \frac{n_0 (V^{(\alpha)})}{(\omega - \mathbf{k}\mathbf{V}^{(\alpha)})^2}. \tag{2.11'}$$

This result is not affected if the summation over a finite number of streams is extended to an infinite set $\mathbf{V}^{(\alpha)}$. We shall assume that the velocities of neighboring (in the sense of the velocity $\mathbf{V}$) streams of this infinite set differ by the same amount $\Delta V \ll V^{(\alpha)}$. Denoting the number of particles per unit velocity interval by

$$f_0 (\mathbf{V}^{(\alpha)}) = \frac{n_0 (\mathbf{V}^{(\alpha)})}{\Delta V},$$

we can write Eq. (2.11') in the form

$$\varepsilon_0 = 1 - \frac{4\pi e^2}{m} \sum_{V^{(\alpha)}=-\infty}^{\infty} \frac{f_0 (\mathbf{V}^{(\alpha)}) \Delta V}{(\omega - \mathbf{k}\mathbf{V}^{(\alpha)})^2}. \tag{2.11"}$$

Now it follows from the analysis of the problem with initial conditions (see §2.1) that the frequency must be assumed to be complex with $\operatorname{Im} \omega > 0$ in the determination of the permittivity; we shall therefore assume $\mathbf{k}\Delta \mathbf{V} < \operatorname{Im} \omega$. This enables us to go over in (2.11") from summation to integration, after which (2.11") reduces to (2.11).

§ 2.2. Collisionless Dissipation of the Energy of Oscillations

The integral in (2.10) can be calculated by assuming that ω is real and replacing ω in the denominator by $\omega + i\Delta$, where Δ is an infinitesimally small positive number. Then, using the relation

$$\frac{1}{\omega - \mathbf{k}\mathbf{v} + i\Delta} = \frac{\mathcal{P}}{\omega - \mathbf{k}\mathbf{v}} - i\pi\delta(\omega - \mathbf{k}\mathbf{v}), \tag{2.12}$$

we can represent the permittivity in the form

$$\varepsilon_0 (\omega, \mathbf{k}) = \operatorname{Re} \varepsilon_0 (\omega, \mathbf{k}) + i \operatorname{Im} \varepsilon_0 (\omega, \mathbf{k}), \tag{2.13}$$

where

$$\operatorname{Re} \varepsilon_0 = 1 + \frac{4\pi e^2}{mk^2} \mathscr{P} \int \frac{\mathbf{k}\, \partial f_0/\partial \mathbf{v}}{\omega - \mathbf{k}\mathbf{v}}\, d\mathbf{v}, \tag{2.14}$$

$$\operatorname{Im} \varepsilon_0 = -\frac{4\pi^2 e^2}{mk^2} \int \mathbf{k} \frac{\partial f_0}{\partial \mathbf{v}} \delta(\omega - \mathbf{k}\mathbf{v})\, d\mathbf{v}. \tag{2.15}$$

In (2.12) and (2.14) the symbol $\mathscr{P}$ denotes the principal value of the integral.

It follows from these relations that ε_0 is complex if the distribution function has a nonvanishing derivative at velocities near the phase velocity of the perturbation. If ε_0 calculated for real ω is complex, the frequency of the plasma oscillations is also complex. The imaginary part of the frequency can be readily found if it is small compared with the real part; this is possible if $\operatorname{Im} \varepsilon_0$ is small. Let us consider this case in more detail.

If $\operatorname{Im} \omega \ll \operatorname{Re} \omega$, it is sufficient to take into account the imaginary correction to the frequency in only $\operatorname{Re} \varepsilon_0 (\omega)$. Expanding the latter in a series in $\operatorname{Im} \omega$, substituting the result into the equation $\varepsilon_0 = 0$, and separating the real and imaginary parts in the latter, we obtain

$$\operatorname{Re} \varepsilon_0 (\omega_k, \mathbf{k}) = 0, \tag{2.16}$$

$$\gamma_k = -\frac{\operatorname{Im} \varepsilon_0 (\omega_k, \mathbf{k})}{\dfrac{\partial \operatorname{Re} \varepsilon_0 (\omega_k, \mathbf{k})}{\partial \omega_k}}. \tag{2.17}$$

Here

$$\omega_k \equiv \operatorname{Re} \omega (\mathbf{k}); \quad \gamma_k \equiv \operatorname{Im} \omega (\mathbf{k}).$$

Equation (2.17) can be re-expressed in a form that admits an energy interpretation of the process of growth or decay of oscillations:

$$\frac{\partial}{\partial t} \left\{ \omega_k \frac{\partial \operatorname{Re} \varepsilon_0 (\omega_k, \mathbf{k})}{\partial \omega_k} \cdot \frac{|\mathbf{E}(t)|^2}{8\pi} \right\} = -\frac{\omega_k \operatorname{Im} \varepsilon_0}{4\pi} |\mathbf{E}|^2 \equiv -\operatorname{Re} \sigma |\mathbf{E}|^2. \tag{2.18}$$

Here $\mathbf{E}_{k, \omega_k}(t)$ stands for $\mathbf{E}_{k, \omega_k} e^{\gamma t}$, and σ is the conductivity.

The quantity $\operatorname{Re}\sigma$ corresponds to the resistance, and the expression $\operatorname{Re}\sigma|\mathbf{E}|^2$ gives the amount of oscillatory energy dissipated in unit time. It follows that the expression to which the symbol $\partial/\partial t$ is applied can be interpreted as the energy of the oscillations:

$$W_k = \omega_k \frac{\partial \operatorname{Re} \varepsilon_0(\omega_k)}{\partial \omega_k} \frac{|\mathbf{E}|^2}{8\pi}, \tag{2.19}$$

and Eq. (2.18) can be regarded as the equation of the energy balance of the oscillations.

Problem. Show that the oscillation energy defined by Eq. (2.19) is equal to the sum of the energy of the nonresonant particles and the electrostatic energy.

Solution. The motion of a particle in an electric field is described by the equations

$$\frac{dv}{dt} = \frac{e}{m} E(x(t), t), \tag{2.20a}$$

$$\frac{dx}{dt} = v(t). \tag{2.20b}$$

In the first of these equations we set

$$E(x, t) = \bar{E} \cos(\omega t - kx)\, e^{\gamma t}, \tag{2.21}$$

where $\bar{E}$ is independent of the time and $\gamma > 0$, and we represent v and x in series in powers of E,

$$\left.\begin{aligned} v &= v_0 + v_1 + v_2 + \ldots, \\ x &= x_0 + x_1 + x_2 + \ldots. \end{aligned}\right\} \tag{2.22}$$

We then find

$$v_1 = \frac{e\bar{E}}{m} \frac{e^{\gamma t}}{\omega'^2 + \gamma^2} [\gamma \cos(\omega' t - kx_0) + \omega' \sin(\omega' t - kx_0)],$$

$$x_1 = \frac{e\bar{E}}{m} \cdot \frac{e^{\gamma t}}{\omega'^2 + \gamma^2} [(\gamma^2 - \omega'^2) \cos(\omega' t - kx_0) + 2\gamma\omega' \sin(\omega' t - kx_0)], \tag{2.23}$$

$$\langle v_2 \rangle = \frac{1}{2} \left(\frac{e\bar{E}}{m}\right)^2 \frac{k\omega'}{(\omega'^2 + \gamma^2)^2} e^{2\gamma t}.$$

Here, $\omega' = \omega - kv_0$ and only the slowly varying part – all that is required in what follows – is taken into account in v_2.

Using (2.23), we obtain an expression for the oscillatory energy of a particle that is not in resonance with the wave (i.e., such that $\omega' \gg \gamma$):

$$\left\langle \left(\frac{mv^2}{2}\right) \right\rangle_{\text{oscil}} \equiv \left\langle \frac{m(v^2 - v_0^2)}{2} \right\rangle = \frac{(e\bar{E})^2 e^{2\gamma t}}{4m\omega'^2} \left(2\frac{\omega}{\omega'} - 1\right). \tag{2.24}$$

The oscillatory energy of all the nonresonant particles is

$$\int \left(\frac{mv^2}{2}\right)_{\text{oscil}} f_0(v_0)dv_0 = \frac{e^2 |\bar{E}|^2 e^{2\gamma t}}{4m} \int \left(2\frac{\omega}{\omega'} - 1\right) \frac{f_0(v_0)}{\omega'^2} dv_0 =$$
$$= \frac{|\bar{E}|^2 e^{2\gamma t}}{16\pi} \left(\omega \frac{\partial \operatorname{Re} \varepsilon_0}{\partial \omega} - 1\right). \tag{2.25}$$

In this equation we have used the definition of ε_0 and the equation $\varepsilon_0 = 0$. Adding the energy of the particles to the electrostatic energy, which is equal to $\frac{|\bar{E}|^2}{16\pi} \equiv \frac{|E_k|^2}{8\pi}$ if the field is given in the form (2.21), we obtain (2.19).

§ 2.3. Energy Classification of Weakly Growing Perturbations

If the dissipative part of the permittivity is sufficiently small and the oscillations neither grow nor decay if the dissipation is neglected, the level of the oscillations can be characterized by their energy, which is defined by (2.19). The problem in § 2.2 shows that the energy introduced in this manner has a clear physical meaning – it is equal to the sum of the electrostatic energy of the field of the oscillations and the perturbed energy of the nonresonant particles.

For example, the energy of the plasma oscillations of a cold plasma at rest ($\operatorname{Re} \varepsilon_0 = 1 - \omega_p^2 / \omega^2$, $\omega \partial \operatorname{Re} \varepsilon_0 / \partial \omega = 2$) is equal to twice the electrostatic energy averaged over an oscillation period:

$$W_k = \frac{|E_k|^2}{4\pi}. \tag{2.26}$$

This result can be immediately understood by noting that the energy in the plasma oscillations is continuously being passed from the field to the charges and back to the field again.

The expression (2.19) shows that the energy of the oscillations may be not only positive, as in the above example of the plasma oscillations, but also negative. The simplest example of a medium in which the oscillations can have a negative energy is a moving cold plasma, for which

$$\operatorname{Re} \varepsilon_0 = 1 - \frac{\omega_p^2}{(\omega - kV)^2}, \tag{2.27}$$

$$\omega_k = kV \pm \omega_p, \tag{2.28}$$

so that

$$\omega_k \frac{\partial \operatorname{Re} \varepsilon_0}{\partial \omega_k} = 2\left(1 \pm \frac{kV}{\omega_p}\right). \quad (2.29)$$

If $kV/\omega_p > 1$, this shows that for solutions with the minus sign

$$\omega_k \frac{\partial \operatorname{Re} \varepsilon_0}{\partial \omega_k} < 0 \quad (2.30)$$

and, in accordance with (2.19)

$$W_k < 0. \quad (2.31)$$

It follows from (2.18) that waves with positive energy grow if the conductivity is negative, $\operatorname{Re}\sigma < 0$. If $W_k < 0$, the wave amplitude grows if the conductivity has the usual meaning ($\operatorname{Re}\sigma > 0$).

It is clear that the sign of the energy of the oscillations is not invariant if the frame of reference is changed. The same applies to the sign of the conductivity. However, it is easily verified that the quantities $\partial \operatorname{Re} \varepsilon(\omega_k)/\partial\omega_k$ and $\operatorname{Im} \varepsilon(\omega_k)$ do not depend on the frame of reference and that the same is therefore true of the growth rate. To see this we note that under the coordinate transformation $x \to x' + Vt$ the frequency ω in $\varepsilon(\omega_k)$ is replaced by $\omega' - kV$; but the frequency of the plasma oscillations changes in accordance with the same law, $\omega_k \to \omega_k' - kV$. As a result, $\partial \operatorname{Re} \varepsilon_0/\partial\omega$ and $\operatorname{Im} \varepsilon_0$ in the expression for γ are always taken at the value $\omega = \omega_k$ and γ therefore remains unchanged.

In a number of cases a particular frame of reference may be physically distinguished from the others. For example, if a low-density beam passes through a plasma, the system of rest of the plasma is a frame of reference of this kind; in the more general case, the laboratory system is distinguished. In a frame of reference distinguished in this manner, all the dissipative processes of growth and decay of oscillations can be reduced to the four following types:

A. Damping of positive-energy waves as a result of normal dissipation:

$$W_k > 0, \quad \sigma_k > 0; \quad (2.32)$$

B. Growth of positive-energy waves as a result of anomalous dissipation:

$$W_k > 0, \quad \sigma_k < 0; \tag{2.33}$$

C. Growth of negative-energy waves as a result of normal dissipation:

$$W_k < 0, \quad \sigma_k > 0; \tag{2.34}$$

D. Damping of negative-energy waves as a result of anomalous dissipation:

$$W_k < 0, \quad \sigma_k < 0. \tag{2.35}$$

The expression for the dissipative part (2.15) of the permittivity shows that normal dissipation occurs if the derivative of the distribution function of the resonant particles is negative (for $\omega/k > 0$):

$$\left(\frac{\partial f_0}{\partial v}\right)_{v=\frac{\omega}{k}} < 0, \tag{2.36}$$

and is anomalous if

$$\left(\frac{\partial f_0}{\partial v}\right)_{v=\frac{\omega}{k}} > 0. \tag{2.37}$$

A typical distribution function of a plasma whose oscillations possess positive energy is shown in Figs. 2.1a and 2.1b. In the first case (Fig. 2.1a), only normal dissipation is possible; in the second case (Fig. 2.1b), anomalous dissipation is also possible.

The relations (2.36) and (2.37) uniquely characterize the energy balance of the resonant particles. The physical meaning of these relations is clear: the interaction of a wave with the resonant particles increases the total energy of the latter if there is an excess of particles that lag behind the wave ($\partial f_0/\partial v < 0$); conversely, if $\partial f_0/\partial v > 0$, their total energy is decreased. The wave accelerates the slower particles – giving them energy – and slows down the faster particles – taking energy from them. The total effect is therefore determined by the sign of $\partial f_0/\partial v$.

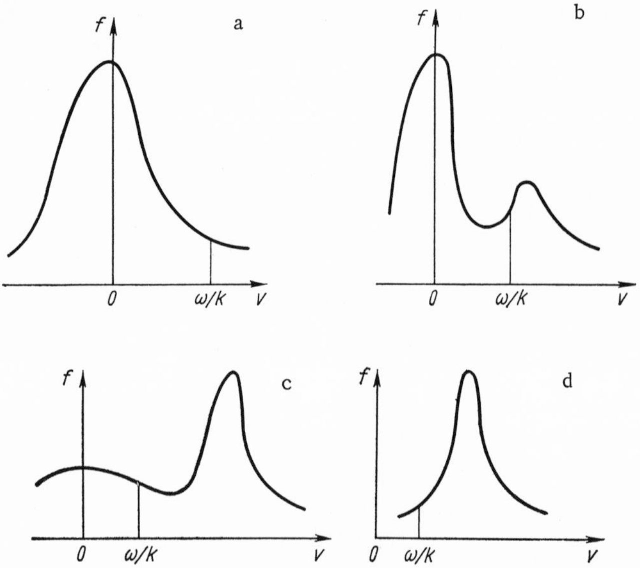

Fig. 2.1. Particle velocity distributions corresponding to different types of dissipative processes: a) $W > 0$, $\sigma > 0$; b) $W > 0$, $\sigma < 0$; c) $W < 0$, $\sigma > 0$; d) $W < 0$, $\sigma < 0$.

An example of a medium in which negative-energy waves can exist is the system considered in § 3.3 consisting of a cold beam and a hot plasma at rest. In this case the distribution function has the form shown in Fig. 2.1c. The instability is due to the interaction between the oscillations and the resonant particles of the hot plasma, for which $\partial f_0/\partial v < 0$. The resonant particles of the plasma therefore take energy (normal dissipation, $\sigma > 0$) and cause a decrease in the energy of the oscillations, $dW/dt < 0$. Since W is negative, this corresponds to an increase in the absolute magnitude of W, the result being growth of the electric field.

The actual reservoir of the energy which is given partly to the resonant particles of the hot plasma and partly to the electric field is the total energy of the moving beam, which is equal to the sum of the oscillatory energy of the particles and the energy of their equilibrium motion:

$$W_{tot} = W_{oscil} + W_{equi} \tag{2.38}$$

In accordance with the assumption that the oscillations have a small amplitude, the oscillatory energy is small compared with W_{equi}, i.e.,

$$W_{tot} > 0. \tag{2.39}$$

Using the concept of the total energy, we can write the energy balance for the case of growth of waves with W < 0 in the form

$$\delta (W_{el.f} + W_{res.\,part}) = -\delta W_{tot} \tag{2.40}$$

Here $W_{el.f}$, $W_{res.part}$, and W_{tot} are positive.

When waves with W < 0 grow, the resonant particles act as an "intermediary," transforming the ordered energy of the beam into the oscillatory energy. However, not all of the energy taken from the beam is transformed into oscillatory energy – part of it is needed to accelerate the resonant particles.

§ 2.4. Damping of Electron Plasma Oscillations in a Maxwellian Plasma

1. Permittivity of a Maxwellian Plasma. Suppose the electrons have a Maxwellian velocity distribution:

$$f_0 = n_0 \left(\frac{m}{2\pi T}\right)^{3/2} \exp\left(-\frac{mv^2}{2T}\right). \tag{2.41}$$

In this case the expressions (2.14) and (2.16) take the form*

$$\operatorname{Re} \varepsilon_0 - 1 = \frac{1}{(kd)^2}\left[1 - \frac{2\omega}{|k| v_T} e^{-\left(\frac{\omega}{kv_T}\right)^2} \int_0^{\omega/|k| v_T} e^{t^2} dt\right], \tag{2.42}$$

$$\operatorname{Im} \varepsilon_0 = \frac{1}{(dk)^2} \cdot \frac{\sqrt{\pi}\,\omega}{|k| v_T} e^{-(\omega/kv_T)^2}. \tag{2.43}$$

*To perform the integration with respect to the velocities it is convenient to use the transformation

$$(\omega + i\Delta - \mathbf{kv})^{-1} = -i \int_0^{\infty} \exp[i(\omega + i\Delta - \mathbf{kv})\tau]\, d\tau.$$

Here $v_T = (2T/m)^{1/2}$ is the thermal velocity and $d = (T/4\pi e^2 n_0)^{1/2}$ is the Debye radius. The expressions (2.42) and (2.43) are derived for real ω. The analytic continuation of these expressions into the region of complex ω reduces to the simple replacement of the real ω by a complex ω in the various expressions. Combining (2.42) and (2.43), we find that the permittivity for arbitrary complex ω has the form

$$\varepsilon_0 = 1 + \frac{1}{k^2 d^2}\left[1 + i\sqrt{\pi}\,\frac{\omega}{|k|v_T}\,W\left(\frac{\omega}{|k|v_T}\right)\right], \qquad (2.44)$$

where

$$W(x) = e^{-x^2}\left(1 + \frac{2i}{\sqrt{\pi}}\int_0^x e^{t^2}\,dt\right). \qquad (2.45)$$

The function W is known as the probability integral (or error function) for complex argument or the Kramp function. It is tabulated in the book by Faddeeva and Terent'ev. The function

$$Z(x) = i\sqrt{\pi}\,xW(x), \qquad (2.46)$$

is also tabulated. (In the book by Fried and Conte, which contains tables of the function Z, the latter is denoted by W.)

If the argument of W is large or small compared with unity, this function can be approximated by means of expressions that follow from (2.45):

a) $|x| \gg 1$

$$W(x) = \frac{i}{\sqrt{\pi}\,x}\left(1 + \frac{1}{2x^2} + \frac{3}{4x^4} + \dots\right) + e^{-x^2}, \qquad (2.47)$$

b) $|x| \ll 1$

$$W(x) = 1 + \frac{2ix}{\sqrt{\pi}} + \quad . \qquad (2.48)$$

2. Damping of Plasma Oscillations when $\omega > kv_T$. The "cold-plasma" approximation discussed in Chapter 1 corresponds to the case

$$\omega \gg kv_T. \qquad (2.49)$$

Using Eqs. (2.44), (2.45), and (2.47), we find that the dispersion equation $\varepsilon_0 = 0$ reduces to Eq. (1.14) in the limit $\omega/kv_T \to \infty$. At large, but finite values of ω/kv_T, the permittivity is complex, its imaginary part being exponentially small:

$$\varepsilon_0 = 1 - \frac{\omega_p^2}{\omega^2}\left(1 + \frac{3k^2T}{m\omega^2}\right) + \frac{i\sqrt{\pi}}{k^2d^2}\cdot\frac{\omega}{|k|v_T} e^{-\omega^2/k^2v_T^2}. \tag{2.50}$$

In this case, Eqs. (2.16) and (2.17) can be used to find the real and imaginary parts of ω. If (2.50) is substituted into (2.16), we obtain

$$\omega_k^2 = \omega_p^2(1 + 3k^2d^2). \tag{2.51}$$

The condition $\omega \gg kv_T$ leads to a lower bound on the wavelength of the plasma oscillations; in accordance with (2.51) it has the form

$$k^2d^2 \ll 1. \tag{2.52}$$

It follows from Eq. (2.17) with allowance for (2.50) and (2.51) that

$$\gamma_k = -\frac{1}{2}\sqrt{\frac{\pi}{2}}\cdot\frac{\omega}{(|k|d)^3}\exp\left(-\frac{3}{2} - \frac{1}{2d^2k^2}\right). \tag{2.53}$$

The growth rate is negative – the oscillations are damped. The collisionless damping of oscillations was established by Landau and is usually called Landau damping.

§ 2.5. Ion-Acoustic and Ion Plasma Oscillations

If $\omega/kv_{T_e} \lesssim 1$, the permittivity is almost independent of the frequency,

$$\left(\varepsilon_0^{(e)}\right)_{\omega \ll kv_{T_e}} \approx \frac{1}{k^2d_e^2}. \tag{2.54}$$

If the ion temperature is less than the electron temperature, the relative contribution of the ions to the permittivity increases as the frequency decreases and when

$$\left(\frac{\omega}{kv_{T_e}}\right)^2 \lesssim \frac{m_e}{m_i} \tag{2.55}$$

the ions are as important as the electrons. The expression for the permittivity of the ions is identical with that for the electrons if the electron subscripts are replaced by the ion subscripts. If both the electrons and ions have a Maxwellian velocity distribution, Eqs. (2.44) and (2.48) yield the following dispersion equation for oscillations with $\omega \ll kv_{T_e}$:

$$\varepsilon_0 = 1 + \frac{1}{k^2 d_e^2}\left(1 + \frac{i\sqrt{\pi}\,\omega}{|k|\,v_{T_e}}\right) + \frac{1}{k^2 d_i^2}\left[1 + i\sqrt{\pi}\,xW(x)\right] = 0. \quad (2.56)$$

Here $d_i = (T_i/4\pi e^2 n_0)^{1/2}$;

$$x = \frac{\omega}{|k|\,v_{T_i}}, \qquad v_{T_i} = \sqrt{\frac{2T_i}{m_i}}.$$

The structure of Eq. (2.56) shows that the oscillations it describes are strongly damped unless the phase velocity of the oscillations greatly exceeds the thermal velocity of the ions:

$$\omega/k \gg v_{T_i}. \quad (2.57)$$

If this condition is satisfied, the real part of the frequency is determined, in accordance with (2.16) and (2.47), by the expression

$$\mathrm{Re}\,\varepsilon_0(\omega_k) = 1 + \frac{1}{k^2 d_e^2} - \frac{\omega_{pi}^2}{\omega_k^2}\left(1 + \frac{3T_i k^2}{m_i \omega_k^2}\right) = 0, \quad (2.58)$$

and the decay rate has the form (2.17), where $\mathrm{Re}\,\varepsilon_0(\omega_k)$ is given by Eq. (2.58) and

$$\mathrm{Im}\,\varepsilon_0(\omega_k) = \frac{1}{k^2 d_e^2}\cdot\frac{\sqrt{\pi}\,\omega_k}{|k|\,v_{T_e}}\left[1 + \left(\frac{m_i}{m_e}\right)^{1/2}\left(\frac{T_e}{T_i}\right)^{3/2} e^{-\frac{\omega_k^2}{k^2 v_{T_i}^2}}\right]. \quad (2.59)$$

The frequency of the oscillations described by Eq. (2.58) is given by

$$\omega_k^2 = \frac{\omega_{p_i}^2}{1 + \frac{1}{k^2 d_e^2}}\left[1 + 3k^2 d_i^2\left(1 + \frac{1}{k^2 d_e^2}\right)\right],$$

$$\omega_{p_i}^2 = 4\pi n_0 e^2/m_i. \quad (2.60)$$

It follows that the condition (2.57) is satisfied if

$$T_i/T_e + k^2 d_i^2 \ll 1. \tag{2.61}$$

This means, first, that weakly damped oscillations are possible only in a nonisothermal plasma,

$$T_i \ll T_e, \tag{2.62}$$

and, secondly, that they have wavelengths that are not too short:

$$(kd_i)^2 \ll 1. \tag{2.63}$$

The wavelength of the perturbations may be both longer or shorter than the electron Debye radius d_e. For long-wavelength perturbations,

$$(kd_e)^2 \ll 1, \tag{2.64}$$

it follows from (2.60) that

$$\omega_k^2 = \frac{k^2 T_e}{m_i}\left(1 + \frac{3T_i}{T_e}\right). \tag{2.65}$$

These oscillations, like the acoustic oscillations of an ordinary fluid, have a linear dispersion law, $\omega \sim k$. By analogy, they are therefore called i o n - a c o u s t i c oscillations.

For oscillations for which

$$(kd_e)^2 \gg 1, \tag{2.66}$$

the dispersion equation (2.60) reduces to

$$\omega_k^2 = \omega_{p_i}^2 (1 + 3k^2 d_i^2). \tag{2.67}$$

Except for the notation, this equation is identical with Eq. (2.51) for the electron plasma oscillations. We shall therefore call oscillations of the type (2.67) i o n p l a s m a o s c i l l a t i o n s.

Using (2.17), (2.58), and (2.59), we find the following expressions for the decay rates of the ion-acoustic and ion plasma oscillations:

$$\gamma_{\text{ion-ac}} = -\sqrt{\frac{\pi}{8}}\, k \left(\frac{T_e}{m_i}\right)^{1/2} \left[\left(\frac{m_e}{m_i}\right)^{1/2} + \left(\frac{T_e}{T_i}\right)^{3/2} e^{-\frac{1}{2}\left(\frac{T_e}{T_i} + 3\right)}\right]. \tag{2.68}$$

$$\gamma_{\text{ion-plasma}} = -\sqrt{\frac{\pi}{8}}\,\omega_{p_i}\left(\frac{1}{|k|\,d_e}\right)^3\left[\left(\frac{m_e}{m_i}\right)^{1/2} + \left(\frac{T_e}{T_i}\right)^{3/2} e^{-\frac{\omega_{p_i}^2}{k^2 v_{T_i}^2} - \frac{3}{2}}\right]. \tag{2.69}$$

Problem. Obtain the dispersion equation for low-frequency plasma oscillations assuming that the ions are cold and the electrons are distributed in space in accordance with the Boltzmann law $n_e = n_0 \, {}^{-e_e\psi/T_e}$ (Tonks and Langmuir).

Solution. Linearizing the equation for n_e, we find

$$n'_e = -\frac{e_e n_0}{T_e}\psi. \tag{2.70}$$

The perturbed ion density can be found as in § 1.3:

$$n'_i = \frac{e_i k^2 n_0}{m_i \omega^2}\psi. \tag{2.71}$$

Substituting (2.70) and (2.71) into Poisson's equation, we arrive at Eq. (2.60) with $T_i = 0$.

§ 2.6. Sufficient Condition for the Stability of a Homogeneous Collisionless Plasma

Investigation of the dispersion equation reveals whether a plasma is stable against any chosen class of perturbations. An example of such perturbations is afforded by the electrostatic oscillations described by Eq. (2.9) with ε_0 in the form (2.10). The dispersion equation becomes much more complicated if allowance is made for violation of the electrostatic approximation or a static magnetic field; in this case new simplifying assumptions must be made in order to determine the oscillation frequency.

An alternative to the dispersion equation method is to obtain general stability conditions without specifying the class of perturbations. This approach makes it possible to establish the stationary states that are definitely stable and thereby restrict the set of stationary states for which a detailed analysis of the dispersion equation must be made in order to investigate their stability.

Proceeding from the basic equations (I) and (II) for the plasma and the electromagnetic field, one can show (the Bernstein–

Newcomb–Rosenbluth theorem) that a homogeneous collisionless plasma is stable against all forms of perturbation if the distribution function of each species of particle depends only on the square of the velocity in such a way that for all v^2

$$\frac{df_0}{dv^2} \leqslant 0. \tag{2.72}$$

To see this, consider the functional

$$L = \int d\mathbf{r} \left\{ \frac{E^2}{8\pi} + \frac{(\mathbf{B} - \mathbf{B}_0)^2}{8\pi} + \sum \int d\mathbf{v} \left[\frac{mv^2}{2} (f - f_0) + G(f) \right] \right\}. \tag{2.73}$$

Here, G is a function of f which we shall define later; B_0 is a constant magnetic field; f_0 is the equilibrium distribution function, which is independent of the coordinates and the time. The summation in (2.73) is over the species of charges and the integration is over all **r** and **v**. The functional L is independent of the time – it is a constant of the motion – as is readily verified by taking the time derivative of both sides of Eq. (2.73) and using Eqs. (I) and (II) (with $C_\alpha = 0$).

In the perturbed state, f, **E**, **B**, and G have the form of series in powers of the amplitude of the perturbations:

$$\left.\begin{aligned} f &= f_0 + f_1 + f_2 + \ldots \\ \mathbf{E} &= 0 + \mathbf{E}_1 + \mathbf{E}_2 + \ldots \\ \mathbf{B} &= \mathbf{B}_0 + \mathbf{B}_1 + \mathbf{B}_2 + \ldots \\ G(f) &= G(f_0) + (f_1 + f_2) G'(f_0) + \frac{1}{2} f_1^2 G''(f_0) + \ldots \end{aligned}\right\} \tag{2.74}$$

Substituting (2.74) into (2.73) and neglecting terms of higher than quadratic order, we obtain

$$L = \int d\mathbf{r} \left\{ \frac{E^2 + B^2}{8\pi} + \sum \int d\mathbf{v} \left[\left(\frac{mv^2}{2} + G'(f_0) \right)(f_1 + f_2) + \frac{1}{2} f_1^2 G''(f_0) \right] \right\}. \tag{2.75}$$

It can be seen that L is a quadratic functional if G satisfies the condition

$$\frac{mv^2}{2} + G'(f_0) \equiv 0, \tag{2.76}$$

which is possible if f_0 depends only on the square of the velocity,

$$f_0 = f_0(v^2). \tag{2.77}$$

If the condition (2.76) is satisfied, the right-hand side of (2.75) is certainly equal to a sum of nonnegative terms if

$$G''(f_0) \geqslant 0. \tag{2.78}$$

On the other hand, the condition (2.76) shows that the second derivative is equal to

$$G''(f_0) = -\frac{m}{2 df_0\, dv^2}. \tag{2.79}$$

The condition (2.78) therefore reduces to the requirement that f_0 decrease monotonically with increasing v^2, i.e., it reduces to (2.72). Thus, the theorem formulated above has been proved. (A growth of a perturbation would correspond to an increase in the value of L, but L is a constant of the motion.)

The condition (2.72) shows that the Maxwellian distribution is included among the definitely stable distributions, since for this distribution

$$G'(f_0) = T \ln \left\{ f_0 / n_0 \left(\frac{m}{2\pi T} \right)^{3/2} \right\}. \tag{2.80}$$

It follows from the proof that the theorem does not presuppose the absence of a static magnetic field or even its homogeneity. The result obtained can therefore be used in problems of plasma stability in a magnetic field, which can be either homogeneous, $\nabla B_0 = 0$, or inhomogeneous, $\nabla B_0 \neq 0$.

§ 2.7. Conditions for Stability and Instability of a Plasma Against Electrostatic Perturbations when $B_0 = 0$

Instability of a plasma against electrostatic perturbations means that the dispersion equation $\varepsilon_0(\omega, k) = 0$ has solutions with $\mathrm{Im}\,\omega > 0$. In other words, the function $\varepsilon_0(\omega, k)$ vanishes in the

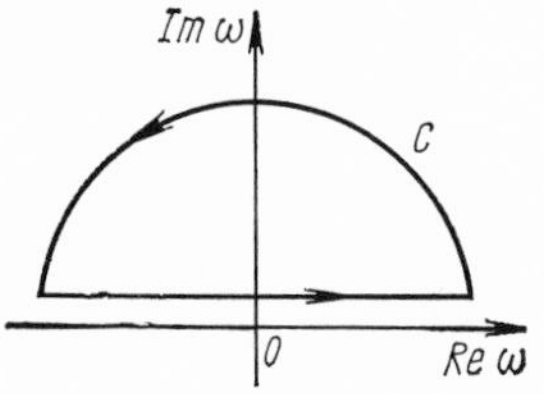

Fig. 2.2. Integration contour in Eq. (2.81).

upper half-plane of ω, that is, somewhere in the region bounded by the contour C in Fig. 2.2, if the plasma is unstable. Since, in accordance with the definition (2.6), the function $\varepsilon_0(\omega)$ is analytic in the upper half-plane and does not have singularities as $|\omega| \to \infty$ [$\varepsilon(\infty) = 1$], the number of unstable roots $\omega_n = \omega_n(k)$, ($n = 1, 2, \ldots, N$), which is equal to the number of zeros of ε_0 in the upper half-plane of ω, is equal to the number of poles of the function $1/\varepsilon_0$ and this (by the residue theorem) is

$$N = \frac{1}{2\pi i}\int_C \frac{d\omega \frac{d\varepsilon}{d\omega}}{\varepsilon(\omega)} \qquad (2.81)$$

(we have omitted the subscript zero of ε_0). The sense of the contour C is indicated by the arrows in Fig. 2.2.

Going over in (2.81) from the variable of integration ω to the variable ε, we can write

$$N = \frac{1}{2\pi i}\int_D \frac{d\varepsilon}{\varepsilon}, \qquad (2.82)$$

where the contour D is the image of the contour C on the plane of the complex ε. This relation is usually known as the Nyquist formula. An example of the contour D for a plasma with a Maxwellian velocity distribution is shown in Fig. 2.3.

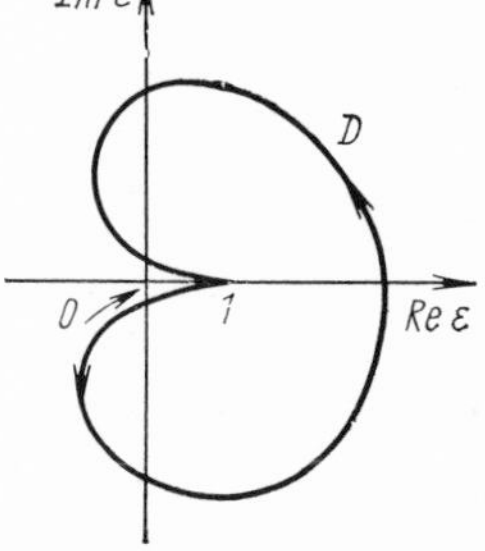

Fig. 2.3. Integration contour in Eq. (2.82) in the case of a Maxwellian plasma.

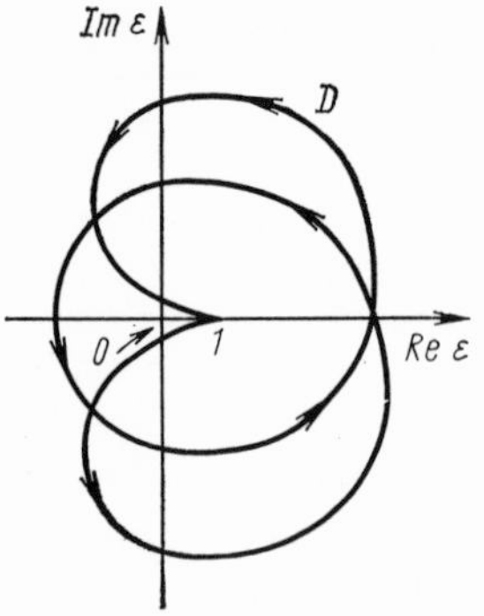

Fig. 2.4. Integration contour in Eq. (2.82) in the case of an unstable plasma.

On the transition from ω to ε, the inifinite semicircle in Fig. 2.2 is transformed into the point $\varepsilon = 1$ of the contour D. All the remaining points of D correspond to the horizontal part of the contour C, on which ω is real. To plot the contour D it is therefore sufficient to have the expression for ε for real ω. In the case of a Maxwellian plasma, this expression is given by Eq. (2.44).

The contour D depicted in Fig. 2.3 does not surround the point $\varepsilon_0 = 0$ and it therefore follows from (2.82) that there are no unstable solutions in the case of a Maxwellian plasma. This agrees with the general theorem of the foregoing section.

A contour of the type shown in Fig. 2.4 corresponds to an instability. For D to have such a form it is necessary that there be a real $\omega = \omega_0$ on the contour C at which $\operatorname{Im} \varepsilon(\omega_0) = 0$, $\operatorname{Im} \varepsilon'(\omega_0) < 0$, $\operatorname{Re} \varepsilon(\omega_0) \leqslant 0$. It is in just such a case that the contour D has a section that passes vertically downwards to the left of the point $\varepsilon = 0$. This yields a sufficient condition for stability: if for all $\omega = \omega_0$ such that

$$\begin{aligned} &\operatorname{Im} \varepsilon_0(\omega_0) = 0, \\ &\operatorname{Im} \varepsilon'(\omega_0) < 0, \end{aligned} \tag{2.83}$$

the real part of ε is positive,

$$\operatorname{Re} \varepsilon(\omega_0) \geqslant 0, \tag{2.84}$$

the plasma is stable.

This assertion can be reformulated if the explicit form of ε_0 [see (2.13)-(2.15)] is used.

A plasma is stable if for all $v = v_0$ (where v is the projection of **v** onto the direction of **k**) satisfying the conditions

$$\left.\begin{array}{l} f'_0(v_0) = 0, \\ f''_0(v_0) > 0, \end{array}\right\} \qquad (2.85)$$

the following inequality holds:

$$\mathcal{P}\int_{-\infty}^{\infty} \frac{f'_0(v)\,dv}{v - v_0} \leqslant \frac{mk^2}{4\pi e^2}. \qquad (2.86)$$

This is the condition of stability of a plasma against perturbations with the given k. The plasma is stable against perturbations with all k (including the case when $k \to 0$) if

$$\mathcal{P}\int_{-\infty}^{\infty} \frac{f'_0(v)\,dv}{v - v_0} \leqslant 0. \qquad (2.87)$$

Integrating by parts, we can represent this stability condition in the form

$$\int_{-\infty}^{\infty} \frac{f_0(v) - f_0(v_0)}{(v - v_0)^2}\,dv \leqslant 0. \qquad (2.88)$$

We shall now show that the condition (2.87) or (2.88) is not only a sufficient but also a necessary condition for stability. In other words, we shall prove the following theorem.

If the opposite inequality to (2.87) holds at even one of the points $v = v_0$ satisfying the conditions (2.85), one can find a perturbation with some value of k against which the plasma is unstable.

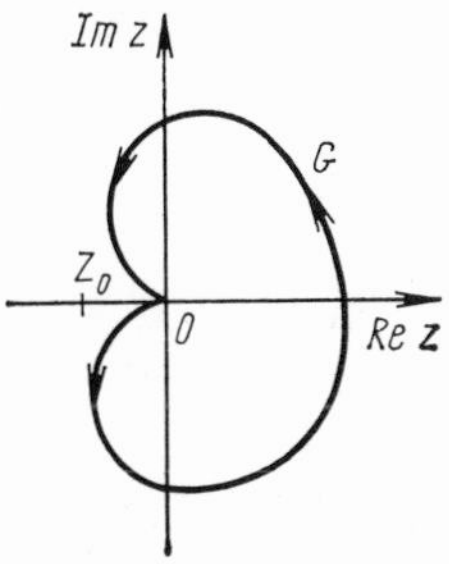

Fig. 2.5. Integration contour in Eq. (2.91) in the case of a Maxwellian plasma.

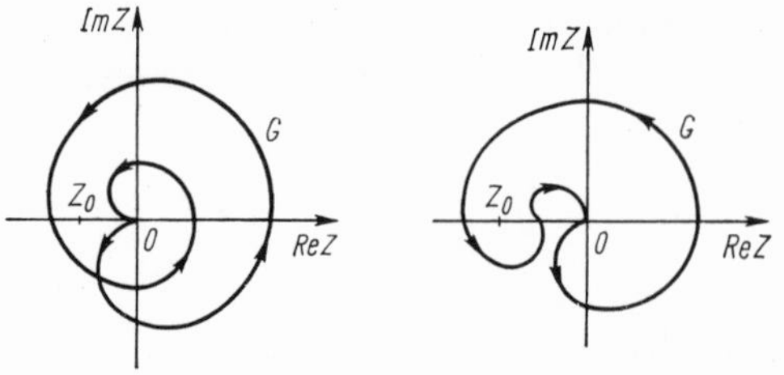

Fig. 2.6. Integration contours in Eq. (2.91) in the case of an unstable plasma.

We replace the variable $\varepsilon(\omega, k)$ by $Z(\omega/k)$:

$$Z\left(\frac{\omega}{k}\right)=(\varepsilon-1)\,k^2. \tag{2.89}$$

In accordance with (2.89) and (2.13)-(2.15), the function $Z(\xi)$ has the form

$$Z(\xi)=-\mathscr{P}\int_{-\infty}^{\infty}\frac{f_0(v)\,dv}{v-\xi}-i\,\pi f'_0(\xi). \tag{2.90}$$

Under the substitution (2.89), the relation (2.82) is transformed into

$$N=\frac{1}{2\pi i}\int_G\frac{dZ}{Z-Z_0}, \tag{2.91}$$

where $Z_0 \equiv -k^2$ and the contour G is the image of C (or D) on the Z plane. The contours G corresponding to Figs. 2.2 and 2.3 are shown in Figs. 2.5 and 2.6.

By making the transformation (2.89), we explicitly separated the parameter $k^2 \equiv -Z_0$, which occurs in the problem. This parameter can take arbitrary positive values. Therefore, if G intersects the negative half-axis of Z, one can always find a Z_0 that lies within G (see Fig. 2.6). It follows that the condition for such an intersection to exist is also a sufficient condition of instability.

Now G intersects Re Z < 0 if $\mathrm{Re}\,Z(v_0) \leq 0$, where v_0 is any one of the ξ_0 that satisfies the conditions $\mathrm{Im}\,Z(v_0) = 0$ and $\mathrm{Im}\,Z'(v_0) < 0$. Recalling the relationship (2.90) between Z and $f_0(v)$, we see that the above theorem is valid.

§ 2.8. Some Examples of Stable and Unstable Distributions (in the Approximation of Electrostatic Oscillations and $B_0 = 0$)

The stability condition (2.87) contains a "one-dimensional" distribution function — that is, one integrated over the velocity components perpendicular to the wave vector. This condition shows that any one-dimensional distribution with a single maximum is stable; for there is then no single point $v = v_0$ that satisfies the conditions (2.85).

A one-dimensional distribution function with a single maximum (situated at $v = 0$) is obtained, in particular, for any spherically symmetric distribution function,

$$f_0(v) = F(v_x^2 + v_y^2 + v_z^2). \tag{2.92}$$

To see this, integrate (2.92) over, say v_x and v_y; the result is

$$\left.\begin{aligned} f_0(v_z) &= \int_{-\infty}^{\infty} dv_x dv_y F(v_x^2 + v_y^2 + v_z^2) = \\ &= 2\pi \int_0^{\infty} F(v_\perp^2 + v_z^2)\, v_\perp dv_\perp = \pi \int_{v_z^2}^{\infty} F(\xi)\, d\xi, \\ v_\perp^2 &= v_x^2 + v_y^2 . \end{aligned}\right\} \tag{2.93}$$

Therefore

$$\frac{\partial f_0(v_z)}{\partial v_z} = -\pi v_z F(v_z^2). \tag{2.94}$$

Since F is nonnegative, it follows from (2.94) that the only extremum of $f_0(v_z)$ occurs at $v_z = 0$. This extremum corresponds to a maximum since

$$\left(\frac{\partial^2 f_0(v_z)}{\partial v_z^2}\right)_{v_z=0} = -\pi F(0) \leqslant 0. \tag{2.95}$$

Thus, any spherically symmetric distribution (with an arbitrary number of maxima and minima) is stable. This enlarges the class of stable, spherically symmetric distributions found in § 2.6. However, in contrast to the stability against all kinds of perturbations in an arbitrary magnetic field B_0 obtained in § 2.6, the stability in this case refers to only electrostatic perturbations and the approximation $B_0 = 0$.

A one-dimensional distribution with a single maximum is also obtained as a result of integration over two velocity components of an anisotropic distribution function that decreases monotonically with the square of the velocity components, i.e., in the case of a function of the form

$$f_0(\mathbf{v}) = F(v_x^2,\ v_y^2,\ v_z^2), \tag{2.96}$$

which is such that for all v_α

$$\frac{\partial F}{\partial v_\alpha^2} \leqslant 0 \quad (\alpha = x, y, z). \tag{2.97}$$

This class of functions includes, for example, the two-temperature, or bi-Maxwellian distribution

$$f_0(\mathbf{v}) = n_0\left(\frac{m}{2\pi T_{\parallel}}\right)^{1/2}\frac{m}{2\pi T_{\perp}}\exp\left[-\frac{mv_z^2}{2T_{\parallel}} - \frac{m\left(v_x^2+v_y^2\right)}{2T_{\perp}}\right]. \tag{2.98}$$

The one-dimensional distribution corresponding to (2.98) has the form

$$f_0(u) = n_0\left(\frac{m}{2\pi T_{\text{eff}}}\right)^{1/2}\exp\left[-\frac{mu^2}{2T_{\text{eff}}}\right], \tag{2.99}$$

where

$$T_{\text{eff}} = T_{\perp}\sin^2\theta + T_{\parallel}\cos^2\theta;$$
$$u = \frac{(\mathbf{kv})}{|k|},\qquad \theta = \tan^{-1}\frac{k_x}{k_z},\qquad \mathbf{k} = (k_x, 0, k_z). \tag{2.100}$$

Thus, the class of stable distributions includes an anisotropic distribution that decreases monotonically with the square of the velocity components. It must be remembered that this result is valid only in the adopted approximation of electrostatic oscillations $(\mathbf{E} = -\nabla\psi)$ and vanishing magnetic field $(\mathbf{B}_0 = 0)$. An anisotropic distribution may be

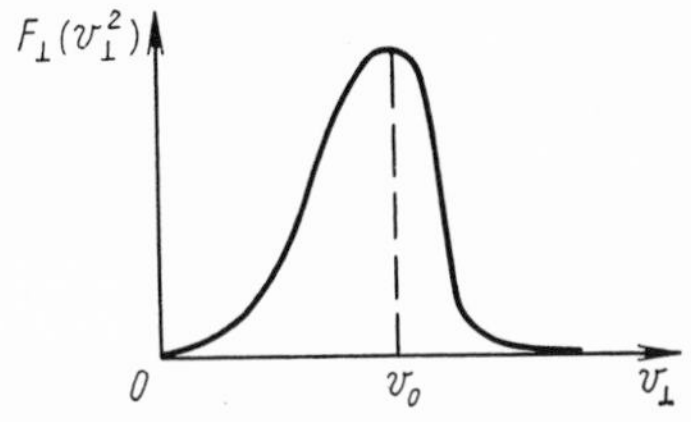

Fig. 2.7. Two-dimensional distribution with a single maximum.

unstable against more complicated perturbations when the electrostatic approximation is violated or a static magnetic field plays a significant role.

Let us now consider functions that possess a symmetry with respect to two given velocity components, for example, v_x and v_y. This could be a function of the form

$$f_0(\mathbf{v}) = F(v_x^2 + v_y^2, v_z). \tag{2.101}$$

In the case of perturbations with wave vector in the symmetry plane ($\mathbf{k} \perp \mathbf{z}$), only the "two-dimensional" distribution function

$$F_\perp(v_\perp^2) = \int F(v_x^2 + v_y^2, v_z)\, dv_z \tag{2.102}$$

is important.

A two-dimensional distribution with one maximum is stable against such perturbations even if this maximum occurs at $v_\perp^2 \neq 0$. This assertion is not *a priori* obvious, since a two-dimensional function with a single maximum could correspond to a one-dimensional distribution function with a minimum (see Figs. 2.7 and 2.8). For example, the function

$$F_\perp(v_\perp^2) = n_0 \frac{\delta(v^2 - v_0^2)}{\pi} \tag{2.103}$$

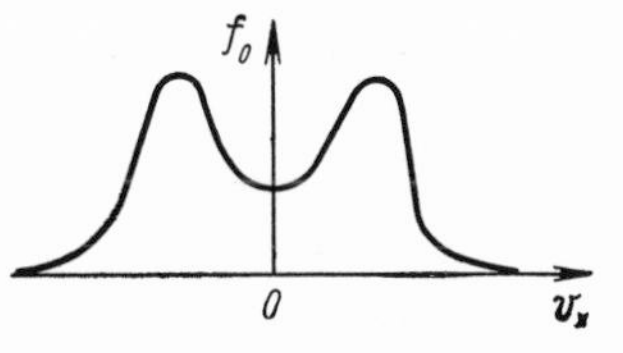

Fig. 2.8. One-dimensional distribution function corresponding to a two-dimensional distribution with a single maximum.

corresponds to the one-dimensional function

$$f_0(v_x) = \begin{cases} \dfrac{1}{\pi\sqrt{v_0^2 - v_x^2}}, & v_x < v_0, \\ 0, & v_x > v_0, \end{cases} \tag{2.104}$$

which has a minimum at $v_x = 0$.

Before we prove this assertion, we shall show that a one-dimensional function that corresponds to a two-dimensional function with a single maximum cannot possess a minimum at any point except $v_x = 0$.

To do this we calculate the number of extrema of the one-dimensional function, i.e., the number of points at which

$$\frac{\partial f_0(v_x)}{\partial v_x} = 0, \tag{2.105}$$

assuming in doing so that $F(v_\perp^2)$ attains a maximum at only a single point $v_\perp = v_0$, i.e., that

$$\left.\begin{aligned} \frac{\partial F}{\partial v_\perp^2} &> 0 \quad \text{for} \quad v_\perp < v_0, \\ \frac{\partial F}{\partial v_\perp^2} &= 0 \quad \text{for} \quad v_\perp = v_0, \\ \frac{\partial F}{\partial v_\perp^2} &< 0 \quad \text{for} \quad v_\perp > v_0. \end{aligned}\right\} \tag{2.106}$$

Since

$$f_0(v_x) = \int_{-\infty}^{\infty} F(v_\perp^2)\, dv_y, \tag{2.107}$$

it follows that

$$\frac{\partial f_0(v_x)}{\partial v_x} = \int_{-\infty}^{\infty} \frac{\partial F(v_\perp^2)}{\partial v_x}\, dv_y =$$

$$= 2v_x \int_{-\infty}^{\infty} \frac{dF(v_\perp^2)}{dv_\perp^2}\, dv_y = 2v_x \int_{v_x^2}^{\infty} \frac{dF(v_\perp^2)}{dv_\perp^2} \cdot \frac{dv_\perp^2}{\sqrt{v_\perp^2 - v_x^2}}. \tag{2.108}$$

This means that one of the extrema is attained at $v_x = 0$. All the other extrema satisfy the condition

$$\int_{v_x^2}^{\infty} \frac{dF(v_\perp^2)}{dv_\perp^2} \cdot \frac{dv_\perp^2}{\sqrt{v_\perp^2 - v_x^2}} = 0. \tag{2.109}$$

It is clear that this equation can be satisfied only for $|v_x| < v_0$, since otherwise, as follows from (2.106), the integrand is everywhere negative. Equation (2.109) can therefore be represented in the form

$$\int_{v_x^2}^{v_0^2} \frac{dF(v_\perp^2)}{dv_\perp^2} \cdot \frac{dv_\perp^2}{\sqrt{v_\perp^2 - v_x^2}} - \int_{v_0^2}^{\infty} \left| \frac{dF(\xi)}{d\xi} \right| \frac{d\xi}{\sqrt{\xi - v_x^2}} = 0. \tag{2.110}$$

Each of these integrals is a monotonic function of $|v_x|$ and their difference can therefore vanish at only a single point v_x. This point can correspond only to the maximum of $f_0(v_x)$, since otherwise the condition $\lim_{v\to\infty} f_0(v_x) = 0$ cannot be satisfied. If such a maximum exists, the point $v_x = 0$ corresponds to a minimum of $f_0(v_x)$ (see Figs. 2.7 and 2.8, which illustrate a two-dimensional distribution with a maximum at $v_\perp \neq 0$ and the corresponding one-dimensional distribution).

For a function that is minimal at $v_x = 0$, the stability condition (2.87) can be written in the form

$$\mathcal{P}\int \frac{f'_0(v_x)\,dv_x}{v_x} \leq 0. \tag{2.111}$$

Substituting the expression (2.107) into the left-hand side and going over from the variables of integration v_x and v_y to $v_\perp$ and $\alpha = \tan^{-1}(v_x/v_y)$, we find

$$\mathcal{P}\int \frac{f'_0(v_x)\,dv_x}{v_x} = -\pi F(0) \leq 0. \tag{2.112}$$

This completes the proof of our above assertion that a two-dimensional distribution with a single maximum is stable.

Bibliography

1. L. Tonks and I. Langmuir, Phys. Rev., 33:195 (1929). The dispersion equation for low-frequency plasma oscillations [Eq. (2.60)] for $T_i = 0$ is obtained.
2. A. A. Vlasov, Zh. Eksp. Teor. Fiz., 8:291 (1938). The kinetic description of plasma oscillations is developed. An expression is obtained for the thermal correction to the frequency of the plasma oscillations [Eq. (2.51)].
3. L. D. Landau, Zh. Eksp. Teor. Fiz., 16:574 (1946) [J. Phys. (Moscow), USSR, 10:25 (1946)]. The Boltzmann equation is solved by the Laplace method (§ 2.1). Collisionless damping of the plasma oscillations is established [Eq. (2.53)].
4. D. Bohm and E. P. Gross, Phys. Rev., 75:1851 (1949). A method is given for obtaining a kinetic dispersion equation by integrating over an infinite number of streams (Problem § 2.1).
5. D. Bohm and E. P. Gross, Phys. Rev., 75:1864 (1949). The physical meaning of collisionless dissipation (Problem § 2.2) is discussed. It is shown that a kinetic instability (§ 2.7) can arise if there is an additional peak in the distribution function. Equation (2.17) is obtained for the specific case of the plasma oscillations of a cold plasma and a low-density beam with a large thermal spread.
6. A. I. Akhiezer and Ya. B. Fainberg, Zh. Eksp. Teor. Fiz., 21:1262 (1951). The kinetic and hydrodynamic effects due to a finite plasma temperature are discussed in connection with the problem of the interaction of a low-density beam with a plasma. It is shown that a kinetic instability can arise as a result of the resonant particles of the plasma.
7. M. E. Gertsenshtein, Zh. Eksp. Teor. Fiz., 23:669 (1952). The diagram method explained in § 2.7 is used to investigate kinetic instabilities. It is shown that a necessary condition for the growth of oscillations is that the sum of the electron and ion distribution functions taken with an appropriate weight has at least two maxima (§ 2.7). The stability condition (2.87) is obtained. For a medium with small dissipation Eq. (2.17) with $\mathrm{Im}\,\varepsilon$ of the form (2.15) is obtained.
8. G. V. Gordeev, Zh. Eksp. Teor. Fiz., 27:19 (1954). Low-frequency oscillations of a Maxwellian plasma (§ 2.5) are investigated by a kinetic equation method. The damping of these oscillations is obtained [Eqs. (2.68) and (2.69)].
9. O. Penrose, Phys. Fluids, 3:258 (1960).
10. P. D. Noerdlinger, Phys. Rev., 118:879 (1960).
11. A. I. Akhiezer, G. Ya. Lyabarskii, and R. V. Polovin, Zh. Eksp. Teor. Fiz., 40:963 (1961).
12. Y. Ozawa, I. Kaji, and M. Koto, Nuclear Fusion, Suppl. Part 3, 1123 (1962). In [9-12] a method similar to that used in [7] is employed to investigate general conditions of stability of a plasma against electrostatic perturbations. The results of these papers are given in § 2.7.
13. V. L. Ginzburg, Usp. Fiz. Nauk, 69:537 (1958) [Soviet Phys. 2:874 (1960)].
14. L. M. Kovrizhnykh and A. A. Rukhadze, Zh. Eksp. Teor. Fiz., 38:850 (1960) [Soviet Phys. — JETP, 11:615 (1960)]. The stability of a spherically symmetric velocity distribution (§ 2.8) is proved in [13, 14].
15. I. B. Bernstein, S. K. Trehan, and M. P. H. Weenink, Nucl. Fusion, 4:61 (1964). The general theorem on plasma stability (§ 2.6) is proved.

16. M. N. Rosenbluth and R. F. Post, Phys. Fluids, 8:547 (1965). The stability of a two-dimensional distribution (§ 2.8) is discussed.
17. V. N. Faddeeva and N. M. Terent'ev, Tables of Values of the Function W(x) for Complex Argument. Pergamon Press, Oxford (1961).
18. B. D. Fried and S. D. Conte, The Plasma Dispersion Function. The Hilbert Transform of the Gaussian. Academic Press, New York (1961). The function Z(x) is tabulated [formula (2.46)].
19. P. A. Sturrock, J. Appl. Phys., 31:2052 (1960).
20. J. R. Pierce, J. Appl. Phys., 32:2580 (1961).
21. R. J. Briggs, J. Appl. Phys., 35:3268 (1964).
22. B. B. Kadomtsev, A. B. Mikhailovskii, and A. V. Timofeev, Zh. Eksp. Teor. Fiz., 47:2266 (1964) [Soviet Phys. — JETP, 20:1517 (1965)].
23. A. Bers and S. Gruber, Appl. Phys. Lett., 6:27 (1965).
24. L. S. Hall and W. Heckrotte, Phys. Fluids, 9:1496 (1966).
25. V. M. Dikasov, L. I. Rudakov, and D. D. Ryutov, Zh. Eksp. Teor. Fiz., 48:913 (1965) [Soviet Phys. — JETP, 21:608 (1965)]. The energy of oscillations (§ 2.3) is discussed in [19-25].

Chapter 3

Kinetic Instabilities

§ 3.1. Influence of the Thermal Motion of the Particles on the Hydrodynamic Two-Stream Instability

In solving the Boltzmann equation (I) in Chapter 1, we neglected the thermal spread of the particles. The distribution function of each group of particles was assumed to be equal to a sum of several δ functions [see Eq. (1.1)]. This made it possible to reduce the Boltzmann equation to a sum of two equations, (1.2) and (1.3), for the stream density and velocity and to reduce the dispersion equation to the rather simple form (1.35) with $\varepsilon_0^{(\alpha)}$ given by (1.36'). Let us now consider the limits of applicability of these relations.

We shall adopt the kinetic description of a plasma, assuming that its equilibrium distribution function has the form

$$f_0 = \sum_\alpha f_0^{(\alpha)}, \tag{3.1}$$

where

$$f_0^{(\alpha)} = n_0^{(\alpha)} \left(\frac{m_\alpha}{2\pi T_\alpha} \right)^{3/2} \exp\left(-\frac{m_\alpha (\mathbf{v} - \mathbf{V}^{(\alpha)})^2}{2T_\alpha} \right). \tag{3.2}$$

As $T_\alpha \to 0$, the expression (3.2) goes over into $n_0^{(\alpha)} \delta(\mathbf{v} - \mathbf{V}^{(\alpha)})$ and corresponds to the approximation of a cold plasma.

Substituting (3.1) and (3.2) into the expression for the permittivity (2.10) and equating the result to zero, we obtain by analogy with (2.44)

$$\varepsilon_0 = 1 + \sum_\alpha \frac{1}{k^2 d_\alpha^2}\left[1 + i\sqrt{\pi}\,\frac{\omega - kV^{(\alpha)}}{|k|\,v_T^{(\alpha)}}\,W\left(\frac{\omega - kV^{(\alpha)}}{|k|\,v_T^{(\alpha)}}\right)\right] = 0, \qquad (3.3)$$

where

$$d_\alpha^2 = \frac{T_\alpha}{4\pi e^2 n_0^{(\alpha)}}, \qquad v_T^{(\alpha)} = \left(\frac{2T_\alpha}{m_\alpha}\right)^{1/2};$$

for simplicity we assume $\mathbf{k} \parallel \mathbf{V}$ in this case.

As we have noted in § 2.4, large arguments of the W function correspond to the cold-plasma approximation. It follows that the general condition of applicability of the cold-plasma approximation in the problem of the hydrodynamic two-stream instability is

$$|\omega - kV^{(\alpha)}| \gg kv_T^{(\alpha)}. \qquad (3.4)$$

In the case of two electron streams of equal density (see § 1.5.1), when $\gamma \approx kV$, the condition (3.4) leads to the requirement that the velocity of the relative motion of the streams should be much greater than the thermal velocity of each of the streams:

$$V \gg (v_T^{(1)},\ v_T^{(2)}). \qquad (3.5)$$

If the density of one of the electron streams is low, $n_1/n_0 \ll 1$, the restriction on the thermal spread of the more dense component is obtained in the same way as in (3.5) but this restriction is more stringent for the less dense component:

$$\left(\frac{n_1}{n_0}\right)^{1/3} V \gg v_{T1}. \qquad (3.6)$$

In the problem of the stability of relative motion of electrons and ions, in which the ratio of the "effective densities" is (m_e/m_i), the restriction on the thermal spread of the ions is

$$v_{T_i} \ll \left(\frac{m_e}{m_i}\right)^{1/3} V. \qquad (3.7)$$

The relations (3.6) and (3.7) refer to the case of the "resonant" instability, when $\gamma \approx \mathrm{Re}\,\omega(n_1/n_0)^{1/3}$ or $\gamma \approx \mathrm{Re}\,\omega(m_e/m_i)^{1/3}$ (see §§ 1.5 and 1.6). The instability of beam oscillations, whose growth rate is proportional to the square root of the ratio of the densities, is described by the hydrodynamic equations only if there is a smaller thermal spread of the "less dense" component, namely, if

$$v_{T1} \ll \left(\frac{n_1}{n_0}\right)^{1/2} V, \tag{3.8}$$

$$v_{T_i} \ll \left(\frac{m_e}{m_i}\right)^{1/2} V. \tag{3.9}$$

To obtain conditions that are more precise than the strong inequalities (3.4)-(3.9), we must consider Eq. (3.3) for large but finite arguments of the W function. Then, to terms of order v_T^2 inclusive, this equation yields

$$1-\sum\frac{\omega_{p\alpha}^2}{(\omega-kV^{(\alpha)})^2}\left(1+\frac{3kT_\alpha}{m_\alpha(\omega-kV^{(\alpha)})^2}\right)=0. \tag{3.10}$$

It is now easy to show that the maximal growth rate of oscillations excited by a "slightly heated" low-density beam in a "slightly heated" dense plasma is

$$\gamma_{\max}=\frac{\sqrt{3}}{2}\omega_{p0}\left(\frac{\alpha}{2}\right)^{1/3}\left[1-\frac{T_0}{2mV^2}-\left(\frac{2}{\alpha}\right)^{2/3}\frac{T_1}{mV^2}\right]. \tag{3.11}$$

It can be seen that the approximation of a cold beam and plasma gives results that are of the correct order even if the inequalities (3.4)-(3.9) are not strong.

The conditions (3.4)-(3.9) have a clear physical meaning: the mean thermal displacement of the particles during an oscillation period (or during the time defined by the reciprocal of the growth rate) measured in a coordinate system that moves with the directed velocity of the particles must not exceed the wavelength of the oscillations.

§ 3.2. Instability of a Beam with a Large Thermal Spread in a Dense Cold Plasma

The hydrodynamic two-stream instability in § 1.5 does not develop if the thermal spread of the beam is sufficiently large –

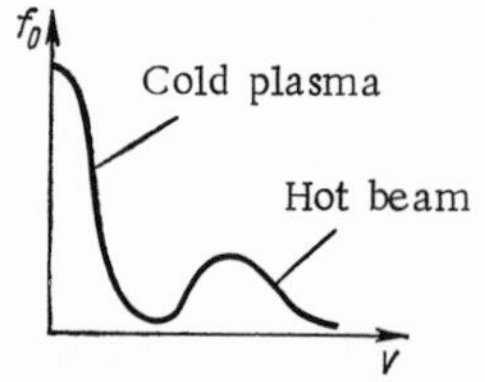

Fig. 3.1. Distribution function of a cold plasma-hot beam system.

that is, the opposite inequality to (3.6) is satisfied:

$$\frac{v_T^{(1)}}{V} > \left(\frac{n_1}{n_0}\right)^{1/3}. \qquad (3.12)$$

Under these conditions a kinetic instability can arise because of the effect of collisionless dissipation (see §§ 2.2 and 2.3). That this possibility exists follows from the general treatment of plasma instability given in § 2.7. This showed that a velocity distribution is unstable if a sufficiently deep minimum is situated between two maxima in the distribution function. In the case in which we are now interested the distribution function of the cold plasma-hot beam system is depicted in Fig. 3.1. It is evident that this distribution satisfies the instability condition.

To investigate the kinetic instability we proceed from Eqs. (2.10), (2.16), and (2.17). Substituting a distribution function of the form

$$f_0 = f_{00} + f_{01} = n_0\delta(\mathbf{v}) + n_1\left(\frac{m}{2\pi T_1}\right)^{3/2}\exp\left(-\frac{m(\mathbf{v}-\mathbf{V})^2}{2T_1}\right) \qquad (3.13)$$

into these equations, we find the frequency and growth rate of the oscillations

$$\left.\begin{aligned} &\omega \approx \omega_{p_e}, \\ &\gamma = \frac{1}{2}\left(\frac{kV}{\omega_{p_e}} - 1\right)\frac{\omega_{p_e}^2}{|k|^3 d_1^2 v_{T1}}\exp\left[-\frac{(\omega_{p_e} - kV)^2}{(kv_{T1})^2}\right], \end{aligned}\right\} \qquad (3.14)$$

where

$$d_1^2 = \frac{T_1}{4\pi e^2 n_1}, \quad n_1 \ll n_0.$$

If the thermal spread of the beam is small compared with the directed velocity, $v_{T1} \ll V$, the maximum of γ is attained at

$$k \approx \frac{\omega_{p_e}}{V}. \qquad (3.15)$$

The order of magnitude of the growth rate is

$$\gamma_{max} \simeq \frac{\omega_{p_e}}{2} \cdot \frac{n_1}{n_0} \left(\frac{V}{v_{T1}} \right)^2 . \tag{3.16}$$

Comparing (3.16) and (1.52) and taking into account (3.12), we conclude that the growth rate of the instability of a beam with a large thermal spread is less than the corresponding growth rate for a cold beam. Extrapolating these expressions to the region of intermediate values of the thermal spread of the beam,

$$v_{T_1} \simeq \alpha^{1/3} V, \tag{3.17}$$

we see that in this region they are of the same order of magnitude. This means that the hydrodynamic instability goes over relatively smoothly into the kinetic instability as the parameter $v_{T1}/(\alpha^{1/3} V)$ is increased to values of order unity or greater. The relationship between these types of instability is depicted schematically in Fig. 3.2. In the same figure we have plotted the growth rate (deduced from the estimates of § 3.1) of the nonresonant hydrodynamic instability [see also Eq. (1.48)] as a function of the thermal spread of the beam.

A kinetic instability is also possible in the case of a beam with small thermal spread, $v_{T1} < \alpha^{1/3}V$, but in this case the growth rate is exponentially small (see Fig. 3.2).

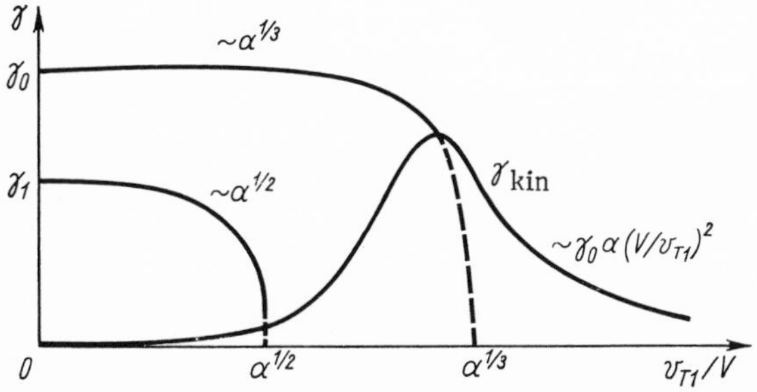

Fig. 3.2. Schematic dependence of the growth rates on v_{T1}. (Here γ_0 and γ_1 are the resonant and nonresonant hydrodynamic growth rates; γ_{kin} is the kinetic growth rate.)

§ 3.3. Instability of a Cold Beam in a Dense Hot Plasma

Let us now consider the excitation of oscillations by a beam whose velocity is small compared with the thermal velocity of the plasma

$$V < v_{T0}. \tag{3.18}$$

We assume that the density of the beam is low compared with the plasma density, $\alpha \equiv n_1/n_0 \ll 1$. We shall neglect the thermal spread of the beam. We assume that $\omega \approx k_z V$ ($\mathbf{V} \| \mathbf{z}$) and, by virtue of (3.18), $\omega \ll k v_{T0}$. Then Eq. (3.3) yields the dispersion equation

$$1 + \frac{1}{k^2 d_0^2}\left(1 + \frac{i\sqrt{\pi}\,\omega}{|k|\,v_{T0}}\right) - \frac{\alpha\omega_p^2}{(\omega - k_z V)^2} = 0. \tag{3.19}$$

Here, $\omega_p^2 = 4\pi e^2 n_0/m$, $d_0^2 = T_0/4\pi e^2 n_0$. In deriving (3.19) we have used the expansions (2.47) and (2.48). It follows from (3.19) that

$$\omega_k = k_z V - \frac{\alpha^{1/2}\,\omega_p}{\left(1 + \frac{1}{k^2 d_0^2}\right)^{1/2}}; \tag{3.20}$$

$$\gamma_k = \left(\frac{\pi\alpha}{8}\right)^{1/2} \frac{1}{\left(1 + k^2 d_0^2\right)^{3/2}} \left\{ k_z V - |k|\, v_{T0} \left\{\frac{\alpha}{2\,(1 + k^2 d_0^2)}\right\}^{1/2} \right\}. \tag{3.21}$$

The growth rate is positive (there is instability) if $V > V_{cr}$, where

$$V_{cr} = \frac{v_{T0}}{\cos\theta}\left(\frac{\alpha}{2}\right)^{1/2} \frac{1}{\left(1 + k^2 d_0^2\right)^{1/2}}, \quad \cos\theta = k_z/k. \tag{3.22}$$

If $kd_0 \approx 1$, then

$$\gamma_k \simeq \alpha^{1/2} \frac{V}{v_{T0}}\,\omega_p, \tag{3.23}$$

$$V_{cr} \simeq \alpha^{1/2} v_{T0}. \tag{3.24}$$

The instability of a beam with $\alpha^{1/2} < V/v_{T0} < 1$ is due to the interaction between the oscillations (3.20) and the resonant electrons of the plasma and is therefore essentially a kinetic effect.

At the limits of applicability of the hydrodynamic and kinetic treatments, i.e., at $V \approx v_{T0}$, the kinetic growth rate (3.23) is small compared with the resonant hydrodynamic growth rate (1.52) but is of the same order of magnitude as the nonresonant hydrodynamic growth rate (1.48). The transition from the hydrodynamic to the kinetic description can be made by leaving the ratio ω/kv_{T0} in (3.3) arbitrary:

$$1+\frac{1}{k^2 d_0^2}\left[1+\frac{i\sqrt{\pi}\,\omega}{|k|\,v_{T0}}\,W\left(\frac{\omega}{kv_{T0}}\right)\right]-\frac{\alpha\omega_p^2}{(\omega-k_zV)^2}=0. \tag{3.25}$$

As with (1.47), we find ($\cos\theta = 1$)

$$\omega=k_zV\pm\frac{\alpha^{1/2}\,\omega_p}{\left\{1+\frac{1}{k^2d_0^2}\left[1+i\sqrt{\pi}\,\frac{V}{v_{T1}}\,W\left(\frac{V}{v_{T1}}\right)\right]\right\}}. \tag{3.26}$$

If k_z is not too near ω_p/V, the imaginary part of the right-hand side of (3.26) determines $\gamma = \gamma(v_{T0}/V)$ for arbitrary values of the parameter v_{T0}/V. If $V > v_{T0}$, the dependence $\gamma(v_{T0}/V)$ is identical with (1.48); if $V < v_{T1}$, with (3.23) (see also Fig. 3.3).

An instability with the growth rate (1.52) is possible only if the denominator of the right-hand side of (3.26) is sufficiently small:

$$\left|1+2\frac{V^2}{v_{T0}^2}\left[1+i\sqrt{\pi}\,\frac{V}{v_{T0}}\,W\left(\frac{V}{v_{T0}}\right)\right]\right|\leqslant\alpha^{1/3}. \tag{3.27}$$

This condition can be satisfied only if $v_{T0}/V < 1$. Thus, the boundary of the instability with the growth rate (1.52) lies at smaller

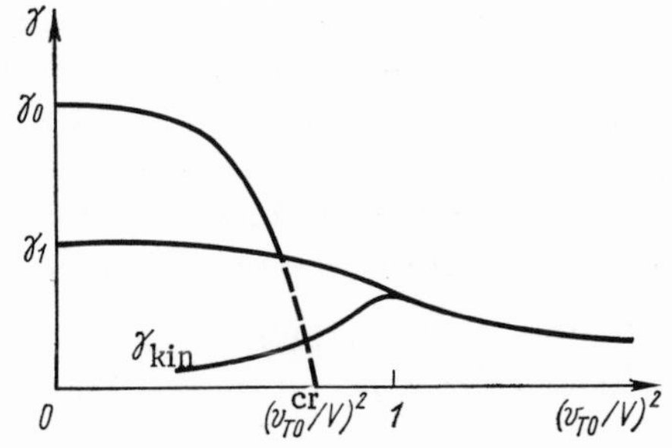

Fig. 3.3. Schematic dependence of the growth rates of the hydrodynamic and kinetic instabilities on the plasma temperature.

values of v_{T0}/V than the boundary between the hydrodynamic and the kinetic instability. This is illustrated in Fig. 3.3.

A kinetic instability is also possible in a hydrodynamically unstable plasma, $V > v_{T0}$ (compare this with the similar effect in § 3.2). For very small v_{T1}/V, the growth rate is exponentially small. At the limits of applicability of the hydrodynamic description the growth rate is comparable with the hydrodynamic growth rate (1.48) (see Fig. 3.3).

Equation (3.18) and the results that can be deduced from it are valid provided $|\omega - kV| \gg kv_{T1}$. With allowance for (3.20) and (3.24) this yields the condition

$$v_{T1} < (V, \alpha^{1/2} v_{T0}). \tag{3.28}$$

In the above treatment we have neglected the role of the ion component of the plasma. This is justified if $\omega \gg \omega_{ion}$, where ω_{ion} [defined by Eq. (2.60)] is the frequency of the ion-acoustic and ion plasma oscillations. This and the approximate equation $\omega \approx k_z V$ yields the condition under which the contribution of the ions can be neglected.

$$V \gg \frac{k}{k_z} \cdot \frac{(T_e/m_i)^{1/2}}{\left(1 + k^2 d_e^2\right)^{1/2}}. \tag{3.29}$$

For $k_z \approx k$ this condition is violated only for very slow beams. The stability of such beams will be considered in § 3.5.

3.4. Instability of a Plasma with a Relative Motion of the Electrons and Ions

The stability of a plasma whose electrons move relative to the ions was considered in § 1.6 in the hydrodynamic approximation. The conditions of applicability of the hydrodynamic treatment for this purpose were discussed in § 3.1. In the same section we established the influence of the real thermal corrections on the growth rate of the oscillations.

We shall now take into account kinetic effects. Since the expressions for $\varepsilon_0(k, \omega)$ for the electron-ion and electron oscillations

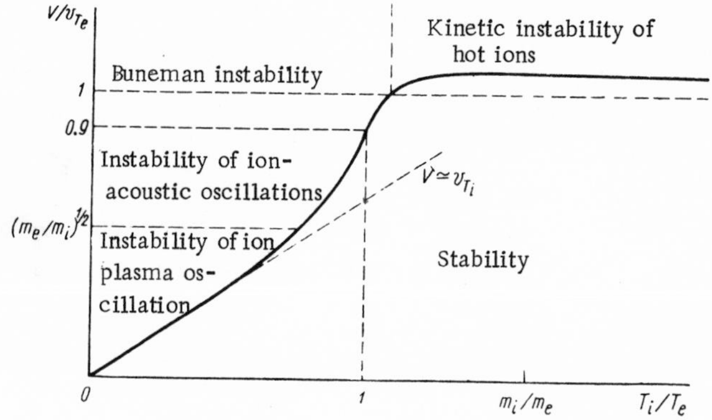

Fig. 3.4. Instability diagram of a plasma in which the electrons and ions have a relative velocity.

differ only in the notation, we can exploit the analogy with §§ 3.2 and 3.3.

The possible types of instability of an electron-ion plasma in which there is a relative motion between the components are illustrated in Fig. 3.4. They depend essentially on the ratio T_i/T_e. In a plasma with $T_e \approx T_i$, only the hydrodynamic instability can arise; see § 1.6 – this is the Buneman instability. For this it is necessary that $V \gtrless v_{T_e}$ (see § 3.1).

In a plasma with $T_e \gg T_i$ we have the branch of ion-acoustic and ion plasma oscillations (§ 2.5). A relative motion of the components may lead to their excitation. In this case the dispersion equation has a form similar to (3.19):

$$1+\frac{1}{k^2d_e^2}\left[1+\frac{i\sqrt{\pi}\,(\omega-k_zV)}{|k|\,v_{T_e}}\right]-\frac{\omega_{p_i}^2}{\omega^2}\left(1+3\frac{k^2T_i}{m_i\omega^2}\right)+$$

$$+\frac{i\sqrt{\pi}}{k^2d_i^2}\cdot\frac{\omega}{|k|\,v_{T_i}}\exp\left[-\frac{\omega^2}{k^2v_{T_i}^2}\right]=0. \qquad (3.30)$$

It follows from (3.30) that the plasma is unstable if $V > V_{cr}$, where

$$V_{cr}=\frac{(T_e/m_i)^{1/2}}{(1+k^2d_e^2)^{1/2}}\left[1+\left(\frac{T_e}{T_i}\right)^{3/2}\left(\frac{m_i}{m_e}\right)^{1/2}\exp\left(-\frac{3}{2}-\frac{T_e}{2T_i\,(1+k^2d_e^2)}\right)\right].$$

(3.31)

The growth rate of the oscillations is

$$\gamma \simeq \left(\frac{m_e}{m_i}\right)^{1/2} \omega_{p_i}. \tag{3.32}$$

For not too large T_e/T_i, when the damping of the oscillations by the resonant ions is important, the minimum of V_{cr} is attained for $(kd_e)^2 < 1$:

$$V_{cr} = \left(\frac{T_e}{m_i}\right)^{1/2} \left[1 + \left(\frac{T_e}{T_i}\right)^{1/2} \left(\frac{m_i}{m_e}\right)^{1/2} \exp\left(-\frac{3}{2} - \frac{T_e}{2T_i}\right)\right]. \tag{3.33}$$

In this case the oscillations that are excited have the frequency (2.65). If

$$T_e/T_i > \ln\left[\left(\frac{T_e}{T_i}\right)^3 \frac{m_i}{m_e}\right], \tag{3.34}$$

the minimum of V_{cr} occurs at perturbations with $(kd_e)^2 \gtrless 1$, i.e., with the frequency (2.67). Then

$$V_{cr.min} \simeq \left(\frac{T_i}{m_i}\right)^{1/2} \ln^{1/2}\left[\left(\frac{T_e}{T_i}\right)^3 \frac{m_i}{m_e}\right]. \tag{3.35}$$

Jackson has calculated $V_{cr.min}$ numerically as a function of T_e/T_i for a hydrogen plasma. The result is shown in Fig. 3.5. For $T_e = T_i$ the value of $V_{cr.min}$ has been found by Buneman:

$$V_{cr.min} = 0.9 V_{T_e}. \tag{3.36}$$

If $V > v_{T_e}$ and $T_i/T_e > (m_i/m_e)^{1/2}$, an instability arises that is similar to that considered in § 3.2. Its maximal growth rate [cf. (3.16)] is

$$\gamma_{max} \simeq \left(\frac{V}{v_{T_i}}\right)^2 \frac{m_e}{m_i} \omega_{p_e}. \tag{3.37}$$

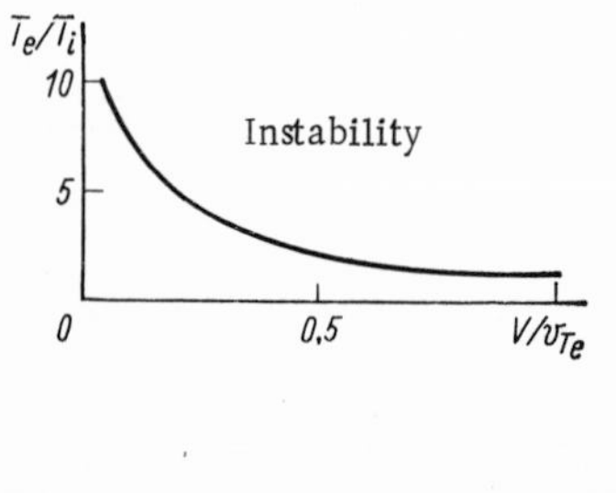

Fig. 3.5. Critical velocity of the relative motion of the electrons and ions as a function of T_e/T_i).

This instability is situated in the upper right-hand corner of Fig. 3.4.

§ 3.5. Excitation of Ion-Electron Oscillations of a Plasma by an Ion Beam

In § 3.3 we have considered the excitation of plasma oscillations by a beam whose velocity is small but nevertheless bounded below [by the condition (3.29)]. A beam of this kind excites oscillations in which only the electron component of the plasma participates. Let us now consider the instabilities of a slower beam by assuming

$$V \lesssim \left(\frac{T_e}{m_i}\right)^{1/2}. \quad (3.38)$$

It will be shown that a beam that is as slow as this can excite oscillations in which the ion component also plays a significant role.

Ion-electron oscillations can be excited by either an ion or an electron beam. For an ion beam the condition $V \lesssim (T_e/m_i)^{1/2}$ means that the energy of the ion beam must not appreciably exceed the temperature of the electron component of the plasma:

$$W_i \lesssim T_e. \quad (3.39)$$

In the case of an electron beam, the condition $V < (T_e/m_i)^{1/2}$ is satisfied provided the energy of the beam particles is very low compared with the mean energy of the plasma electrons:

$$W_e \lesssim \frac{m_e}{m_i} T_e. \quad (3.40)$$

This condition seriously restricts the possibilities of practical realization of an experiment on the excitation of ion oscillations by an electron beam. Suppose for the sake of an estimate we take the energy of the beam electrons to be of the order of a few electron-volts; we then find that to carry out an experiment of this kind we must first heat the plasma electrons to a temperature of a few kiloelectronvolts.

In view of this observation, we shall concentrate our attention in the following exposition on the stability of an ion beam. Formally, the only difference between the excitation of ion oscillations by an electron beam and the case we shall consider resides in the notation.

We shall assume $|\omega - k_z V| \gg kv_{T1}$. This approximation will be justified if it is shown that $\gamma > kv_{T1}$. We shall assume that the phase velocity of the oscillations lies in the interval

$$v_{T_i} \ll \frac{\omega}{k} \ll v_{T_e}. \tag{3.41}$$

We shall neglect the small thermal corrections and imaginary terms in ε_0 which arise because the parameters kv_{T_i}/ω, $kv_{T_i}/(\omega - k_z V)$ and ω/kv_{T_e}, are finite. Under these assumptions, Eq. (3.3) yields the dispersion equation

$$1 + \frac{1}{k^2 d_e^2} - \frac{\omega_{p_i}^2}{\omega^2} - \frac{\alpha \omega_{p_i}^2}{(\omega - k_z V)^2} = 0, \quad \alpha = \frac{n_1}{n_0}. \tag{3.42}$$

It differs from (1.46) by the transformation

$$\omega_{p_e}^2 \to \omega_{p_i}^2 \left(1 + \frac{1}{k^2 d_e^2}\right)^{-1}. \tag{3.43}$$

Like Eq. (1.46), Eq. (3.42) has a complex solution with $\gamma > 0$. The maximum of the growth rate,

$$\gamma = \frac{\sqrt{3}}{2^{4/3}} \alpha^{1/3} \omega_k, \tag{3.44}$$

is attained when ω_k, $k_\perp$, and k_z satisfy:

$$\omega_k^2 \approx \frac{\omega_{p_i}^2}{1 + \frac{1}{k^2 d_e^2}} \approx (k_z V)^2. \tag{3.45}$$

The first relation in (3.45) means that the beam excites the branches of ion-acoustic and ion plasma oscillations (see § 2.5). Such oscillations are possible only in a plasma with $T_e \gg T_i$ since the first inequality in (3.41) is not satisfied if $T_i \gtrless T_e$.

The right-hand relation in (3.45) yields an equation for the resonant wave number k_z:

$$k_z^4 - k_z^2 \frac{1}{d_e^2}\left[\frac{T_e}{m_i V^2} - (1 + k_\perp^2 d_e^2)\right] - k_\perp^2 \frac{\omega_{p_i}^2}{V^2} = 0. \quad (3.46)$$

It can be seen that oscillations with $k_\perp < k_z$ can be excited only if

$$V < \left(\frac{T_e}{m_i}\right)^{1/2}. \quad (3.47)$$

In this case, (3.46) yields

$$k_z = \frac{\omega_{p_i}}{V}\left(1 - \frac{m_i V^2}{T_e}\right)^{1/2}. \quad (3.48)$$

A faster beam excites only oblique perturbations with $k_\perp \gtrsim k_z$. In the limit $V \gg (T_e/m_i)^{1/2}$, Eq. (3.46) gives

$$k_z^2 = k_\perp^2 \frac{T_e}{m_i V^2} \cdot \frac{1}{1 + k_\perp^2 d_e^2}. \quad (3.49)$$

If the temperature of the plasma ions T_i is sufficiently low, the total wave number $k = (k_\perp^2 + k_z^2)^{1/2}$ for both small and large $V (m_i/T_e)^{1/2}$ can be large compared with $1/d_e$. To such k there corresponds the maximal $\omega_k = \omega_{p_i}$ and, in accordance with (3.44), the maximal growth rate:

$$\gamma_{max} = \frac{\sqrt{3}}{2^{1/3}} \alpha^{1/3} \omega_{p_i}. \quad (3.50)$$

Let us now consider the limits of applicability of these results. In accordance with (3.50), the assumption $\gamma > kv_{T1}$ is justified if the beam has a sufficiently small thermal spread [cf. (3.6)],

$$\frac{v_{T1}}{V} < \alpha^{1/3} \cos\theta. \quad (3.51)$$

If this inequality is not satisfied, a kinetic instability similar to that considered in § 3.2 can arise.

If $V \gg (T_e/m_i)^{1/2}$, we must assume in (3.51) that $\cos\theta \approx (T_e/m_i V^2)^{1/2}$. In this case (3.51) entails

$$\frac{v_{T1}}{v_{T_e}} < \left(\frac{m_e}{m_i}\right)^{1/2} \alpha^{1/3}. \quad (3.52)$$

This is a more stringent restriction on the thermal spread of the beam than in the case $V \lessgtr (T_e/m_i)^{1/2}$, when $\cos\theta \approx 1$.

If the distribution of the beam particles over the "thermal" velocities is anisotropic, i.e., $T_{\perp 1} \neq T_{\parallel 1}$, the velocity v_{T1} in (3.51) must be replaced by v_{T1}^{eff}, where $v_{T1}^{eff} = (2T_1^{eff}/m_i)^{1/2}$ and T_{eff} is defined by Eq. (2.100). If $\cos\theta \approx 1$, then $v_{T1}^{eff} \approx v_{T\parallel 1}$, if $\cos\theta \ll 1$, then $v_{T1}^{eff} \approx v_{T\perp 1}$.

The relations $\omega \gg kv_{T_i}$, and $\omega \approx k_z V$ entail

$$V \gg v_{T_i}. \tag{3.53}$$

In accordance with § 3.3, the beam can also be unstable if $V \lesssim v_{T_i}$.

The interaction of the oscillations with the resonant electrons is not important if the growth rate (3.50) is large compared with the electron decay rate (2.68) and (2.69). If $kd_e \approx 1$, the electron decay rate is $|\gamma_{el}| \simeq (m_e/m_i)^{1/2}\omega_{p_i}$. Taking into account (3.50), we find that the condition $\gamma > |\gamma_{el}|$ entails

$$\alpha > \left(\frac{m_e}{m_i}\right)^{3/2}. \tag{3.54}$$

An instability is also possible if $\alpha \lessgtr (m_e/m_i)^{3/2}$, but the growth rate is smaller in this case.

Bibliography

1. D. Bohm and E. P. Gross, Phys. Rev., 75:1864 (1949). The kinetic instability of a beam with a large thermal spread is established (§ 3.2).
2. A. I. Akhiezer and Ya. B. Fainberg, Zh. Eksp. Teor. Fiz., 21:1262 (1951). The possibility of a kinetic instability of a beam in a plasma with a large thermal spread is pointed out and investigated in detail (§ 3.3).
3. M. E. Gertsenshtein, Zh. Eksp. Teor. Fiz., 23:669 (1952). The possibility of excitation of low-frequency oscillations when there is a relative motion of the electrons and ions (§ 3.4) is established.
4. G. V. Gordeev, Zh. Eksp. Teor. Fiz., 27:19 (1954). The kinetic growth rate of the ion-acoustic instability is found [Eq. (3.32)].
5. G. V. Gordeev, Zh. Eksp. Teor. Fiz., 27:24 (1954). A study is made of the excitation of low-frequency perturbations by a beam of particles (§ 3.5).
6. O. Buneman, Phy. Rev., 115:503 (1959). A condition of instability in the case of a relative motion of the electrons and ions in an isothermal plasma is found [Eq. (3.35)].
7. P. L. Auer, Phys. Rev. Lett., 1:411 (1958).

8. I. B. Bernstein et al., Phys. Fluids, 3:136 (1960).
9. E. A. Jackson, Phys. Fluids, 3:786 (1960).
10. I. B. Bernstein and R. M. Kulsrud, Phys. Fluids, 4:1037 (1961).
11. I. B. Bernstein and R. M. Kulsrud, Phys. Fluids, 5:210 (1962). Detailed analyses of the conditions under which the ion-acoustic instability (§ 3.4) can arise are made in [7-11].
12. P. J. Kellogg and H. Liemohn, Phys. Fluids, 3:40 (1960). A numerical analysis of the instability that arises when there is a relative motion of streams with a vanishing temperature.
13. A. A. Vedenov, E. P. Velikhov, and R. Z. Sagdeev, Usp. Fiz. Nauk, 73:701 (1961) [Soviet Phys. — Uspekhi 4:332 (1961)]. A study of the instability of two ion streams of comparable density in a plasma with hot electrons.
14. V. S. Imshennik and Yu. I. Morozov, Zh. Tekh. Fiz., 31:640 (1961) [Soviet Phys. — Tech. Phys., 6:464 (1961)]. The effects that occur at the boundary of applicability of the hydrodynamic and the kinetic description of a beam (§§ 3.1 and 3.2) are analyzed.
15. A. A. Rukhadze, Zh. Tekh. Fiz., 31:1236 (1961) [Soviet Phys. — Tech. Phys., 6:900 (1962)]. A discussion of the instability of a beam in a plasma when allowance is made for thermal (§§ 3.2-3.4) and relativistic effects.

Chapter 4

Spatially Localized Perturbations in an Unstable Plasma and the Amplification of Waves

§ 4.1. Introductory Remarks

The plane-wave type perturbations discussed hitherto are, in reality, an idealization of spatially localized perturbations. We obtain a more realistic description of such perturbations if we associate them with a set of plane waves. This approach is adopted in the present chapter.

It turns out that the properties of spatially localized perturbations are not identical in all respects with those of an individual plane wave. In particular, the field of a spatially localized perturbation at a given point of space may decrease in time although the amplitudes of the plane waves that make up this perturbation increase in time. Whether this or the opposite situation obtains depends on the properties of the dispersion equation that describes the instability.

Plasma oscillations may be excited by fluctuations that exist in the plasma or by external factors. The latter possibility is exploited in application to generate or amplify waves. In a theoretical analysis of these questions it is necessary to consider perturbations that are excited during a certain interval of time and not instantaneously.

If the initial state of the perturbed distribution function $g(\mathbf{r}, \mathbf{v})$ is a plane wave, the potential of the perturbation has the form (2.8). In the general case, $g(\mathbf{r}, \mathbf{v})$ can be represented as a superposition of plane waves:

$$g(\mathbf{r}, \mathbf{v}) = \int g_k(\mathbf{v}) e^{i\mathbf{kr}} d\mathbf{k}. \tag{4.1}$$

The potential of the perturbation corresponding to arbitrary initial g (**r**, **v**) is therefore an integral over k of the right-hand side of (2.8):

$$\psi(\mathbf{r}, t) = \frac{1}{2\pi i} \int_{-\infty}^{\infty} d\mathbf{k} \int_{-\infty + i\sigma}^{\infty + i\sigma} d\omega \frac{h(\mathbf{k}, \omega) e^{i(\mathbf{kr} - \omega t)}}{\varepsilon_0(k, \omega)}. \tag{4.2}$$

Here

$$h(\mathbf{k}, \omega) = \frac{4\pi e}{k^2} \int \frac{g_k(\mathbf{v})\, d\mathbf{v}}{\omega - \mathbf{kv}}. \tag{4.3}$$

Assuming that the dispersion equation ε_0 (k, ω) = 0 does not have multiple roots, i.e., all the $\omega_n = \omega_n$ (k) are distinct, we represent the function ε_0 (k, ω) near $\omega = \omega_n$ (k) as a series

$$\varepsilon_0(\omega, \mathbf{k}) = (\omega - \omega_n(\mathbf{k})) \left(\frac{\partial \varepsilon_0(\omega, \mathbf{k})}{\partial \omega} \right)_{\omega = \omega_n(\mathbf{k})} + \cdots. \tag{4.4}$$

Using (4.4), we shall calculate the integral over ω in (4.2); this expression then takes the form

$$\psi(\mathbf{r}, t) = \sum_n \int d\mathbf{k} e^{i\mathbf{kr} - i\omega_n(\mathbf{k})t} \frac{h(\mathbf{k}, \omega_n(\mathbf{k}))}{(\partial \varepsilon_0 / \partial \omega)_{\omega = \omega_n}(\mathbf{k})}. \tag{4.5}$$

The summation in (4.5) is extended over the roots of the dispersion equation.

It follows from (4.5) that the behavior of the perturbation is determined not only by the properties of the functions $\omega_n = \omega_n$ (k) but also by the original form of the perturbation, i.e., by the form of g (**r**).

The function g (k) can be found from a given g (**r**) by means of the relation

$$g_k(\mathbf{v}) = \frac{1}{(2\pi)^3} \int g(\mathbf{r}, \mathbf{v}) e^{-i\mathbf{kr}} d\mathbf{r}. \tag{4.6}$$

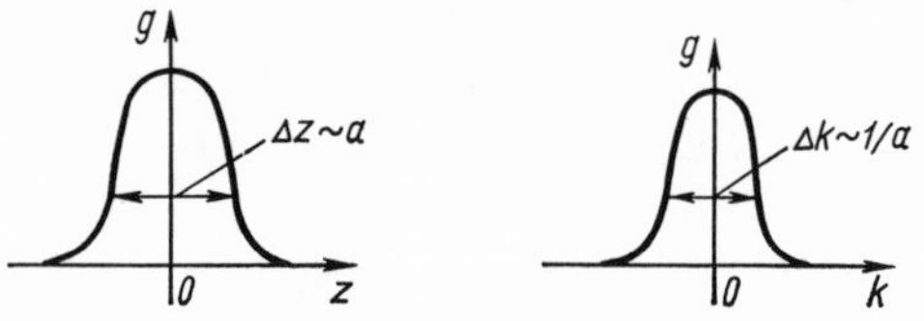

Fig. 4.1. Form of the functions g(z) and g(k) in the case of a smooth, spatially localized distribution.

To a plane wave with wave vector $\mathbf{k}_0$ ($g(\mathbf{r}) \propto e^{i\mathbf{k}_0\mathbf{r}}$) there corresponds the Fourier component $g_{\mathbf{k}} \propto \delta(\mathbf{k}-\mathbf{k}_0)$. To a function localized in a region with dimension $\Delta\mathbf{r}$ there corresponds a superposition of plane waves whose wave numbers differ from zero in an interval $\Delta\mathbf{k}$ such that

$$\Delta\mathbf{r}\Delta\mathbf{k} \simeq 1. \quad (4.7)$$

The mean wave number $\mathbf{k}_0$ in this interval may be of the order of $\Delta\mathbf{k}$ or greater. The first possibility means that the scale of the spatial variation of the function $g(\mathbf{r})$ is equal to the dimension of the localization region. An example of a function of this kind is the Gaussian function

$$g(\mathbf{r}) \propto \exp[-(\mathbf{r}-\mathbf{r}_0)^2/2a^2]. \quad (4.8)$$

Its Fourier transform is

$$g_{\mathbf{k}} \propto \exp\left[-\frac{k^2a^2}{2} - i\mathbf{k}\mathbf{r}_0\right]. \quad (4.9)$$

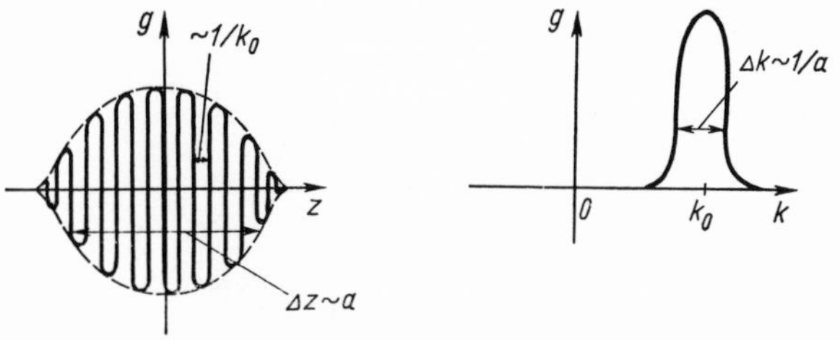

Fig. 4.2. Form of the functions g(z) and g(k) in the case of a wave packet.

The case $k_0 \gg \Delta k$ corresponds to a function g ($\mathbf{r}$) that varies over a distance that is small compared with $\Delta \mathbf{r}$. This class of functions includes, in particular, the function

$$g(\mathbf{r}) \propto \exp[-(\mathbf{r}-\mathbf{r}_0)^2/2a^2 + i\mathbf{k}_0(\mathbf{r}-\mathbf{r}_0)], \tag{4.10}$$

where $k_0 a \gg 1$. In this case, the function g_k has the form

$$g_k \propto \exp[-(\mathbf{k}-\mathbf{k}_0)^2 a^2/2 - i\mathbf{k}\mathbf{r}_0]. \tag{4.11}$$

For $r_0 = 0$, the dependences (4.8)-(4.11) for the case of one-dimensional functions are given in Fig. 4.1 (for a smooth, spatially localized perturbation) and in Fig. 4.2 (for spatially localized perturbations with a large number of oscillations).

In § § 4.2 and 4.3 we shall investigate the behavior of perturbations of the type (4.8) and in § 4.4 perturbations of the type (4.10). In § 4.5 we shall consider perturbations excited by spatially localized external sources of the type (4.8). Throughout this chapter we shall assume that the perturbations are one-dimensional, $\mathbf{k} = (0, 0, k)$, g ($\mathbf{r}$) = g (z).

§ 4.2. Perturbations with Many Wave Numbers in an Unstable Plasma

We shall assume that the plasma is unstable. Suppose the growth rate is maximal at $k = k^*$, $\gamma(k^*) = \gamma_{max}$. If the initial perturbation h (k) is a fairly smooth function of k that is appreciably different from zero at $k = k^*$, the largest amplitude after a long time, $t > 1/\gamma^*$, will be associated with the Fourier harmonics with $k \approx k^*$. In the neighborhood of $k = k^*$

$$\omega(k) \approx \omega^* + i\gamma_{max} + V^*(k-k^*) - \frac{\beta + i\alpha}{2}(k-k^*)^2. \tag{4.12}$$

Here

$$\omega^* = \operatorname{Re}\omega(k^*), \qquad V^* = \left(\frac{\partial\omega}{\partial k}\right)_{k=k^*},$$

$$\alpha = -\left(\frac{\partial^2\gamma}{\partial k^2}\right)_{k=k^*}, \qquad \beta = -\left(\frac{\partial^2 \operatorname{Re}\omega}{\partial k^2}\right)_{k=k^*}.$$

Setting $h(k) = \bar{h}(k)\exp(-ikz_0)$, where $\bar{h}(k)$ is a smooth function [cf. (4.8)] and taking into account (4.12), we reduce (4.5) to the form

$$\psi(z, t) \underset{t\gamma_{\max}>1}{\simeq} \frac{\bar{h}(k^*)\, e^{i(k^*z'-\omega^* t)+\gamma_{\max} t}}{\left(\frac{\partial \varepsilon_0}{\partial \omega}\right)\Big|_{\substack{\omega=\omega(k^*)\\ k=k^*}}}\, e^{-\frac{(z'-V^*t)^2}{2t(\alpha-i\beta)}} \times$$

$$\times \int dk \exp\left\{-\frac{\alpha-i\beta}{2} t \left[k - k^* - \frac{i(z'-V^*t)}{t(\alpha-i\beta)}\right]^2\right\}, \qquad (4.13)$$

$$z' = z - z_0.$$

The interval of wave numbers whose integration gives an appreciable contribution to (4.13) is

$$\Delta k \simeq \min\left\{\left(\frac{1}{\alpha t}\right)^{1/2}, \left(\frac{1}{\beta t}\right)^{1/2}\right\}. \qquad (4.14)$$

Over this interval the function $\bar{h}(k)$ can be assumed to be constant if the localization scale of the initial perturbation is sufficiently small,

$$a^2 < t \max(\alpha, \beta). \qquad (4.15)$$

The integrand in (4.13) attains a maximum at a value of k that differs slightly from k^*. This difference can be neglected if $|z-V^*t|$ is sufficiently small:

$$|z - V^*t| \lesssim \sqrt{t}\, |\alpha - i\beta|^{1/2}. \qquad (4.16)$$

If this is not the case, the function $\omega(k)$ would have to be expanded in a series for $k \neq k^*$ in order to find the maximum of the amplitude of the potential.

However, one can also regard (4.13) as an approximate (model) equation and use it to describe the field not only for $|z'-V^*t|$ satisfying the condition (4.16) but also for $|z'-V^*t| \gtrsim \sqrt{t}\,|\alpha-\beta i|^{1/2}$. In such an approach, Eq. (4.13) yields

$$\psi(z, t) \simeq \frac{\bar{\psi}}{\sqrt{t\gamma^*}} \exp\left\{i(k^*z' - \omega^* t) + \gamma_{\max} t - \frac{(z'-V^*t)^2(\alpha+i\beta)}{2t(\alpha^2+\beta^2)}\right\}, \qquad (4.17)$$

$$t\gamma_{\max} > 1,$$

where

$$\psi = \left(2\pi \frac{\gamma^*}{\alpha - i\beta}\right)^{1/2} \left\{ \frac{\bar{n}(k, \omega(k))}{\left(\frac{\partial \varepsilon_0}{\partial \omega}\right)_{\omega_n(k)}} \right\}_{k=k^*}$$

In accordance with (4.17), the point $z = z_{opt}$ corresponding to the maximal amplutide of the potential is displaced in space with the group velocity corresponding to the maximal growth rate:

$$z_{opt}(t) = z_0 + V^* t. \tag{4.18}$$

This effect may be called c o n v e c t i o n of the perturbation. Near $z = z_{opt}$, the amplitude of the potential increases in time as $\exp(\gamma_{max} t)$. This time dependence of the amplitude holds provided (4.16) is satisfied, i.e., precisely when Eq. (4.18) is most exact.

In the framework of the model (4.17), $\Delta z \approx \sqrt{t}\,|\alpha - i\beta|^{1/2}$ characterizes the size of the localization region of the major part of the perturbed field. It can be seen that the effective size of the perturbation increases with the time (the perturbation s p r e a d s in space). The rate of spreading is determined by the second derivatives of the frequency and the growth rate at $k = k^*$.

It also follows from (4.17) that with the passage of time a group of Fourier components with wave numbers $k \approx k^*$ is separated out from the original broad k spectrum; the amplitudes of this group grow faster than for values of k that differ appreciably from k^*. Physically, this corresponds to the fact that a perturbation of the type shown in Fig. 4.1 takes on the form depicted in Fig. 4.2.

Thus, the behavior of the field described by Eq. (4.17) can be thought of as the result of competition between the effects of convection and spreading, this being accompanied by the growth in time. Convection leads to a displacement of the amplitude maximum of the potential from the fixed point z to other points of the space. In the case of a point-like perturbation and a nonvanishing group velocity V^*, this would result in the disappearance of the field at the point z. However, the increase in the size of the perturbation due to spreading means that the field is only partly carried away from this point. The remaining part grows in time as exp ($\gamma_{max}t$). This growth is compensated by the field that leaves this point if, in accordance with (4.17),

$$\gamma_{max} > \frac{(V^*)^2}{2}\left(\frac{|\gamma''|}{(\gamma'')^2 + (\mathrm{Re}\,\omega'')^2}\right)_{k=k^*} \tag{4.19}$$

This result can also be represented as smallness of the group velocity V^* compared with a certain critical V_{cr}, where

$$V_{cr}^2 = 2\gamma_{max}\left(\frac{\gamma''^2 + \mathrm{Re}\,\omega''^2}{|\gamma''|}\right)_{k=k^*}. \tag{4.20}$$

One can also use V_{cr} and γ_{max} to express Δz (t):

$$\Delta z(t) \simeq V_{cr}\left(\frac{t}{\gamma_{max}}\right)^{1/2}. \tag{4.21}$$

Thus, to obtain an approximate picture of the behavior of the originally localized perturbation in the unstable plasma it is sufficient to know only three quantities:

$$\gamma_{max},\ V^* \text{ and } V_{cr}$$

By way of an example let us consider the instability of two cold beams of equal density. In § 1.5.1 the corresponding dispersion equation was investigated for $V_2 = -V_1 \equiv V$. We found that

$$\left.\begin{aligned} &\gamma_{max} = \frac{\omega_p}{2}, \quad V^* = 0, \\ &V_{cr} = \sqrt{\frac{3}{2}}\,V; \quad \beta = 0; \quad |\alpha| = \frac{3}{2}\cdot\frac{V^2}{\omega_p}. \end{aligned}\right\} \tag{4.22}$$

In this case there is no convection, so that the field increases in the region of the original localization as exp ($\gamma_{max}t$).

For arbitrary V_1 and V_2, the relations (4.22) are replaced by

$$\left.\begin{aligned} &\gamma_{max} = \frac{\omega_p}{2}, \quad V^* = \frac{V_1 + V_2}{2}, \\ &V_{cr} = \sqrt{\frac{3}{2}}\,\frac{|V_1 - V_2|}{2}, \quad \beta = 0, \quad \alpha = \frac{3}{8}\cdot\frac{(V_1 - V_2)^2}{\omega_p}. \end{aligned}\right\} \tag{4.22'}$$

It follows that the field in the originally specified region increases if

$$|V_1 + V_2| < \frac{\sqrt{3}}{2}\,|V_1 - V_2|. \tag{4.23}$$

Like the original equation (4.17), this condition is only approximate, i.e., true in order of magnitude.

§ 4.3. Absolute and Convective Instability

As in § 4.2, we shall investigate the field of a perturbation that is originally localized in the neighborhood of the point z_0 and contains a large number of Fourier harmonics. In contrast to § 4.2, in which the main attention was devoted to the field near its maximum point, we shall now investigate the behavior of the field for fixed z and, in particular, near the point at which it is originally localized.

1. Conditions of Absolute and Convective Instability. For simplicity, we shall assume that only one particular branch of oscillations $\omega_\alpha = \omega_\alpha(k)$ corresponds to instability. Denoting by k_1 and k_2 the limits of the interval of wave numbers for which $\gamma(k) > 0$ and letting t tend to infinity, we can replace the infinite limits of integration in the integral (4.5) by finite limits – from k_1 to k_2. At these limits, $\omega(k)$ is real and equal to $\omega_1 = \omega(k_1)$ and $\omega_2 = \omega(k_2)$, respectively. If the integration over k is replaced by integration over $\omega = \omega(k)$, the asymptotic value of (4.5) can be represented in the form

$$\psi(z, t) \underset{t\to\infty}{\propto} \int\limits_C d\omega \frac{dk(\omega)}{d\omega} \frac{\exp[ik(\omega)z - i\omega t]}{\left[\frac{\partial\varepsilon(\omega, k)}{\partial\omega}\right]_{k=k(\omega)}}. \tag{4.24}$$

The path of integration C is shown in Fig. 4.3. It can be plotted given the known functions $\mathrm{Re}\,\omega = \mathrm{Re}\,\omega(k)$ and $\gamma = \gamma(k)$. (In the case of a cold low-density beam that passes through a cold plasma, these functions are shown in Fig. 1.1.) Obviously, the function $k = k(\omega)$ is real at all points of C.

We shall calculate the integral (4.24) by displacing the contour C into the lower half-plane of complex ω. If $k(\omega)$ is analytic

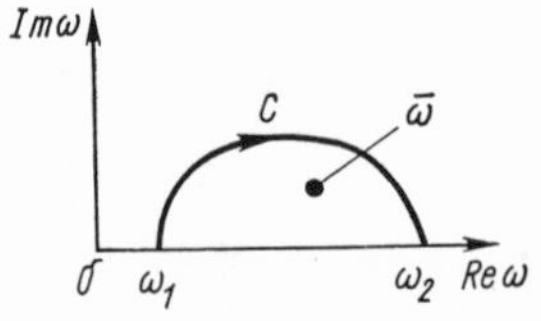

Fig. 4.3. Path of integration in Eq. (4.24).

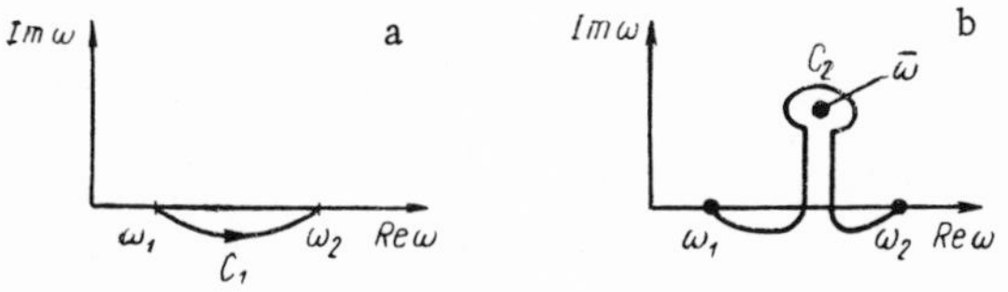

Fig. 4.4. Integration paths for convective (a) and absolute (b) instabilities.

in the region between the curve C and the abscissa (see Fig. 4.3), integration along the contour C reduces to integration along the contour C_1 shown in Fig. 4.4a. On C_1 the imaginary part of ω is negative and the asymptotic value of (4.24) for all finite z (including for $z = z_0$) therefore vanishes in this case and we have convective instability.

In the region between C and the abscissa (see Fig. 4.3) the function $\omega = \omega(k)$ may have saddle points. These values, $\omega \equiv \bar{\omega}$, correspond to branch points of the reciprocal function $k = k(\omega)$. In this case it is impossible to deform the contour C into a contour C_1 that lies entirely in the lower half-plane of ω. However, one can transform C into the contour C_2 shown in Fig. 4.4b in such a way that C_2 does not intersect the cut of the function $k = k(\omega)$. The integration along the sections of the contour C_2 that surround the cut lead to an asymptotic expression for (4.24) of the type

$$\psi(z, t) \underset{t\to\infty}{\propto} \frac{1}{\sqrt{t}} e^{-i\bar{\omega}t} (\mathrm{Im}\,\bar{\omega} > 0), \qquad (4.25)$$

and the limit of this expression is infinite (corresponding to absolute instability).

The result (4.25) can be obtained as follows. We note that in limit $t \to \infty$ only the integration around the small section of C_2 near $\omega = \bar{\omega}_0$ is important. For $\omega = \bar{\omega}$ and $k \approx k(\bar{\omega}) \equiv \bar{k}$ the equation $\varepsilon_0(\mathbf{k}, \omega) = 0$ yields

$$\varepsilon_0(\mathbf{k}, \omega) \approx \left(\frac{\partial \varepsilon_0}{\partial \omega}\right)_{\substack{\omega=\bar{\omega}\\ k=\bar{k}}} (\omega - \bar{\omega}) + \frac{1}{2}\left(\frac{\partial^2 \varepsilon_0}{\partial k^2}\right)_{\substack{\omega=\bar{\omega}\\ k=\bar{k}}} (k - \bar{k})^2 = 0. \qquad (4.26)$$

Hence

$$\left(\frac{dk}{d\omega}\right)_{\omega=\bar{\omega}} \propto \frac{1}{\sqrt{\omega - \bar{\omega}}}. \qquad (4.27)$$

Substituting (4.27) into Eq. (4.24), in which the contour C is replaced by C_2, and integrating over ω, we arrive at (4.25).

Our investigation yields the following condition for distinguishing absolute and convective instability (the Fainberg–Kurilko–Shapiro condition). The instability is absolute if the function $\omega_\alpha = \omega_\alpha(k)$ has a saddle point $(d\omega/dk = 0)$ in the upper half-plane between the contour $\omega_\alpha(k)$ corresponding to real k and the real ω axis (see Fig. 4.3). If this is not the case, the instability is convective.

In obtaining this result we have assumed that there is only one branch of unstable oscillations, $\omega_\alpha = \omega_\alpha(k)$. The more complicated case of several branches of oscillations with $\gamma > 0$ has been considered by Briggs.

Let us apply the above method to establish the nature of the instability in a system consisting of a cold beam and a cold plasma at rest [the dispersion equation (1.46)]. The saddle points can be found by differentiating (1.46) with respect to k with $d\omega/dk = 0$. As a result we find that in this case the function $\omega(k)$ has no saddle points at all, i.e., the corresponding instability is convective.

2. Brigg's Graphical Method for Determining Absolute and Convective Instabilities. If there is a complicated dispersion equation, the graphical method proposed by Briggs can be used to find the saddle points in the complex k region and to establish whether these points belong to the investigated branch of oscillations. We take several points on the contour C in Fig. 4.3 and describe through them straight lines with $\mathrm{Re}\,\omega = \text{const}$ to their intersection with the real ω axis (as is shown in Fig. 4.5).

We use the dispersion equation to find the function $k_\beta = k_\beta(\omega)$ for values of ω that lie on these lines; the subscript β means that the solution which is real for ω on the contour C is to be taken.

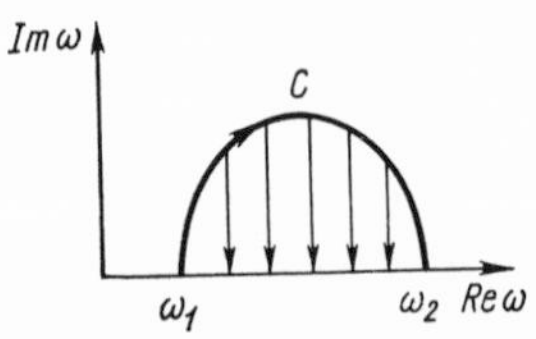

Fig. 4.5. Method for displacing the contour C on the transition to real ω.

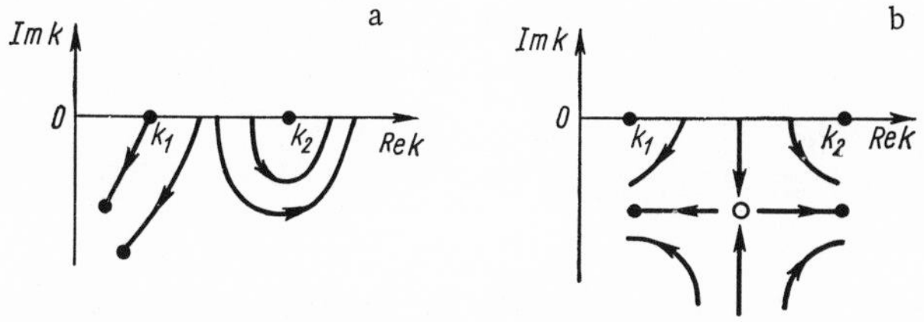

Fig. 4.6. The curves $k = k(\omega)$ corresponding to values of ω on the straight lines in Fig. 4.5 for convective (a) and absolute (b) instability.

As we move along the lines $\mathrm{Re}\,\omega = \text{const}$ (shown in Fig. 4.5 by the arrows) the corresponding curves $k_\beta = k_\beta(\omega)$ either have the form in Fig. 4.6a [when $\omega = \omega(k)$ has no saddle points] or the form in Fig. 4.6b (if a saddle point is present). Figure 4.6a corresponds to convective instability; Fig. 4.6b, to absolute instability.

§ 4.4. Wave Packets

1. Field of a Wave Packet. Suppose that the initial perturbation is such that the function $h(k)$ is nonvanishing only in a small neighborhood of $k = k_0$ (see Fig. 4.2). We shall say that such a perturbation is a wave packet. Let us consider the behavior in time of a packet of the Gaussian type (4.10). Substituting (4.10) into (4.5) and assuming that the most sharply peaked function is a Gaussian exponential function, we find the following expression for $\psi(z, t)$:

$$\psi(z, t)_0 \exp\{i(k_0 z' - \omega_0 t)\} \exp[-(z' - V_0 t)^2/2a^2]. \tag{4.28}$$

Here

$$\omega_0 = \omega(k_0), \quad V_0 = \left(\frac{\partial\, \mathrm{Re}\,\omega_0(k)}{\partial k}\right)_{k=k_0}$$
$$z' = z - z_0.$$

The expression (4.28) is derived under the assumptions

$$a^2 > t \max\left(\frac{\partial^2 \gamma}{\partial k^2}, \frac{\partial^2\, \mathrm{Re}\,\omega}{\partial k^2}\right)_{k=k_0}, \tag{4.29}$$

$$\frac{t}{a}\left(\frac{\partial \gamma}{\partial k}\right)_{k=k_0} < 1. \tag{4.30}$$

The condition (4.29) is the opposite of (4.15). It corresponds to the neglect of the effect of spreading of the packet. If the condition (4.30) is satisfied, the field grows at all points of the packet with the same growth rate. The term with V_0 in Eq. (4.28) should be retained only if $V_0 \gg (\partial\gamma/\partial k)_{k=k_0}$.

It follows from (4.28) that the maximum of the wave packet amplitude is displaced in space with a group velocity corresponding to the mean wave number of the packet:

$$z_{\text{opt}}(t) = z_0 + V_0 t. \tag{4.31}$$

The maximum increases with the growth rate $\gamma_0 = \gamma(k_0)$:

$$|\psi|_{z'=V_0 t} \propto e^{\gamma_0 t}. \tag{4.32}$$

2. Energy Balance of a Wave Packet in a Plasma with Small Dissipation. If the dissipative part of the permittivity is small and the perturbation has the form of a plane wave, one can introduce the concept of an energy of the oscillations [see Eq. (2.19)] and characterize the process of growth or decay of the oscillations by means of the energy balance equation (2.18). We shall show that similar considerations can be applied to perturbations in the form of a wave packet.

As in § 2.2, we shall assume that the perturbations are electrostatic. For such perturbations Maxwell's equations (II) yield

$$\frac{\partial}{\partial t}\cdot\frac{E^2}{8\pi} + \mathbf{j}\mathbf{E} = 0. \tag{4.33}$$

The electric field of the wave packet can be represented in the form

$$\mathbf{E} = e^{i\mathbf{k}_0\mathbf{r} - i\omega_0 t}\,\mathbf{E}_0(\mathbf{r}, t), \tag{4.34}$$

where the part $\mathbf{E}_0(\mathbf{r}, t)$, which depends weakly on the coordinates, is given by

$$\mathbf{E}_0(\mathbf{r}, t) = \int \mathbf{E}(\mathbf{k})\, e^{i(\mathbf{k}-\mathbf{k}_0)\mathbf{r} - i(\omega_k - \omega_0)t}\, d\mathbf{k}. \tag{4.35}$$

The current $\mathbf{j}$ in (4.33) is given by

$$\mathbf{j}(\mathbf{r}, t) = \int d\mathbf{k}\,\mathbf{E}(\mathbf{k})\,\sigma(\mathbf{k}, \omega_k)\, e^{i\mathbf{k}\mathbf{r} - i\omega_k t}. \tag{4.36}$$

In the case of wave-packet perturbations the current **j** can be expressed in terms of $\mathbf{E}_0(\mathbf{r}, t)$ and its derivatives with respect to the coordinates and the time. To this end we expand $\sigma(\mathbf{k}, \omega_k)$ in a series near the point $\mathbf{k} = \mathbf{k}_0$:

$$\sigma(\mathbf{k}, \omega_k) = \sigma(\mathbf{k}_0, \omega_0) + (\mathbf{k} - \mathbf{k}_0)\frac{\partial\sigma}{\partial\mathbf{k}_0} + (\omega_k - \omega_0)\frac{\partial\sigma}{\partial\omega_0} + \cdots. \tag{4.37}$$

This series can be terminated, since $\mathbf{E}(\mathbf{k})$ is nonvanishing only in a small neighborhood of $\mathbf{k}_0$.

Using the relations

$$\left.\begin{aligned} \frac{\partial\mathbf{E}_0}{\partial r_\alpha} &= i\int(\mathbf{k}-\mathbf{k}_0)_\alpha e^{i(\mathbf{k}-\mathbf{k}_0)\mathbf{r}-i(\omega_k-\omega_0)t}\,\mathbf{E}(\mathbf{k})\,d\mathbf{k}, \\ \frac{\partial\mathbf{E}_0}{\partial t} &= -\int(\omega_k-\omega_0)\,e^{i(\mathbf{k}-\mathbf{k}_0)\mathbf{r}-i(\omega_k-\omega_0)t}\,\mathbf{E}(\mathbf{k})\,d\mathbf{k}, \end{aligned}\right\} \tag{4.38}$$

we obtain the following equation from (4.36) and (4.37):

$$\mathbf{j}(\mathbf{r}, t) = \sigma(\mathbf{k}_0, \omega_0)\,\mathbf{E}(\mathbf{r}, t) - i\frac{\partial\sigma}{\partial k}e^{i\mathbf{k}_0\mathbf{r}-i\omega_0 t}\frac{\partial\mathbf{E}_0}{\partial r_\alpha} + i\frac{\partial\sigma}{\partial\omega_0}e^{i\mathbf{k}_0\mathbf{r}-i\omega_0 t}\frac{\partial\mathbf{E}_0}{\partial t}. \tag{4.39}$$

If the dissipation is small, the function $\sigma(\omega_0)$ can be expanded in a finite series in γ_0:

$$\sigma(\mathbf{k}_0, \omega_0) = \sigma(\mathbf{k}_0, \operatorname{Re}\omega_0) + i\gamma_0\frac{\partial\sigma(\mathbf{k}_0, \operatorname{Re}\omega_0)}{\partial\operatorname{Re}\omega_0}. \tag{4.40}$$

Taking into account (4.39) and (4.40), we can reduce Eq. (4.33), averaged over a period of the oscillations, to the form

$$\frac{\partial W}{\partial t} + \operatorname{div}\mathbf{S} = 2\gamma_0 W. \tag{4.41}$$

Here

$$\left.\begin{aligned} W &= \operatorname{Re}\omega_0\,\frac{\partial\operatorname{Re}\varepsilon_0(\operatorname{Re}\omega_0, k_0)}{\partial\operatorname{Re}\omega_0}\cdot\frac{|\mathbf{E}_0(\mathbf{r}, t)|^2}{8\pi}, \\ \mathbf{S} &= -\omega_k\frac{\partial\varepsilon_0}{\partial\mathbf{k}}\cdot\frac{|E_0|^2}{8\pi} \equiv \mathbf{V}_0 W, \\ \mathbf{V}_0 &= \left.\frac{\partial\omega_k}{\partial\mathbf{k}}\right|_{\mathbf{k}=\mathbf{k}_0}. \end{aligned}\right\} \tag{4.42}$$

Equation (4.41) is the energy balance of the wave packet. In it W and S are the energy and energy flux of the oscillations at the point **r** at the time t.

§ 4.5. Excitation of Oscillations of an Unstable Plasma by External Sources and the Amplification of Waves

1. Statement of the Problem. In all the foregoing (in Chapters 1-3 and in § § 4.1-4.4) we have assumed that all the charges of the plasma (the electrons and ions) move in a self-consistent manner, as follows from the Boltzmann equation. We now allow for the fact that there may be other charges in the plasma whose motion is not determined by a self-consistent electric field but by some external factors. We shall call charges of this kind *external sources*.

As an example of external sources we may take conductors in or on the surface of the plasma whose charge is regulated artificially. It should also be remembered that the Boltzmann equation describes only the average behavior of the charges and does not take into account thermal fluctuations. These fluctuating charges can also act as external sources for the self-consistent field.

If external sources are present, Poisson's equation has the form

$$-\Delta\psi(z, t) = 4\pi\rho(z, t) + 4\pi\rho_{ex}(z, t). \tag{4.43}$$

Here, ρ is the density of the self-consistent charges and ρ_{ex} is the density of the external charges (the density of the sources).

We shall calculate the field obtained as a result of excitation of plasma oscillations by these external charges.

2. Excitation of Oscillations by an Instantaneous Source. Let us first consider the excitation of oscillations by an external source that acts for only a very short interval of time, i.e., such that

$$\rho_{ex}(z, t) = q(z, t)\,\delta(t - t_0)$$
$$t_0 > 0. \tag{4.44}$$

We shall assume that for $t < t_0$ the plasma is in an unperturbed state and that its distribution function at $t = t_0$ is an equilibrium function, $g(\mathbf{r}, \mathbf{v}) = 0$. In this case Eq. (2.3) shows that the perturbed distribution function is related to the Fourier–Laplace component of the perturbed potential ψ by the equation

$$f_{kp}(v) = \frac{1}{p + ikv} \cdot \frac{ike\psi_{kp}}{m} \cdot \frac{\partial f_0}{\partial v}. \tag{4.45}$$

Using (4.45) to calculate the self-consistent charge and substituting the result into (4.43), we obtain the analog of (2.4):

$$\psi_{kp} = \frac{4\pi q_k(t_0)\, e^{-pt_0}}{k^2 \varepsilon_0(ip, k)}. \tag{4.46}$$

This shows that the excitation of oscillations by an instantaneous external source leads to the same field behavior as in the problem with initial conditions; the behavior of the field after the instant of excitation is entirely determined by the self-consistent motion of the charges. The oscillations that are excited are those whose frequency satisfy the dispersion equation $\varepsilon_0(\omega, k) = 0$ (characteristic oscillations). The space-time dependence of the potential in this case is determined by an expression that is similar to (2.8):

$$\psi(z, t, t_0) \propto 4\pi \int_{-\infty}^{\infty} dk\, e^{ikz} \int_{-\infty + i\sigma}^{\infty + i\sigma} \frac{q_k(t_0)\, e^{-i\omega(t - t_0)}}{k^2 \varepsilon_0(\omega, k)}\, d\omega. \tag{4.47}$$

3. Excitation of Oscillations by a Source That Depends in an Arbitrary Manner on the Time. To emphasize that the perturbation is due to a source that acts at the instant of time $t = t_0$, the potential ψ in Eq. (4.47) has t_0 included among its arguments. This notation is convenient for the generalization of the results to the case of noninstantaneous sources, to which we now turn.

Any noninstantaneous source $\rho_{ex}(z, t)$ can be represented as a set of instantaneous sources that act at each moment of time:

$$\rho_{ex}(z, t) = \int_0^t q(z, t_0)\, \delta(t - t_0)\, dt_0 \tag{4.48}$$

In this equation it is assumed that there are no sources for $t < 0$.

Each of the instantaneous sources leads to a "response," determined by Eq. (4.47). The resulting potential, which is equal to the sum of the responses of all the sources, is therefore obtained by integrating (4.47) over all $t_0 < t$:

$$\psi(z, t) \propto \int_0^t dt_0 \int_{-\infty}^{\infty} dk e^{ikz} \int_{-\infty+i\sigma}^{\infty+i\sigma} \frac{q_k(t_0) e^{-i\omega(t-t_0)}}{k^2 \varepsilon_0(\omega, k)} d\omega. \tag{4.49}$$

The integral over ω in (4.49) can be reduced to a sum over the branches of oscillations with frequencies $\omega_\alpha = \omega_\alpha(k)$ [cf. (4.5)]. In the case of an unstable plasma the main contribution to the asymptotic limit ($t \to \infty$) of this expression is due entirely to branches with $\operatorname{Im}\omega(k) > 0$. The infinite limits of integration with respect to k can be replaced in this case by finite limits corresponding to the range of complex $\omega(k)$ with $\operatorname{Im}\omega(k) > 0$ (cf. § 4.3) and the integral over the real k can be replaced by an integral along the complex ω lying on the contour C in Fig. 4.3. As a result, Eq. (4.49) is reduced to a form similar to that of (4.24):

$$\psi_{t\to\infty}(z, t) \propto \int_C d\omega \frac{dk(\omega)}{d\omega} \cdot \frac{e^{ik(\omega)z}}{k^2\left(\dfrac{\partial\varepsilon(\omega, k)}{\partial\omega}\right)_{k=k(\omega)}} \int_0^t dt_0 q_k(t_0) e^{-i\omega(t-t_0)}. \tag{4.50}$$

4. Amplification of Waves Excited by a Monochromatic Source. In the case of a monochromatic source of frequency Ω, the function $q_k(t_0)$ has the form

$$q(t_0) \propto e^{-i\Omega t_0}. \tag{4.51}$$

It then follows from (4.50) that

$$\psi_{t\to\infty}(z, t) \propto \int_C d\omega \frac{dk(\omega)}{d\omega} \cdot \frac{e^{ik(\omega)z - i\omega t}}{k^2(\omega)\left(\dfrac{\partial\varepsilon}{\partial\omega}\right)_{k=k(\omega)}(\omega - \Omega)}. \tag{4.52}$$

If the instability is convective and the frequency Ω lies in the interval between ω_1 and ω_2 (which are defined § 4.3), the integration around C reduces to the residue around the point $\omega = \Omega$. In this case the potential has the asymptotic form

$$\psi_{t\to\infty}(z, t) \propto e^{ik(\Omega)z - i\Omega t}. \tag{4.53}$$

The meaning of the function $k = k(\Omega)$ can be established by means of Figs. 4.5 and 4.6. This function satisfies the following conditions.

1. The function $k = k(\Omega)$ is the solution of the dispersion equation calculated for real Ω. This solution is, in general, complex, $\mathrm{Im}\, k \neq 0$.

2. $k(\Omega)$ is not any solution of the dispersion equation but only the one obtained from the real solution $k = k(\omega)$ for complex ω in the upper half-plane (on the contour C in Fig. 4.5) when the contour C is deformed onto the real axis of the ω plane.

If $\mathrm{Im}\, k(\Omega) < 0$ for certain Ω in the interval (ω_1, ω_2), it follows from (4.53) that the perturbed field increases exponentially in space (for $z > 0$). This effect is called wave amplification.

The exponential law of the spatial growth of the field [for $\mathrm{Im}\, k(\Omega) < 0$] is obtained for very large times t and finite z [this was assumed on the transition from (4.52) to (4.53)]. If it is assumed that both t and z are large, the law (4.53) is distorted.

Let us consider the amplification of waves in a cold beam–cold plasma system [the dispersion equation (1.46)]. Solving (1.46) for the wave number, we have (see also Fig. 4.6a):

$$k = \frac{\omega}{V} \pm \frac{\sqrt{\alpha}\,\frac{\omega_p}{V}}{\sqrt{1 - \frac{\omega_p^2}{\omega^2}}}. \tag{4.54}$$

It is shown in § 1.5 that the real part of the frequency of unstable oscillations lies in the interval

$$0 < \mathrm{Re}\,\omega < \omega_p \sqrt{1 + \alpha^{1/3}}. \tag{4.55}$$

In accordance with (4.54), this interval of frequencies is larger than the interval in which $\mathrm{Im}\, k \neq 0$. It follows that only waves with the frequencies

$$0 < \omega < \omega_p \tag{4.56}$$

can be amplified.

In order to establish whether the solutions (4.54) with $\mathrm{Im}\,k < 0$ really do correspond to amplification, it is necessary to consider the order in which the roots of $k = k(\omega)$ follow one another as ω moves from the contour C to the real axis. These roots behave in the manner shown in Fig. 4.6a (Briggs). It can be seen that all the complex k determined by Eq. (4.54) satisfy the conditions 1 and 2 noted above and therefore correspond to amplification.

As an example of waves whose dispersion equation has complex k that do not correspond to amplification we may mention electromagnetic waves [Eq. (1.5)]. In this case

$$k = \pm \frac{1}{c}\sqrt{\omega^2 - \omega_p^2}\,. \tag{4.57}$$

It can be seen that k is complex if $\omega < \omega_p$. These complex roots cannot become real roots for any values of the complex ω with $\mathrm{Im}\,\omega > 0$ and $|\mathrm{Re}\,\omega| < \omega_p$. They therefore do not satisfy the condition 2 formulated above. An additional analysis shows that these complex k correspond to surface waves (Briggs).

If the instability is due to small dissipative effects, so that the real and imaginary parts of the oscillation frequency are determined by Eqs. (2.16) and (2.17), it is readily verified that the real and imaginary parts of the wave number of the amplified waves satisfy the equations

$$\left.\begin{aligned} &\mathrm{Re}\,\varepsilon_0(\mathrm{Re}\,k, \omega) = 0,\\ &\mathrm{Im}\,k(\omega) = -\frac{\gamma(\mathrm{Re}\,k)}{V_{gr}(\mathrm{Re}\,k)}. \end{aligned}\right\} \tag{4.58}$$

To find the gain in this case there is no need to solve the dispersion equation anew if the functions $\gamma(k)$ and $\mathrm{Re}\,\omega(k)$ have already been found in the stability investigation.

Let us now mention some features of wave amplification in a cold plasma through which a low-density beam passes. If the beam is hot (kinetic instability, § 3.2) and if only the dissipative part of the beam contribution is taken into account in the dispersion equation, the value of $\mathrm{Re}\,\omega$ of unstable oscillations does not depend on the wave number ($\mathrm{Re}\,\omega = \omega_{p0}$). In this case $V_{gr} = 0$, and it follows from (4.58) that the gain of the waves of frequency $\omega = \omega_{p0}$ is infinite. To find the correct expression for $\mathrm{Im}\,k(\omega)$ it is necessary to include in the dispersion equation the small real terms due to the thermal motion of the plasma and beam. In this case, Eq. (3.3) shows that the group velocity is finite and for $k^2 d_1^2 \gg 1$ is

$$V_{gr} = \frac{3kT}{m\omega_{po}} + \frac{\omega_{po}}{k} \cdot \frac{1}{k^2 d_1^2}. \tag{4.59}$$

A gain that is formally infinite is obtained for the amplification of waves with $\omega = \omega_p$ in a cold beam–cold plasma system [Eq. (4.54)]. Allowance for the thermal motion of the plasma leads to the following refinement of this result:

$$\operatorname{Im} k(\omega_p) = -\left(\frac{\alpha}{2}\right)^{1/3} \frac{\sqrt{3}}{2} \cdot \frac{\omega_p}{V} \left(\frac{V}{v_{T0}}\right)^{2/3}. \tag{4.60}$$

Bibliography

1. A. V. Haeff, Phys. Rev., 74:1532 (1948).
2. J. R. Pierce, J. Appl. Phys., 19:231 (1948).
3. A. V. Haeff, Proc. IRE, 37:4 (1949).
4. J. R. Pierce, J. Appl. Phys., 20:1060 (1949). In [1-4] a study is made of the wave amplification due to the hydrodynamic two-stream instability, § 4.5. Equation (4.54) and other similar equations are obtained.
5. A. I. Akhiezer and Ya. B. Fainberg, Zh. Eksp. Teor. Fiz., 21:1262 (1951). The maximal gain of waves in a beam-plasma system [Eq. (4.60)] is calculated.
6. M. E. Gertsenshtein, Zh. Eksp. Teor. Fiz., 23:669 (1952). Kinetic amplification is considered. Equation (4.58) for Im K is obtained.
7. L. D. Landau and E. M. Lifshitz, Mechanics of Continuous Media [in Russian], Moscow, Fizmatgiz (1954), § 29. Definition of absolute and convective instability (§ 4.3). Translator's note: This does not seem to have appeared in the English translation. The subject is adequately covered in [10] and [17].
8. G. D. Boyd, L. M. Field, and R. W. Gould, Phys. Rev., 109:1393 (1958).
9. M. Sumi, J. Phys. Soc. Japan, 14:653 (1959). The amplification of waves is calculated numerically in [8, 9] with allowance for the thermal motion of a plasma.
10. P. A. Sturrock, Phys. Rev., 112:1488 (1958). A discussion of the difference between: a) absolute and convective instability (§ 4.3); b) amplification and damping (§ 4.5). In [10] it is emphasized that waves can be amplified only in an unstable plasma.
11. J. E. Drummond and D. B. Chang, Bull. Amer. Phys. Soc., 3:411 (1958). It is shown that the instability of a cold beam moving in a cold plasma is convective (§ 4.3).
12. Ya. B. Fainberg, V. I. Kurilko, and V. D. Shapiro, Zh. Tekh. Fiz., 31:633 (1961) [Soviet Phys. — Technical Physics, 6:459 (1961)]. Exposition of a regular method

for distinguishing absolute and convective instabilities (§ 4.3). A number of applications is considered, in particular, the problem of the nature of instabilities in plasmas.

13. O. Buneman, in: Plasma Physics, ed. J. E. Drummond, McGraw-Hill, New York (1961).
14. R. V. Polovin, Zh. Tekh. Fiz., 31:1220 (1961). The papers [13, 14] concern the questions discussed earlier by Sturrock in [10] (§§ 4.3, 4.5).
15. M. Feix, Nuovo Cimento, 27:1130 (1963). A study of the dynamics of spatially by localized perturbations containing many wave numbers (§ 4.2).
16. B. B. Kadomtsev, in: Reviews of Plasma Physics, Vol. 2, Consultants Bureau, New York (1966), p. 153. The behavior of wave packets (§ 4.4) and perturbations containing many wave numbers (§ 4.2) is discussed. The energy balance equation for a wave packet (§ 4.4) is given.
17. R. J. Briggs, Electron-Stream Interaction with Plasmas, MIT, Cambridge, Massachusetts (1964). This monograph contains a more detailed discussion of the majority of the questions touched on the present chapter and also a critical analysis of the foregoing investigations.
18. V. D. Shafranov, in: Reviews of Plasma Physics, Vol. 3, Consultants Bureau, New York (1967), p. 1. The energy balance of a wave packet (§ 4.4) is considered.

Chapter 5

Influence of Plasma Inhomogeneity on Instability

§ 5.1. Trajectory Integral Method

To study kinetic effects in an inhomogeneous plasma one must know the perturbed distribution function of such a plasma. One method by which it can be found is the trajectory integral method. It was first used by Shafranov and Drummond to investigate oscillations of a homogeneous plasma in a magnetic field and Rosenbluth and Rostoker to study an inhomogeneous plasma in a magnetic field (see [4–6] in the bibliography to Chapter 7). We begin this chapter with an exposition of this method, assuming as hitherto that the role of a magnetic field can be neglected, i.e., $\mathbf{B}_0 = 0$.

In the case of an inhomogeneous plasma, the equilibrium distribution function satisfies the equation

$$\mathbf{v}\nabla f_0 + \mathbf{F}_0 \frac{\partial f_0}{\partial \mathbf{v}} = 0. \tag{5.1}$$

Here, $\mathbf{F}_0$ is the equilibrium force (per unit mass) that acts on a particle. The force $\mathbf{F}_0$ is due to the inhomogeneity of the plasma. In the presence of $\mathbf{F}_0$ the Boltzmann equation for the perturbed distribution function becomes

$$\frac{\partial f}{\partial t} + \mathbf{v}\nabla f + \mathbf{F}_0 \frac{\partial f}{\partial \mathbf{v}} = -\frac{e}{m}\left(\mathbf{E} + \left[\frac{\mathbf{v}}{c}\mathbf{B}\right]\right)\frac{\partial f_0}{\partial \mathbf{v}}. \tag{5.2}$$

Here, f, **E**, and **B** are the perturbed quantities. Since the plasma is inhomogeneous, the function f cannot be assumed to depend on the coordinates as a plane wave (as was assumed in § 2.1). We must therefore have recourse to a different method of solving Eq. (5.2).

We express all the functions of the variables **r**, **v**, and t in (5.2) in terms of new variables $\mathbf{r}_0$, $\mathbf{v}_0$, and τ:

$$\left. \begin{aligned} t &= \tau, \\ \mathbf{r} &= \mathbf{r}(\mathbf{r}_0, \mathbf{v}_0, \tau) = \mathbf{r}_0 + \int_{t_0}^{\tau} \mathbf{v}(\mathbf{r}_0, \mathbf{v}_0, \tau')\, d\tau', \\ \mathbf{v} &= \mathbf{v}(\mathbf{r}_0, \mathbf{v}_0, \tau) = \mathbf{v}_0 + \int_{t_0}^{\tau} \mathbf{F}_0(\mathbf{r}_0, \mathbf{v}_0, \tau')\, d\tau'. \end{aligned} \right\} \tag{5.3}$$

Here, t_0 is an arbitrary instant of time and may, for example, be the time at which a perturbation is applied.

In the variables $\mathbf{r}_0$, $\mathbf{v}_0$, and τ, the functions **r** and **v** are obviously the values measured at the time τ of the coordinates and velocity of the particle that at $\tau = t_0$ was at the point $\mathbf{r}_0$ and had the velocity $\mathbf{v}_0$. With this choide of the new variables

$$\left(\frac{\partial f}{\partial \tau}\right)_{\mathbf{r}_0, \mathbf{v}_0} = \left(\frac{\partial f}{\partial t}\right)_{\mathbf{r}, \mathbf{v}} + \left(\frac{\partial f}{\partial \mathbf{r}}\right)_{t, \mathbf{v}} \left(\frac{\partial \mathbf{r}}{\partial \tau}\right)_{\mathbf{r}_0, \mathbf{v}_0} + \left(\frac{\partial f}{\partial \mathbf{v}}\right)_{t, \mathbf{r}} \left(\frac{\partial \mathbf{v}}{\partial \tau}\right)_{\mathbf{r}_0, \mathbf{v}_0}. \tag{5.4}$$

Since

$$\left(\frac{\partial \mathbf{r}}{\partial \tau}\right)_{\mathbf{r}_0, \mathbf{v}_0} = \mathbf{v}, \quad \left(\frac{\partial \mathbf{v}}{\partial \tau}\right)_{\mathbf{r}_0, \mathbf{v}_0} = \mathbf{F}_0, \tag{5.5}$$

the right-hand side of (5.4) is identical with the left-hand side of (5.2); in the new variables the Boltzmann equation therefore takes the form

$$\left(\frac{\partial f}{\partial \tau}\right)_{\mathbf{r}_0, \mathbf{v}_0} = -\frac{e}{m}\left(\mathbf{E} + \left[\frac{\mathbf{v}}{c}\mathbf{B}\right]\right)\frac{\partial f_0}{\partial \mathbf{v}}. \tag{5.6}$$

From this we find the desired distribution

$$f(\tau, \mathbf{r}_0, \mathbf{v}_0) = f(\tau_0, \mathbf{r}_0, \mathbf{v}_0) - \frac{e}{m}\int_{t_0}^{\tau}\left(\mathbf{E} + \left[\frac{\mathbf{v}}{c}\mathbf{B}\right]\right)\frac{\partial f_0}{\partial \mathbf{v}}\, d\tau'. \tag{5.7}$$

Here $f(\tau_0, \mathbf{r}_0, \mathbf{v}_0)$ is the initial distribution function.

To reduce Eq. (5.7) to a form similar to (2.3), we represent the perturbed electromagnetic field in the form of integrals of the Fourier–Laplace components. In the variables **r** and **t**, these integrals have the form

$$\{\mathbf{E}(\mathbf{r}, t), \mathbf{B}(\mathbf{r}, t)\} = \int d\mathbf{k}\, d\omega \exp\{i\mathbf{kr} - i\omega(t - t_0)\}\{\mathbf{E}_{k\omega}, \mathbf{B}_{k\omega}\}. \tag{5.8}$$

The integration is with respect to all real k and the complex ω with $\operatorname{Im}\omega = \sigma$, where σ lies above all the singularities of the functions $\mathbf{E}_{k\omega}$ and $\mathbf{B}_{k\omega}$ (this is discussed in more detail in § 2.1). Substituting (5.8) into (5.7) and remembering that the coordinates and the time must be expressed in terms of $\mathbf{r}_0$, $\mathbf{v}_0$, and τ, we obtain

$$f = f(t_0, \mathbf{r}_0, \mathbf{v}_0) - \frac{e}{m}\int d\mathbf{k}\, d\omega \exp\{i\mathbf{k}\mathbf{r}(\tau) - i\omega(\tau - t_0)\}\int_{t_0}^{\tau} \exp\left\{i\omega(\tau - \tau') - i\mathbf{k}\int_{\tau'}^{\tau} \mathbf{v}(\tau'')\, d\tau''\right\}\times$$
$$\times\left(\mathbf{E}_{k\omega} + \left[\frac{\mathbf{v}(\tau')}{c}\mathbf{B}_{k\omega}\right]\right)\frac{\partial f_0}{\partial \mathbf{v}(\tau')}\, d\tau'. \tag{5.9}$$

Let us consider how Eq. (2.3) can be deduced from Eq. (5.9). If the equilibrium force is absent, $\mathbf{F}_0 = 0$, Eq. (5.3) shows that $\mathbf{v}(\tau) = \mathbf{v}_0$ and $df_0/d\mathbf{v}$ depends only on $\mathbf{v}_0$. Integrating over τ'' and τ' in (5.9) and going over to the variables **r**, **v**, and **t**, we obtain in this case

$$f = f(t_0, \mathbf{r} - \mathbf{v}(t - t_0), \mathbf{v}) - \frac{ie}{m}\cdot\frac{\partial f_0}{\partial \mathbf{v}}\int d\mathbf{k}\, d\omega \frac{e^{i\mathbf{kr}}}{\omega - \mathbf{kv}}\left(\mathbf{E}_{k\omega} + \right.$$
$$\left. + \left[\frac{\mathbf{v}}{c}\mathbf{B}_{k\omega}\right]\right)[\exp\{-i\omega(t - t_0)\} - \exp\{-i\mathbf{kv}(t - t_0)\}]. \tag{5.10}$$

The contribution to the integral with respect to ω from the second term in the square brackets vanishes since the corresponding integration contour can be closed in this case in the upper half-plane, in which the integrand has no poles. Equation (5.10) can therefore be written in the form

$$f = f(t_0, \mathbf{r} - \mathbf{v}(t - t_0), \mathbf{v}) - \frac{ie}{m}\cdot\frac{\partial f_0}{\partial \mathbf{v}}\int d\mathbf{k}\, d\omega\left(\mathbf{E}_{k\omega} + \right.$$
$$\left. + \left[\frac{\mathbf{v}}{c}\mathbf{B}_{k\omega}\right]\right)\frac{e^{i\mathbf{kr} - i\omega(t - t_0)}}{\omega - \mathbf{kv}}. \tag{5.11}$$

Further, taking into account the relations

$$f(t_0, \mathbf{r} - \mathbf{v}(t - t_0), \mathbf{v}) = \int d\mathbf{k} f_k(t_0, \mathbf{v}) \exp\{i\mathbf{k}[\mathbf{r} - \mathbf{v}(t - t_0)]\} =$$
$$= i \int \frac{d\mathbf{k}\, d\omega}{\omega - \mathbf{kv}} \exp\{i\mathbf{kr} - i\omega(t - t_0)\} f_k(t_0, \mathbf{v}), \tag{5.12}$$

we see that in the case of an electrostatic field the Fourier–Laplace component of (5.11) is identical with (2.3).

Let us assume that the equilibrium force $\mathbf{F}_0$ is independent of the velocity and can be derived from a potential:

$$\mathbf{F}_0 = -\nabla\Phi(\mathbf{r}). \tag{5.13}$$

In this case the solution of (5.1) is a function of the form

$$f_0 = f_0(\varepsilon), \tag{5.14}$$

where $\varepsilon \equiv v^2/2 + \Phi_0$ can be interpreted as the energy of a particle of unit mass and is a constant of the motion.

If the components of $\mathbf{F}_0$ vanish along certain of the directions, the function f_0 may also depend on other integrals of motion. However, in the present treatment we shall restrict the treatment to functions of the form (5.14). For f_0 of this form, the derivative $\partial f_0/\partial \mathbf{v}$ in (5.9) becomes

$$\frac{\partial f_0}{\partial \mathbf{v}} = \mathbf{v}\,\frac{\partial f_0}{\partial \varepsilon}. \tag{5.15}$$

The derivative $df_0/d\varepsilon$ is a constant of the motion and may therefore be taken in front of the integral over t'. Since f_0 is symmetric with respect to the velocities, the term with the perturbed magnetic field disappears. As a result, Eq. (5.9) reduces to

$$f = f(t_0, \mathbf{r}_0, \mathbf{v}_0) - \frac{e}{m}\cdot\frac{\partial f_0}{\partial \varepsilon} \int d\mathbf{k}\, d\omega \exp\{i\,\mathbf{kr} -$$
$$- i\omega(t - t_0)\}\, E_{k\omega} \int_{t_0}^{t} \mathbf{v}(\tau')\exp\left[i\omega(t - \tau') - i\mathbf{k}\int_{\tau'}^{t} \mathbf{v}(\tau'')d\tau''\right]d\tau'. \tag{5.16}$$

§ 5.2. Permittivity of a Weakly Inhomogeneous Plasma

In an inhomogeneous plasma, the velocity of the particles is not constant, but changes under the influence of the force $\mathbf{F}_0$. We shall assume that the change in the particle velocity $\Delta \mathbf{v}$ in the time interval $\Delta t = t - t_0$ in which we are interested is small compared with $\mathbf{v}(t)$. Then

$$\mathbf{v}(\tau') \approx \mathbf{v}(t) - (t - \tau')\mathbf{F}_0(t). \quad (5.17)$$

We shall also assume that $\mathbf{k}\mathbf{F}_0 (t-t_0)^2 \ll 1$. If this condition is satisfied, the term with $\mathbf{F}_0$ in the exponential function of the integrand in (5.16) is small and the exponential function can therefore be expanded in a series in $\mathbf{F}_0$. Equation (5.16) then reduces to [see also the comments made in § 5.1 on the transition from (5.10) to (5.11)]

$$f = f(t_0, \mathbf{r}_0, \mathbf{v}_0) + \int \exp\{-i\omega(t - t_0)\} f_\omega(\mathrm{E})\, d\omega, \quad (5.18)$$

where

$$f_\omega(\mathrm{E}) = -\frac{ie}{m}\cdot\frac{\partial f_0}{\partial \varepsilon}\int d\mathbf{k}\, e^{i\mathbf{k}\mathbf{r}}\, \mathrm{E}_{k\omega}\left[\frac{\mathbf{v}}{\omega - \mathbf{k}\mathbf{v}} - i\mathbf{F}_0\frac{1}{(\omega - \mathbf{k}\mathbf{v})^2} - \frac{i(\mathbf{k}\mathbf{F}_0)\,\mathbf{v}}{(\omega - \mathbf{k}\mathbf{v})^3}\right]. \quad (5.19)$$

In the case of a plasma that is inhomogeneous in one dimension ($F_{0z} \neq 0$, $F_{0x} = F_{0y} = 0$) the force $\mathbf{F}_0$ can be eliminated by means of an equation that follows from (5.1):

$$F_{0z}\frac{\partial f_0}{\partial \varepsilon} = -\frac{\partial f_0}{\partial z}. \quad (5.20)$$

As a result

$$f_\omega(\mathrm{E}) = -\frac{ie}{m}\int d\mathbf{k}\, e^{i\mathbf{k}\mathbf{r}}\, \mathrm{E}_{k\omega}\left\{\mathbf{v}\frac{\frac{\partial f_0}{\partial \varepsilon}}{\omega - \mathbf{k}\mathbf{v}} + \frac{i\mathbf{e}_z\frac{\partial f_0}{\partial z}}{(\omega - \mathbf{k}\mathbf{v})^2} + \frac{i\mathbf{v}k_z\frac{\partial f_0}{\partial v_z}}{(\omega - \mathbf{k}\mathbf{v})^3}\right\} \quad (5.21)$$

($\mathbf{e}_z$ is a unit vector).

We shall use (5.21) to calculate the current and the charge. Let us first consider one-dimensional electrostatic perturbations with $\mathrm{E} \parallel z$, $k_\perp = 0$. The perturbed current associated with $f_\omega(\mathbf{E})$ is equal in this case to

$$j_\omega = \int e^{ikz} E_{k\omega}\left[\sigma^{(0)}(z, k, \omega) - \frac{i}{2}\cdot\frac{d^2\sigma^{(0)}}{dk\,dz}\right] dk. \quad (5.22)$$

Here, $\sigma^{(0)}(k, \omega, z)$ is the "local" conductivity:

$$\sigma^{(0)}(k, \omega, z) = \frac{i e^2 \omega}{m} \int \frac{f_0 d\mathbf{v}}{(\omega - k v_z)^2}. \tag{5.23}$$

The expression (5.22) can also be obtained by proceeding from the general relation between the current and field in an inhomogeneous medium,

$$j_\omega(z) = \int \sigma\left(\omega, z - z', \frac{z+z'}{2}\right) E_\omega(z')\, dz'. \tag{5.24}$$

The inhomogeneity of the plasma is taken into account by the dependence of σ on the argument $(z + z')/2$. Assuming that this dependence is weak, we represent σ as a series:

$$\sigma\left(z - z', \frac{z+z'}{2}\right) = \sigma(z - z', z) + \frac{1}{2}(z' - z) \frac{\partial \sigma(z - z', z)}{\partial z} + \cdots \tag{5.25}$$

where $\partial/\partial z$ denotes differentiation with respect to only the second argument. Substituting (5.25) into (5.24) and expanding $E(z')$ and $\sigma(z-z')$ in Fourier integrals, we arrive at Eq. (5.22), in which $\sigma^{(0)}(z, k, \omega)$ is the Fourier component of $\sigma(z-z')$.

Note that the relationship between j and E need not be expressed in the form (5.24); an alternative is

$$j_\omega(z) = \int \sigma(\omega, z - z', z) E_\omega(z')\, dz'. \tag{5.24'}$$

In this case the Fourier component of $\sigma(\omega, z-z', z)$ with respect to the argument $z-z'$ cannot, however, be interpreted as a local conductivity (5.23), differing from the latter by a term of the type $\partial^2\sigma^{(0)}/\partial k \partial z$.

The expression for the charge density ρ of one-dimensional electrostatic perturbations has the form

$$\rho_\omega(E) = -\frac{i}{4\pi} \int dk e^{ikz} E_k \left\{ k(\varepsilon_0^{(0)} - 1) - i\frac{\partial \varepsilon_0^{(0)}}{\partial z} - i\frac{k}{2} \cdot \frac{\partial^2 \varepsilon_0^{(0)}}{\partial z \partial k} \right\}, \tag{5.26}$$

where

$$\varepsilon_0^{(0)} = 1 - \frac{4\pi e^2}{m} \int \frac{f_0(\mathbf{v}, z)\, d\mathbf{v}}{(\omega - \mathbf{kv})^2}. \tag{5.27}$$

Now suppose $k_\perp \neq 0$. In general, the perturbations cannot then be electrostatic. We shall however assume $E \approx -\nabla\psi$. Using (5.21), we find that the charge density in this case has a form similar to (5.26):

$$\rho_\omega = -\frac{1}{4\pi}\int d\mathbf{k} e^{i\mathbf{kr}} \psi_{k\omega}\left\{k^2(\varepsilon_0^{(0)} - 1) - ik_z\frac{\partial\varepsilon_0^{(0)}}{\partial z} - i\frac{k_z k}{2}\cdot\frac{\partial^2\varepsilon_0^{(0)}}{\partial z\partial k}\right\}. \quad (5.28)$$

The function $\varepsilon_0^{(0)}$ is determined by the previous equation (5.27).

Substituting (5.28) into Poisson's equation, we obtain the equation for the Laplace transform of the potential $\psi_{k\omega}$:

$$\int d\mathbf{k} e^{i\mathbf{kr}} \psi_{k\omega}\left\{k^2\varepsilon_0^{(0)} - ik_z\frac{\partial\varepsilon_0^{(0)}}{\partial z} - i\frac{k_z k}{2}\cdot\frac{\partial^2\varepsilon_0^{(0)}}{\partial z\partial k}\right\} = 4\pi\rho_{ex}(\omega, r). \quad (5.29)$$

In this equation the density ρ_{ex} of the external charges also includes the Laplace transform of the initial perturbed density due tothe term $f(t_0, \mathbf{r}_0, \mathbf{v}_0)$ in Eq. (5.18).

§ 5.3. Limits of Applicability of the Approximation of a Homogeneous Plasma in Stability Problems

Suppose that at some instant of time a wave packet (see § 4.4) arises in the plasma having dimensions that are small compared with the inhomogeneity scale of the plasma. If the inhomogeneity of the plasma is ignored, the behavior of the packet is determined by its energy balance equation (4.41). It follows from this equation that the maximum of the packet amplitude is displaced in space and grows in time with the group velocity V_{gr} and growth rate γ that correspond to the mean wave number of the packet. This applies to a packet that grows slowly because of dissipative effects. A more complicated behavior is obtained for a "hydrodynamically unstable" packet, in which the spreading effect must also be taken into account.

Let us now consider how weak the inhomogeneity must be for the kind of behavior of a wave packet obtained for a homogeneous plasma to remain valid until the packet amplitude increases to values that are appreciably greater than the initial value.

Equation (5.27) shows that the "local" permittivity of an inhomogeneous plasma has the same form as in the case of a homogeneous plasma. The only difference is that the equilibrium distribution function f_0 now depends on both the velocities and coordinates. Inhomogeneity effects can therefore be ignored until the wave packet has moved through some appreciable distance in the direction of the plasma inhomogeneity. This is true only if $B_0 = 0$, since otherwise (when $B_0 \neq 0$) the local ε_0 may also contain terms with $\partial f_0/\partial \mathbf{r}$, which correspond to the gradient effects discussed in Vol. 2.

Let L be the characteristic distance over which the equilibrium parameters in the dispersion equation change appreciably. A packet with group velocity V_{gr} traverses this distance in a time $\tau \approx L/V_{gr}$. This time must be appreciably greater than the reciprocal of the growth rate, $\tau\gamma \gg 1$. This yields the desired condition under which the plasma inhomogeneity can be ignored:

$$\gamma \gg \frac{V_{gr}}{L}. \tag{5.30}$$

The value of L may be less than the characteristic inhomogeneity scale L_0 of the plasma if the instability that leads to the growth of the wave packet is a resonant instability, i.e., can develop only in a narrow range of variation of the equilibrium parameters. An example of an instability of this kind is the resonant hydrodynamic instability of a low-density beam considered in § 1.5.2.

It can arise only if $|\omega - \omega_p|/\omega_p < \alpha^{1/3}$. In this case, therefore, L in (5.30) is not the distance over which the density changes by an order of magnitude but only the distance over which it changes by a small fraction of the order of $\alpha^{1/3}$. The condition (5.30) for this instability therefore becomes

$$L_0 \gg \frac{V}{\omega_p} \alpha^{-2/3}. \tag{5.31}$$

In contrast, $L \approx L_0$ for the nonresonant hydrodynamic instability (see § 1.5.2). For the latter, the condition (5.30) is therefore – despite the smaller growth rate – more stringent than (5.31):

$$L_0 \gg \frac{V}{\omega_p} \alpha^{-1/2}. \tag{5.32}$$

In a plasma–hot beam system (see § 3.2), $\gamma \approx \alpha\omega_p$ and $V_{gr} \approx v_{T0}$, so that (5.30) in this case yields

$$L_0 \gg d_0\alpha^{-1}, \tag{5.33}$$

where $d_0 = (T_0/4\pi e^2 n_0)^{1/2}$ is the Debye radius for the "cold" plasma.

§ 5.4. Wave Packet in an Inhomogeneous Plasma. Quasiclassical Approximation

If the opposite condition to (5.30) holds, a perturbation that has not had sufficient time to grow to large amplitudes is displaced into a region of the plasma with very different equilibrium parameters. In this case the perturbation must be investigated by means of equations that take into account the plasma inhomogeneity.

In the electrostatic approximation, the perturbations of a weakly inhomogeneous plasma are described by Eq. (5.29). Let us use this equation to study the behavior of a wave packet in a cold plasma–hot beam system. We represent $\varepsilon_0^{(0)}$ in a form analogous to (2.13):

$$\varepsilon_0^{(0)} = \mathrm{Re}\,\varepsilon_0^{(0)} + i\,\mathrm{Im}\,\varepsilon_0^{(0)}, \tag{5.34}$$

where

$$\left.\begin{aligned} \mathrm{Re}\,\varepsilon_0^{(0)} &= 1 - \frac{\omega_p^2}{\omega^2}\left(1 + \frac{3k^2T}{m\omega^2}\right), \\ \mathrm{Im}\,\varepsilon_0^{(0)} &= -\frac{4\pi^2 e^2}{mk^2}\int \mathbf{k}\frac{\partial f_0}{\partial \mathbf{v}}\,\delta(\omega - \mathbf{kv})\,d\mathbf{v}. \end{aligned}\right\} \tag{5.35}$$

We replace quantities of the type $k^2\,\mathrm{Re}\,\varepsilon_0^{(0)}$ and $k_z\dfrac{\partial\,\mathrm{Re}\,\varepsilon_0^{(0)}}{\partial z}$ in the integrand in (5.29) by differential operators that act on $\psi(\mathbf{r})$. Then Eq. (5.29) reduces to

$$\widehat{L}\psi_\omega \equiv 3\frac{T}{m}\cdot\frac{\omega_p^2}{\omega^4}\Delta^2\psi_\omega + 6\frac{\partial}{\partial z}\left(\frac{T\omega_p^2}{m\omega^4}\right)\Delta\frac{\partial\psi_\omega}{\partial z} + \bar{\varepsilon}_0\Delta\psi_\omega + \frac{\partial\psi_\omega}{\partial z}\cdot\frac{\partial\bar{\varepsilon}_0}{\partial z} + 4\pi\delta\rho(\psi_\omega) = -4\pi\rho_{ex}(\omega, r). \tag{5.36}$$

Here, $\bar{\varepsilon}_0 = 1 - \omega_p^2/\omega^2$ and $\delta\rho(\psi_\omega)$ takes into account the resonant particles,

$$4\pi\delta\rho(\psi_\omega) = -i\int d\mathbf{k}\,\psi_{k\omega}e^{i\mathbf{kr}}k^2\,\mathrm{Im}\,\varepsilon_0^{(0)}(k, \omega, z). \tag{5.37}$$

We take the expression for the charge density of the "external source" in the form [cf. (4.44)]

$$\rho_{ex}(\omega,\ \mathbf{r}) = \exp\{i\,\omega t_0 + i\mathbf{k}_0(\mathbf{r} - \bar{\mathbf{r}})\}\, q_0(z,\ t_0). \tag{5.38}$$

It is assumed that the function q_0 (z) is nonvanishing over an interval Δz that is large compared with the reciprocal wave number $1/k_{0z}$ but small compared with the characteristic inhomogeneity scale of the plasma. We assume that the source acts only at some instant of time $t = t_0$ (i.e., instantaneously; see § 4.5). The spatial dependence of (5.38) corresponds to a wave packet (see § 4.4) localized in a small region of the plasma.

Our problem is to calculate by means of (5.36) the Laplace transform $\psi_\omega(\mathbf{r})$ and then find the function $\psi(\mathbf{r}, t)$. It is convenient to split this problem into several stages.

1. Solution of the Homogeneous Equation Corresponding to (5.36). Suppose the right-hand side of (5.36) vanishes, so that ψ_ω satisfies the equation

$$\hat{L}\psi_\omega = 0. \tag{5.39}$$

We shall seek a solution of this equation in the form

$$\psi_\omega(\mathbf{r}) = A(z)\exp\left\{ i\mathbf{k}_\perp \mathbf{r}_\perp + i\int^z K(z')dz' \right\}, \tag{5.40}$$

assuming that the characteristic scale of variation of the functions A (z) and K (z) is of the same order as the inhomogeneity scale L_0 of the plasma and is much larger than the characteristic value of the reciprocal "wave number" K (z):

$$\frac{1}{K(z)}\,\frac{d\ln A}{dz} \ll 1,\quad \frac{1}{K(z)}\cdot\frac{d\ln K}{dz} \ll 1,\qquad \frac{1}{K(z)\,L_0} \ll 1. \tag{5.41}$$

The assumption that K (z) and A (z) have a weak spatial dependence corresponds to the quasiclassical approximation in quantum mechanics and to the approximation of geometrical optics in the theory of propagation of light in an inhomogeneous medium. We shall therefore refer to this approach as the *quasiclassical approximation*.

In the zeroth approximation in the parameters (5.41), Eq. (5.39) yields the following equation for K (z):

$$3\frac{T}{m}\cdot\frac{\omega_p^2}{\omega^4}(k_\perp^2+K^2)^2-(k_\perp^2+K^2)\,\varepsilon_0^{(0)}-i\,(k_\perp^2+K^2)\,\operatorname{Im}\varepsilon_0^{(0)}(K,\ z,\ \omega)=0. \tag{5.42}$$

This equation can also be represented as the vanishing of the product of the local scalar of the permittivity (5.27), in which k_z is set equal to K(z), and k^2:

$$(k_\perp^2+K^2(z))\,\varepsilon_0^{(0)}(K(z),\ z,\ \omega)=0. \tag{5.43}$$

We assume that the imaginary terms of Eq. (5.42) that correspond to the resonant particles are small compared with the real terms. In this case the solutions of (5.42) have the form

$$K_{1,2}=K_{1,2}^{(0)}+iK_{1,2}^{(1)}, \tag{5.44}$$

$$K_{3,4}=\pm i|k_\perp|, \tag{5.45}$$

where

$$K_{1,2}^{(0)}=\pm\left[\frac{m\omega^2}{3T}\left(\frac{\omega^2}{\omega_p^2}-1\right)-k_\perp^2\right]^{1/2}, \tag{5.46}$$

$$K_{1,2}^{(1)}=\frac{m\omega^4}{6T\omega_p^2K_{1,2}^{(0)}}\operatorname{Im}\varepsilon_0^{(0)}(K_{1,2}^{(0)},\ z,\ \omega). \tag{5.47}$$

Let us consider the meaning of these solutions, assuming in what follows that $\operatorname{Im}\omega \ll \operatorname{Re}\omega$.

If $(\omega_p^2+3k^2T/m)^{1/2} > \operatorname{Re}\omega$, the roots K_1 and K_2 correspond to spatially oscillating perturbations ($\operatorname{Re}K \neq 0$) with one or other sign of the phase velocity. Equation (5.46) can be interpreted as the law of variation of the wave number of plasma oscillations (§ 1.3) propagating in an inhomogeneous plasma. It can be seen that the wave number K (z) increases with decreasing plasma density and vice versa (for simplicity we assume that the temperature T does not change). The dependence of the wave number on the density is shown schematically in Figs. 5.1 and 5.2.

Equation (5.47) determines the rate of spatial growth of the plasma oscillations (see the problem on the amplification of waves in § 4.3) due to the interaction between the oscillations and the

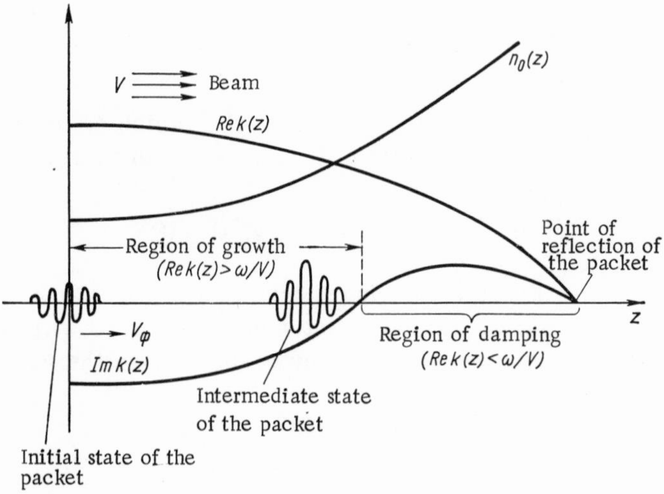

Fig. 5.1. Amplification of a wave packet moving into a region of higher plasma density.

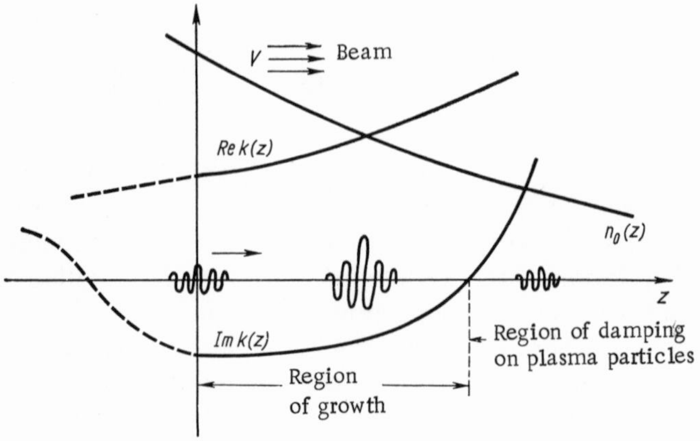

Fig. 5.2. Amplification of a wave packet moving into a region of lower plasma density.

resonant particles. Assuming that all the resonant particles are in the beam (and not the plasma), and the beam moves, say, in the direction of positive z, we find that $\mathrm{Im}\,\varepsilon_0^{(0)} \neq 0$ only for waves with $v_{ph}(z) \equiv \omega/K(z) > 0$. If the velocity distribution in the beam is Maxwellian [see (2.73)], then for such waves

$$(\mathrm{Im}\,\varepsilon_0^{(0)})_{\mathrm{beam}} = \frac{\sqrt{\pi}(\omega - K(z)V)}{(K^2(z)+k_\perp^2)\,d_1^2\,|k(z)|\,v_{T1}} \exp\left[-\left(\frac{\omega - K(z)V}{k(z)\,v_{T1}}\right)^2\right].$$

Here

$$k(z) = \sqrt{K^2(z)+k_\perp^2}. \tag{5.48}$$

It follows from (5.47) and (5.48) that waves with $v_{ph} > 0$ are spatially growing waves for z for which $K(z) > \omega/V$ (in the region of lower plasma density) and are spatially damped waves for $K(z) < \omega/V$ (where the plasma density is greater). This is also shown in Figs. 5.1 and 5.2.

The resonant particles of the "cold" plasma can only be neglected in the parts of space where $K(z) \ll \omega/v_{T0}$. If the opposite condition holds, the contribution of the beam of $\mathrm{Im}\,\varepsilon_0^{(0)}$ is less than the contribution of the particles of the "cold" plasma, which is

$$(\mathrm{Im}\,\varepsilon_0^{(0)})_{\mathrm{pl}} = \frac{\sqrt{\pi}\,\omega\exp\left(-\dfrac{\omega^2}{k^2(z)\,v_{T0}^2}\right)}{(K^2+k_\perp^2)\,d_0^2\,|k|\,v_{T0}}. \tag{5.49}$$

If the wave propagates from the more dense to the less dense part of the plasma (see Fig. 5.2), the spatial growth of the wave due to the beam is replaced by spatial damping due to the interaction with the resonant particles of the plasma. For a wave that propagates from the less dense to the more dense part of the plasma, the situation is reversed: with increasing z the wave is transformed from a damped into a growing wave (see Fig. 5.1). As for waves that propagate against the beam, they can only be spatially damped waves.

For z such that $(\omega_p^2 + 3k_\perp^2 T/m)^{1/2} > \mathrm{Re}\,\omega$, the solutions $K_{1,2}^{(0)}$ are purely imaginary. For real ω these solutions have $\mathrm{Im}\,\varepsilon_0 = 1$. As we shall show below, these solutions describe an effect of nontransmission of waves that propagate in a plasma with a lower density. The coordinate-independent solutions of the type (5.45) also correspond to nontransmitted (evanescent) perturbations.

In the next approximation in the parameters (5.41), Eq. (5.36) yields an expression for A (z):

$$A_\alpha(z)=\frac{C_\alpha}{\sqrt{K_\alpha \varepsilon_0}}, \qquad \alpha=1,\ 2,\ 3,\ 4, \tag{5.50}$$

where C_α are arbitrary constants. In calculating A_1 and A_2, we have neglected the small imaginary terms in K_1 and K_2, assuming $K_{1,2} \approx K_{1,2}^{(0)}$.

Equations (5.40), (5.44)-(5.47), and (5.50) determine the desired solution of the homogeneous equation (5.39). This solution is not valid for all z. It does not hold in regions in which the number of resonant particles is not too low ($\mathrm{Im}\,\varepsilon_0^{(0)} \approx \mathrm{Re}\,\varepsilon_0^{(0)}$) nor near the points $z = z^*$ and $z = z_0$, at which $K_{1,2}^0(z^*)$ and $\varepsilon_0(z_0) = 0$, respectively. If $\mathrm{Im}\,\varepsilon_0^{(0)} \approx \mathrm{Re}\,\varepsilon_0^{(0)}$, Eq. (5.39) itself becomes invalid; for in this case the derivatives of higher orders of ψ ignored in the derivation of the equation make a contribution that is comparable with that of the terms that are retained. For $z = z_0$ and $z = z^*$, Eq. (5.39) remains valid but in this case the limits of applicability of the quasiclassical approximation (5.41) are exceeded. This applies to the roots K_1 and K_2 for $z = z^*$ and all the roots K for $z \approx z_0$.

2. Solution of the Inhomogeneous Equation. We find the solution of (5.36) by the method of variation of constants. We set

$$\psi_\omega(z)=\sum_{\alpha=1}^{4} b_\alpha(z,\ \omega)\exp\left\{i\int^{z} K_\alpha(z,\ \omega)dz\right\}. \tag{5.51}$$

We find the functions $b_\alpha(z)$ by solving the following system of equations, which follows from (5.36) and (5.51):

$$\left.\begin{aligned}
&\sum_{\alpha=1}^{4} K_\alpha^n \frac{\partial b_\alpha(z)}{\partial z} e^{i\int^{z} K_\alpha(z')\,dz'}=0, \qquad n=0,\ 1,\ 2,\\
&\sum_{\alpha=1}^{4} K_\alpha^3 \frac{\partial b_\alpha(z)}{\partial z} e^{i\int^{z} K_\alpha(z')\,dz'}=-\frac{4\pi\rho_{ex}(\omega,\ z)}{\frac{3T}{m}\cdot\frac{\omega_p^2}{\omega^2}}.
\end{aligned}\right\} \tag{5.52}$$

The result is

$$\psi_\omega(z) = \sum_{\alpha=1}^{4} \int^{z} g_\alpha(z', \omega)\, \rho_{ex}(\omega, z') \exp\left\{- i \int_{z}^{z'} K_\alpha(z'')\, dz''\right\} dz', \quad (5.53)$$

where $g_\alpha(z, \omega)$ are certain smooth functions of z.

3. Space-Time Behavior of a Wave Packet. Using (5.53), we find the desired function $\psi(z, t)$:

$$\psi(z, t) = \frac{1}{2\pi} \int_{-\infty+i\sigma}^{+\infty+i\sigma} \psi_\omega(z)\, e^{-i\omega t}\, d\omega = \frac{1}{2\pi} \int_{-\infty+i\sigma}^{+\infty+i\sigma} d\omega \exp\{- i\omega(t - t_0)\} \sum_{\alpha=1}^{4} \int^{z} dz' \times$$

$$\times g_\alpha(z', \omega)\, q_0(z', t_0) \exp\left[ik_0(z' - \bar{z}) - i \int_{z}^{z'} K_\alpha(z'')\, dz''\right]. \quad (5.54)$$

As in (5.53), the lower limits of integration over z' in the terms of the sum over α are not yet determined. We are interested in the behavior of a wave packet excited by an instantaneous source $\rho_{ex}(z)$ that is nonvanishing only in a small interval Δz near $z = \bar{z}$. It was shown in § 4.4 that, if the effects of plasma inhomogeneity are neglected, the wave packet propagates with group velocity corresponding to its mean wave number. We shall assume that a similar situation obtains when the inhomogeneity effects are taken into account until the packet reaches the points z^* or z_0. In (5.54) we must then substitute limits of integration to take into account only waves "emitted" from the region in which the source is localized.

Since a wave with one and the same group velocity can correspond to both positive and negative values of K (z) (depending on the sign of Re ω), it is convenient to replace $K_{1,2}(z)$ by

$$\left.\begin{array}{l} K_+(z) = \begin{cases} K_1, & \mathrm{Re}\,\omega > 0, \\ K_2, & \mathrm{Re}\,\omega < 0; \end{cases} \\ K_-(z) = \begin{cases} K_2, & \mathrm{Re}\,\omega > 0, \\ K_1, & \mathrm{Re}\,\omega < 0. \end{cases} \end{array}\right\} \quad (5.55)$$

The group velocity of the wave K_+ is positive and that of the wave K_- is negative.

In the light of these remarks, the expression for ψ (z, t) becomes

$$\psi(z, t) = \frac{1}{2\pi} \int_{-\infty+i\sigma}^{+\infty+i\sigma} d\omega e^{-i\omega(t-t_0)} \left\{ \int_{-\infty}^{z} dz' g_+ (z', \omega) q_0 \times \right.$$

$$\times (z', t_0) \exp\left\{ i k_0 (z' - \bar{z}) - i \int_{z}^{z'} K_+ dz'' \right\} +$$

$$+ \int_{-\infty}^{z'} dz' g_- (z', \omega) q_0 (z', t_0) \exp\left\{ i k_0 (z' - \bar{z}) - \right.$$

$$\left. - i \int_{z}^{z} K_- dz'' \right\} + \int_{-\infty}^{z} dz' g_3 (z', \omega) q_0 (z', t_0) \exp\{ i k_0 (z' - \bar{z}) -$$

$$- (z - z') |k_\perp| \} + \int_{+\infty}^{z} dz' g_4 (z', \omega) q_0 (z', t_0) \exp\{ i k_0 (z' - z) +$$

$$\left. + (z - z') |k_\perp| \} \right\}. \tag{5.56}$$

Here, the functions $g_\pm$ are related to $g_{1,2}$ by equations similar to (5.55).

4. <u>The Example of a Gaussian Packet.</u> In analyzing the expression (5.56) we shall restrict ourselves to the case

$$q_0 (z) \propto \exp\left\{ -\frac{(z - \bar{z})^2}{2a_0^2} \right\}. \tag{5.57}$$

In accordance with the remark made after Eq. (5.38), $a_0 |K_\alpha| \gg 1$. We shall also assume that $|dK_\alpha/dz| a_1^2 \ll 1$. It then follows from (5.56) and (5.57) that for $a_0 \ll |z - \bar{z}| < L_0$ the potential has the form

$$\psi_\pm (z, t) \propto \exp\left\{ - i\omega_0 (t - t_0) + i \int_{\bar{z}}^{z} K_\pm (\omega_0, z') dz' - \right.$$

$$\left. - \frac{1}{2} \left[\frac{V_{gr}^{(\pm)} (\bar{z}, \omega_0)}{a_0} \left(t - t_0 - \int_{\bar{z}}^{z} \frac{dz'}{V_{gr}^{(\pm)} (z', \omega_0)} \right) \right]^2 \right\}. \tag{5.58}$$

Here, ψ_+ is the solution in the region $z - \bar{z} > 0$ and ψ_- is ψ for $z - \bar{z} < 0$. The value of ω_0 is determined by the equation $K_0 = K_1$

$(\bar{z}, \omega_0)$, i.e., ω_0 is the solution of the local dispersion equation (5.43) with $K(z) = k_0$ and $z = \bar{z}$:

$$\varepsilon_0^{(0)}(K_0, \bar{z}, \omega) = 0. \tag{5.59}$$

The local group velocity, $V_{gr}(z, \omega_0)$, is

$$V_{gr}^{(\pm)}(z, \omega_0) = \left(\frac{\partial K_\pm(\mathrm{Re}\,\omega_0, z)}{\partial\,\mathrm{Re}\,\omega_0}\right)^{-1}. \tag{5.60}$$

It follows from (5.58) that a wave packet of frequency $\omega = \mathrm{Re}\,\omega_0$ is emitted (in both the positive and negative directions) from the localization region of the source and that the position of the wave packet maximum is determined by the equation

$$t - t_0 - \int_{\bar{z}}^{z} \frac{dz'}{V_{gr}^{(\pm)}(z', \mathrm{Re}\,\omega_0)} = 0. \tag{5.61}$$

Differentiating (5.61) with respect to t, we find that the coordinate z (t) of the center of the packet changes in accordance with the law

$$\frac{dz(t)}{dt} = V_{gr}(z(t), \mathrm{Re}\,\omega_0). \tag{5.62}$$

This result is a generalization of the law of motion (4.31) of a wave packet in a homogeneous plasma. Equation (5.62) can also be obtained from the energy balance equation of a packet in an inhomogeneous plasma.

The real and imaginary parts of $K_\pm(\omega_0, z)$ are

$$\begin{aligned} \mathrm{Re}\,K_\pm(\omega_0, z) &= \mathrm{Re}\,K_\pm(\mathrm{Re}\,\omega_0, z), \\ \mathrm{Im}\,K_\pm(\omega_0, z) &= \mathrm{Im}\,K_\pm(\mathrm{Re}\,\omega_0, z) + \frac{\gamma_0}{V_{gr}^{(\pm)}(\omega_0, z)}, \end{aligned} \tag{5.63}$$

where

$$\gamma_0 = \mathrm{Im}\,\omega_0 = -\frac{\mathrm{Im}\,\varepsilon_0^{(0)}(K_0, \bar{z}, \mathrm{Re}\,\omega_0)}{\dfrac{\partial\,\mathrm{Re}\,\varepsilon_0^{(0)}(K_0, \bar{z}, \mathrm{Re}\,\omega_0)}{\partial\,\mathrm{Re}\,\omega_0}}. \tag{5.64}$$

Using (5.58), (5.61), and (5.63), we find the law of growth of the maximum of $|\psi|$:

$$|\psi| \sim \exp\left[-\int_{\bar{z}}^{z(t)} \operatorname{Im} K_{\pm}(z', \quad \operatorname{Re}\omega_0)\, dz'\right]. \tag{5.65}$$

This relation shows that $-\operatorname{Im} K(z)$ can be interpreted as a local gain (cf. § 4.5).

The relation (5.65) can also be represented in the form

$$|\psi| \propto \exp\left[\int_{t_0}^{t} \gamma\,[z(t'), \ \omega_0]\, dt'\right], \tag{5.66}$$

where z (t) is determined by Eq. (5.62) and $\gamma(z, \omega_0)$ is the local growth rate defined as follows:

$$\varepsilon_0^{(0)}\,[\operatorname{Re} K(z), \ z, \ \operatorname{Re}\omega_0 + i\gamma(\bar{z})] = 0. \tag{5.67}$$

The local growth rate is related to the local gain by the equation

$$\gamma(z) = -V_{gr}(z) \operatorname{Im} K(z). \tag{5.68}$$

Bibliography

1. L. D. Landau and E. M. Lifshitz, Quantum Mechanics, Pergamon Press, Oxford (1965), § 46.
2. L. D. Landau and E. M. Lifshitz, Electrodynamics of Continuous Media, Pergamon Press, Oxford (1960), § 65.
3. V. L. Ginzburg, The Propagation of Electromagnetic Waves in Plasmas, Pergamon Press, Oxford (1964), § 23.
4. T. H. Stix, The Theory of Plasma Waves, McGraw-Hill, New York (1962). The monographs [1-4] contain discussions of the propagation of waves in inhomogeneous media and the approximation of geometrical optics (the quasiclassical approximation) (§ 5.4).
5. B. B. Kadomtsev, In: Reviews of Plasma Physics, Vol. 2, Consultants Bureau, New York (1966), p. 153. The behavior of wave packets in an inhomogeneous plasma is discussed (§ 5.4). An expression is given for the permittivity of a weakly inhomogeneous plasma (§ 5.2).
6. P. A. Sturrock, Phys. Rev., 117:1426 (1960).
7. M. F. Gorbatenko, in: Plasma Physics and Problems of Controlled Thermonyclear Fusion, Vol. 1 [in Russian], Naukova Dumka, Kiev (1961). The excitation of plasma oscillations by a radially inhomogeneous beam is considered in [6, 7].

8. Ya. B. Fainberg, V. D. Shapiro, and V. I. Shevchenko, Plasma Physics and the Problems of Controlled Thermonuclear Fusion (Beam-Plasma Interactions) [in Russian], Naukova Dumka, Kiev (1967), p. 31.
9. Ya. B. Fainberg, Czech. J. Phys., 5B:652 (1968). The papers [8, 9] contain a discussion of the instabilities of a longitudinally inhomogeneous stream.

Chapter 6

Instabilities of a Collisional Plasma

§ 6.1. Limits of Applicability of the Collisionless Approximation in Stability Problems

1. Boltzmann Equation for a Collisional, Fully Ionized Plasma. Collisions between particles are taken into account by the right-hand side, C_α, of the Boltzmann equation (I). The particles of each species (electrons or ions) collide with each other, with the charges of the other species, and with neutral atoms. If collisions between charged particles are predominant (fully ionized plasma), then

$$C_\alpha = \sum_{\beta=e,i} C_{\alpha\beta}, \quad \alpha = e, i, \tag{6.1}$$

where the expression for $C_{\alpha\beta}$ has the form (Landau)

$$C_{\alpha\beta} = -\frac{2\pi\lambda e_\alpha^2 e_\beta^2}{m_\alpha} \cdot \frac{\partial}{\partial v_\alpha} \int \left\{ \frac{f_\alpha(\mathbf{v})}{m_\beta} \frac{\partial f_\beta(\mathbf{v}')}{\partial v'_l} - \frac{f_\beta(\mathbf{v}')}{m_\alpha} \cdot \frac{\partial f_\alpha(\mathbf{v})}{\partial v_l} \right\} U_{kl} d\mathbf{v}'. \tag{6.2}$$

Here

$$U_{kl} = \frac{1}{u^3}(u^2\delta_{kl} - u_k u_l), \quad u_k = v_k - v'_k, \tag{6.3}$$

λ is the "Coulomb logarithm" (λ is a weak function of the temperature and plasma density; for the purpose of estimates one can assume $\lambda \approx 10$).

For practical purposes, the electron velocities are always large compared with the ion velocities, since otherwise the ion energy would have to exceed the electron energy by a factor of at least m_i/m_e. If v_i/v_e is small, the crossed collision integrals C_{ei} and C_{ie} have a simple form. They are given in the problem.

2. <u>Characteristic Collision Times.</u> If there is no electromagnetic field (or other fields) in a plasma and its distribution is homogeneous in space, collisions lead to the setting up of a Maxwellian velocity distribution of the particles (Boltzmann's H-theorem). The most rapid process is that of the establishment of a Maxwellian electron distribution. If the plasma contains no very fast groups of electrons (with velocities that greatly exceed the mean values), then the Boltzmann equation (I) and Eqs. (6.1)-(6.3) show that the time of this process satisfies

$$\tau^{(e)}_{\text{Maxw}} \simeq \frac{T_e^{3/2} m_e^{1/2}}{\lambda e^4 n_0}. \tag{6.4}$$

It can be seen that this time is shorter, the denser and colder is the plasma.

The mean velocities of the ions and electrons are also equalized in a term of order $\tau^{(e)}_{\text{Maxw}}$.

The time needed to establish a Maxwellian distribution of the ions satisfies

$$\tau^{(i)}_{\text{Maxw}} \simeq \frac{T_i^{3/2} m_i^{1/2}}{\lambda e^4 n_0}. \tag{6.5}$$

If $T_i \approx T_e$, this time is greater than $\tau^{(e)}_{\text{Maxw}}$ by a factor of $(m_i/m_e)^{1/2}$. The slowest process is that of equalization of the ion and electron temperatures:

$$\tau_{ie} \simeq \frac{m_i}{m_e} \tau^{(e)}_{\text{Maxw}} \tag{6.6}$$

A group of fast particles with velocity $V \gg v_T$ loses its velocity and comes into equilibrium with the remaining particles in a time of order

$$\tau_{\text{fast}} \simeq \left(\frac{V}{v_T}\right)^{3/2} \tau_{\text{Maxw}} \tag{6.7}$$

3. Limits of Applicability of the Collisionless Approximation in Stability Problems. In stability theory one studies a plasma with a non-Maxwellian particle velocity distribution, since a plasma with a Maxwellian distribution is stable (see § 2.6). However, it follows from § 6.1.2 that a non-Maxwellian distribution is not stationary. It can only be assumed to be stationary if its collisional relaxation time τ is sufficiently long compared with the reciprocal of the growth rate γ of perturbations:

$$\tau\gamma \gg 1. \tag{6.8}$$

Even if the equilibrium velocity distribution of a component of the plasma is Maxwellian, the perturbed distribution does not have this property. Collisions will cause the perturbed distribution to become a Maxwellain distribution in a time that is bounded below by $\tau^{(\alpha)}_{\text{Maxw}}$ (α = i, e). (This time may also depend on ω and k.) This yields a further condition which, in addition to (6.8), restricts the applicability of the collisionless approximation. It can also be represented in the form (6.8), in which τ now characterizes the relaxation of the perturbed state.

Let us use (6.8) to estimate the limits of applicability of the results obtained in the foregoing chapters. We shall do this by taking τ in this inequality equal to the shortest of the characteristic times for the establishment of a Maxwellian distribution in each problem.

A. Hydrodynamic Instability of Two Streams of Equal Density (§ 1.5.1). In this case $\gamma \simeq \omega_{p_e}$, and $\tau \simeq \tau^{(e)}_{\text{Maxw}}$. The condition (6.8) takes the form

$$n_0 d_e^3 \gg 1. \tag{6.9}$$

The meaning of the parameter $n_0 d_e^3$ is clear: to within a numerical coefficient it is the number of particles in a sphere of radius d_e. The inequality (6.9) is simultaneously a necessary condition for the applicability of a selfconsistent description of the plasma; thus, in using the Boltzmann equation, it is necessary to assume that (6.9) is satisfied automatically. This requirement is also satisfied under almost all practical experimental conditions. We conclude that the instability of two streams of equal density is not very sensitive to collisional processes.

B. Low-Density Electron Stream. The condition (6.9) is now replaced by

$$\alpha^{1/3}\, n_0 d_e^3 \gg 1. \qquad (6.10)$$

This condition can be violated only in the physically uninteresting cases of extremely small α.

The condition (6.10) applies to a beam with a small thermal spread $v_{T1} < \alpha^{1/3}V$. If the beam has a larger spread, collisions are unimportant if

$$\alpha\left(\frac{V}{v_{T1}}\right)^2 n_0 d_e^3 \gg 1. \qquad (6.11)$$

C. Relative Motion of Electrons and Ions (§ § 1.6 and 3.4). For $V > v_{T_e}$ the growth rate is $\gamma \approx \mu^{1/3}\omega_{p_e}$, where $\mu \equiv m_e/m_i$. The shortest of the $\tau^{(\alpha)}_{\text{Maxw}}$ is $\tau^{(e)}_{\text{Maxw}}$. The condition for neglect of collisions therefore has a form similar to (6.10) with α replaced by μ.

In the case of the ion-acoustic instability ($v_{Ti} < V < v_{T_e}$, $T_e \gg T_i$) the growth rate is relatively small, $\gamma \simeq \mu^{1/2}\omega_{p_i}$ (see § 3.4). If an equilibrium electric field is not present in the plasma, the electrons are slowed down in a time of order $\tau^{(e)}_{\text{Maxw}}$. The condition of applicability of the collisionless approximation for the ion-acoustic instability is therefore

$$d_e^3 n_0 \mu \gg 1. \qquad (6.12)$$

If $kd_e \ll 1$, then $\gamma \approx \mu^{1/2}k\,(T_e/m_i)^{1/2}$ and the condition (6.12) is replaced by

$$kd_e^4 n_0 \mu \gg 1. \qquad (6.13)$$

4. Stationary State of a Fully Ionized Plasma in an Electric Field. A relative motion of the electrons and ions may be maintained by an electric field. The velocity of the stationary motion of the electrons can be found from the condition that the slowing down of the electrons as a result of interaction with the ions is compensated by the acceleration of the electrons in the electric field:

$$e_e n_0 \mathbf{E}_0 + \mathbf{R}^{(e)} = 0. \qquad (6.14)$$

The force of friction $\mathbf{R}^{(e)}$ can be expressed by Eq. (6.31) (see the problem in § 6.1 below). Assuming in this equation that $\mathbf{V}^{(i)} = 0$, $f_e = f_{Maxw}(\mathbf{v}-\mathbf{V}^{(e)})$ and assuming $V^{(e)} \ll v_{T_e}$, we find

$$\mathbf{R}^{(e)} = -\frac{m_e n_e}{\tau_e}\mathbf{V}^{(e)}, \tag{6.15}$$

where

$$\tau_e = \frac{3\sqrt{m_e}T_e^{3/2}}{4\sqrt{2\pi}\lambda e_e^2 e_i^2 n_0}. \tag{6.16}$$

It follows from (6.14) and (6.15) that the desired expression for $\mathbf{V}^{(e)}$ is

$$\mathbf{V}_0^{(e)} = \frac{e_e}{m_e}\tau_e \mathbf{E}_0. \tag{6.17}$$

The assumption $V^{(e)} \ll v_{T_e}$ is satisfied if

$$E_0 \ll \frac{m_e v_{T_e}}{e_e \tau_e} \simeq \frac{e\lambda}{d_e^2}. \tag{6.18}$$

In the opposite case it follows from Eq. (6.31) that

$$R^{(e)} = -\frac{e^2 n_e \lambda}{d_e^2} G\left(\frac{V_0}{v_{T_e}}\right), \tag{6.19}$$

where

$$G(x) = \frac{1}{x^2\sqrt{\pi}}\left(-\int_0^x e^{-y^2}\,dy + x\right). \tag{6.20}$$

The stationary velocity of the electrons can now be determined from the condition

$$E_0 - \frac{e_e\lambda}{d_e^2} G\left(\frac{V_0}{v_{T_e}}\right) = 0. \tag{6.21}$$

The function G (x) has a maximum at $x \approx 1$ approximately equal to 0.2. It follows that a stationary motion of the electrons is possible only if (Dreicer)

$$E < E_{cr} \equiv 0.2\,\frac{e\lambda}{d_e^2}. \tag{6.22}$$

For $\omega\tau_e > 1$, the stability of a stationary state of a plasma in an electric field is determined by the collisionless approximation (§ 3.4). We shall consider the opposite limiting case, $\omega\tau_e \ll 1$, in § 6.3, deriving the equations needed for this purpose in § 6.2.

5. Weakly Ionized Plasma. If the fraction of neutral atoms is sufficiently large, the collisions between charged particles and neutral particles are more important than the collisions between the charged particles. This is the case in a weakly ionized plasma.

Collisions with the neutral particles change the momentum of the charged particles. This effect, averaged over the distribution function, leads to the appearance of a frictional force that acts on the corresponding component of the plasma [cf. (6.15)]:

$$\mathbf{R}^{(\alpha)} = -m_\alpha n_\alpha \mathbf{V}^{(\alpha)} \nu_\alpha. \tag{6.23}$$

We shall call ν_α the collision frequency. It is proportional to the density of the neutrals and depends on the properties of the atoms of the neutral gas. Approximately,

$$\nu \simeq n_{\text{neutr}}\, \bar{v}\bar{\sigma}, \tag{6.24}$$

where $\bar{v}$ and $\bar{\sigma}$ are the characteristic values of the relative velocity of the charged and neutral particles and the effective scattering cross section.

Problem. Simplify the expressions (6.2) for C_{ei} and C_{ie} under the assumption that the ratio of the ion to the electron velocity is small (Braginskii).

Solution 1. Simplification of C_{ei}. Expand $U_{\alpha\beta}(\mathbf{v}-\mathbf{v}')$ in powers of the ion velocity $\mathbf{v}'$. Neglecting terms of higher than quadratic order in $\mathbf{v}'$, we obtain

$$U_{\alpha\beta} = V_{\alpha\beta} - v'_\gamma \frac{\partial V_{\alpha\beta}}{\partial v_\gamma} + \frac{v'_\gamma v'_\delta}{2} \frac{\partial^2 V_{\alpha\beta}}{\partial v_\gamma \partial v_\delta}, \tag{6.25}$$

where $V_{\alpha\beta} = (v^2\delta_{\alpha\beta} - v_\alpha v_\beta)/v^3$. With allowance for (6.25), Eq. (6.2) yields the desired result

$$C_{ei} = \frac{2\pi\lambda e_e^2 e_i^2 n_i}{m_e^2} \cdot \frac{\partial}{\partial v_\alpha} \left\{ V_{\alpha\beta} \frac{\partial f_e}{\partial v_\beta} - V_\gamma^{(i)} \frac{\partial V_{\alpha\beta}}{\partial v_\gamma} \cdot \frac{\partial f_e}{\partial v_\beta} - \mu f_e \frac{\partial V_{\alpha\beta}}{\partial v_\beta} + \mu f_e V_\gamma^{(i)} \frac{\partial^2 V_{\alpha\beta}}{\partial v_\gamma \partial v_\beta} + \right.$$

$$\left. + \frac{1}{2} \frac{\partial f_e}{\partial v_\beta} \cdot \frac{\partial^2 V_{\alpha\beta}}{\partial v_\gamma \partial v_\delta} \left(V_\gamma^{(i)} V_\delta^{(i)} + \frac{T_i}{m_i} \delta_{\gamma\delta} + \frac{\pi_{\alpha\beta}^{(i)}}{m_i n_i} \right) \right\}. \tag{6.26}$$

Here $n_i = \int f_i d\mathbf{v}$ and $\mathbf{V}^{(i)} = \frac{1}{n_i}\int \mathbf{v} f_i d\mathbf{v}$ are the density and mean directed velocity of the ions; $\mu \equiv m_e/m_i$;

$$T_i = \frac{1}{n_i}\int \frac{m_i(\mathbf{v}-\mathbf{V}^{(i)})^2}{3} f_i d\mathbf{v}; \tag{6.27}$$

T_i has the meaning of an ion temperature [T is identical with the true temperature in the case of a Maxwellian distribution $f_i(\mathbf{v}-\mathbf{V}^{(i)})$];

$$\pi^{(i)}_{\alpha\beta} = m_i \int f_i \left\{ (\mathbf{v}-\mathbf{V}^{(i)})_\alpha (\mathbf{v}-\mathbf{V}^{(i)})_\beta - \frac{(\mathbf{v}-\mathbf{V}^{(i)})^2}{3}\delta_{\alpha\beta} \right\} d\mathbf{v}; \tag{6.28}$$

$\pi_{\alpha\beta}$ is the viscosity tensor of the ions.

2. Simplification of C_{ie}. In this case the small parameter is the ratio $|\mathbf{v}|/|\mathbf{v}'|$. Equation (6.25) is replaced by the series

$$U_{\alpha\beta} = V'_{\alpha\beta} - v_\gamma \frac{\partial V'_{\alpha\beta}}{\partial v'_\gamma} + \ldots$$
$$V'_{\alpha\beta} = (v'^2\delta_{\alpha\beta} - v'_\alpha v'_\beta)/v'^3. \tag{6.29}$$

Substituting this into (6.3), we obtain the desired expression for C_{ie}:

$$C_{ie} = \mu \mathbf{R}^{(e)} \frac{\partial f_i}{\partial \mathbf{v}} + \frac{2\pi\lambda e_i^2 e_e^2}{m_i m_e} \cdot \frac{\partial}{\partial v_\alpha} \left\{ (v_\gamma - V^{(i)}_\gamma) f_i \int \frac{\partial f_e(v')}{\partial v'_\beta} \times \right.$$
$$\left. \times \frac{\partial V'_{\alpha\beta}}{\partial v'_\gamma} d\mathbf{v}' + \mu \frac{\partial f_i}{\partial v_\beta} \int f_e(v') V'_{\alpha\beta} d\mathbf{v}' \right\}. \tag{6.30}$$

Here

$$R^{(e)}_\alpha = -\frac{2\pi\lambda e_e^2 e_i^2 n_i}{m_e^2} \int \left\{ V_{\alpha\beta} \frac{\partial f_e}{\partial v_\beta} - V^{(i)}_\gamma \frac{\partial V_{\alpha\beta}}{\partial v_\gamma} \cdot \frac{\partial f_e}{\partial v_\beta} \right\} d\mathbf{v}. \tag{6.31}$$

Using (6.26), we can show that $R^{(e)}_\alpha$ is the frictional force that the ions exert on the electrons; by definition, it is equal to

$$R^{(e)}_\alpha = m_e \int v_\alpha C_{ei} d\mathbf{v}. \tag{6.32}$$

From (6.26) and (6.30), we deduce (as from the general expressions (6.2) for C_{ei} and C_{ie}) the natural result that the collisions do not alter the total momentum of the electrons and ions:

$$R^{(i)}_\alpha \equiv m_i \int v_\alpha C_{ie} dv = -R^{(e)}_\alpha. \tag{6.33}$$

The total energy of the colliding particles also remains constant:

$$W_{ei} \equiv \int \frac{m_e v^2}{2} C_{ei} d\mathbf{v} = -W_{ie} \equiv -\int \frac{m_i v^2}{2} C_{ie} d\mathbf{v}. \tag{6.34}$$

In accordance with (6.26), the energy W_{ei} acquired by the electrons in unit time from the ions is

$$W_{ei} = \mathbf{V}^{(e)}\mathbf{R}^{(e)} + Q_{ei}, \tag{6.35}$$

where $\mathbf{V}^{(e)} = \frac{1}{n_e}\int \mathbf{v} f_e d\mathbf{v}$ is the mean electron velocity,

$$R_{ei} = \mathbf{R}^e\,(\mathbf{V}^{(i)} - \mathbf{V}^{(e)}) - Q_{ie}, \tag{6.36}$$

$$Q_{ie} = \frac{2\pi\lambda e_e^2 e_i^2 n_i}{m_i} 2 \int \left\{ \frac{f_e}{v} + \frac{T^{(i)}}{m_e} \cdot \frac{v_\beta}{v^3} \cdot \frac{\partial f_e}{\partial v_\beta} - \frac{\pi_{\alpha\beta}^{(i)}}{2m_e n_i} \cdot \frac{\partial f_e}{\partial v_\beta} \frac{\partial V_{\beta\gamma}}{\partial v_\beta} \right\} d\mathbf{v}. \tag{6.37}$$

Here, Q_{ei} and Q_{ie} are the increments in unit time of the random energy of the electrons or ions as a result of their mutual collisions:

$$\left.\begin{aligned} Q_{ei} &= \int \frac{m_e}{2} (\mathbf{v} - \mathbf{V}^{(e)})^2\, C_{ei} d\mathbf{v}; \\ Q_{ie} &= \int \frac{m_i}{2} (\mathbf{v} - \mathbf{V}^{(i)})^2\, C_{ie} d\mathbf{v}. \end{aligned}\right\} \tag{6.38}$$

§ 6.2. Hydrodynamics of a Fully Ionized Collisional Plasma

To investigate oscillations with frequencies that are low compared with the frequency of binary collisions, it is above all necessary to solve the Boltzmann equation in the approximation of frequent collisions. To do this we shall follow Braginskii's method.

In the Boltzmann equation (I) we replace the velocity $\mathbf{v}$ by a new variable $\mathbf{v}^{(\alpha)}$, setting

$$\mathbf{v} = \mathbf{v}^{(\alpha)} + \mathbf{V}^{(\alpha)}(\mathbf{r},\ t). \tag{6.39}$$

Here $\mathbf{V}^{(\alpha)}$ is the mean velocity of the corresponding component of the plasma (defined in the problem in § 6.1). As a result of this change of variables, Eq. (I) reduces to

$$\hat{L}^{(\alpha)} f_\alpha(\mathbf{r},\ \mathbf{v}^{(\alpha)},\ t) = \sum_\beta C_{\alpha\beta}, \tag{6.40}$$

where

$$\hat{L}^{(\alpha)} = \frac{d_\alpha}{dt} + \mathbf{v}^{(\alpha)}\nabla + \left(\frac{e_\alpha}{m_\alpha}\mathbf{E} - \frac{d_\alpha \mathbf{V}^{(\alpha)}}{dt}\right)\frac{\partial}{\partial \mathbf{v}^{(\alpha)}} - \frac{\partial V_\gamma^{(\alpha)}}{\partial x_\beta} \mathbf{v}_\beta^{(\alpha)} \frac{\partial}{\partial v_\gamma^{(\alpha)}},$$

$$\frac{d_\alpha}{dt} = \frac{\partial}{\partial t} + \mathbf{V}^{(\alpha)}\nabla. \tag{6.41}$$

It is assumed that the magnetic field vanishes.

The integral $C_{\alpha\alpha}$ depends on $\mathbf{v}^{(\alpha)}$ in the same way as on $\mathbf{v}$ [see Eq. (6.2)]. In the electron–ion collision integral C_{ei} [see Eq. (6.26)], one must make the substitution $\mathbf{V}^{(i)} \to \mathbf{V}^{(i)} - \mathbf{V}^{(e)}$ as well as adding the electron subscript to $\mathbf{v}$. An analogous substitution must also be made in the expression $\mathbf{R}^{(e)}$ in C_{ie} [see Eqs. (6.30) and (6.31)]. These transformations of the crossed collision integrals are valid if the difference $\mathbf{V}^{(i)} - \mathbf{V}^{(e)}$ is small compared with $\mathbf{v}^{(e)}$.

The approximation of frequent collisions means that the principal terms of the right-hand side of (6.40) are large compared with the left-hand side. If the left-hand side of (6.40) and the small terms in C_α are ignored, we obtain

$$C_{ee} = C_{ei}^{(0)} = 0, \quad C_{ii} = 0. \tag{6.42}$$

Here, $C_{ei}^{(0)}$ is the highest term of the right-hand side of (6.26) expressed in the variables $\mathbf{v}^{(e)}$.

Maxwellian functions are a solution of (6.42), so that in the limit of frequent collisions

$$f_\alpha = n_\alpha \left(\frac{m_\alpha}{2\pi T_\alpha}\right)^{3/2} \exp\left(-\frac{m_\alpha v^{(\alpha)\,2}}{2T_\alpha}\right) \equiv F_\alpha. \tag{6.43}$$

The density n_α and temperature T_α introduced here and the previously introduced mean velocity $\mathbf{V}^{(\alpha)}$ are related to each other and the other moments of the distribution function by the equations

$$\left.\begin{aligned} &\frac{\partial n_\alpha}{\partial t} + \operatorname{div}(n_\alpha \mathbf{V}^{(\alpha)}) = 0, \\ &m_\alpha n_\alpha \frac{d_\alpha V_k^{(\alpha)}}{dt} = -\frac{\partial p_\alpha}{\partial x_k} - \frac{\partial \pi_{kl}^{(\alpha)}}{\partial x_l} + e_\alpha n_\alpha E_k + R_k^{(\alpha)}, \\ &\frac{3}{2} n_\alpha \frac{d_\alpha T_\alpha}{dt} + p_\alpha \operatorname{div} \mathbf{V}^{(\alpha)} = -\operatorname{div} \mathbf{q}^{(\alpha)} - \\ &\qquad - \pi_{kl}^{(\alpha)} \frac{\partial V_k^{(\alpha)}}{\partial x_l} + Q_\alpha. \end{aligned}\right\} \tag{6.44}$$

These results are obtained by integrating (I) with respect to the velocities. The notation can be found in the problem in § 6.1.

To find π_{kl}, **q**, **R**, and Q it is necessary to calculate the correction to (6.43) due to the terms of the Boltzmann equation (I) ignored in (6.42). We set

$$f_\alpha = F_\alpha + f_\alpha^{(1)}. \tag{6.45}$$

The equations for $f_e^{(1)}$ and $f_i^{(1)}$ are, respectively,

$$\left.\begin{aligned} C_{ee}[f_e^{(1)}, F_e] + C_{ei}^{(0)}(f_e^{(1)}) &= \hat{L}^{(e)} F_e - C_{ei}^{(1)}(F_e), \\ C_{ii}[f_i^{(1)}, F_i] &= \hat{L}^{(i)} F_i - C_{ie}(F_i). \end{aligned}\right\} \tag{6.46}$$

We now use Eqs. (6.44) to express the derivatives d_α/dt in the operator $\hat{L}^{(\alpha)}$ in terms of derivatives with respect to the coordinates. In Eqs. (6.44) we assume that π_{kl} and **q** vanish; these quantities are determined by the function $f_\alpha^{(1)}$ and their contribution to the right-hand side of (6.46) is therefore small.

As a result, Eqs. (6.46) reduce to

$$\left.\begin{aligned} C_{ee}[f_e^{(1)}, F_e] + C_{ei}^{(0)}(f_e^{(1)}) \,\mathbf{v} F_e \Bigg\{ \left(\frac{m_e v^2}{2T_e} - \frac{5}{2}\right) \nabla \ln T_e + \\ + \frac{m_e}{T_e \tau_e} \left[\frac{3\sqrt{\pi}}{\sqrt{2} v^3} \left(\frac{T_e}{m_e}\right)^{3/2} - 1 \right] \left(\mathbf{V}^{(e)} - \mathbf{V}^{(i)} + \frac{\mathbf{R}_1^{(e)}}{n_e T_e} \right) \Bigg\}, \\ C_{ii}[f_i^{(1)}, F_i] = F_i \Bigg[\left(\frac{m_i v^2}{2T_i} - \frac{5}{2}\right) \mathbf{v} \nabla \ln T_i + \\ + \frac{m_i}{T_i} \left(v_k v_l - \frac{v^2}{3} \delta_{kl} \right) W_{kl}^{(i)} \Bigg]. \end{aligned}\right\} \tag{6.47}$$

For brevity we have here omitted the superscript α in $\mathbf{v}^{(\alpha)}$. The quantities $\mathbf{R}_1^{(e)}$ and $W_{kl}^{(i)}$ are defined by

$$\left.\begin{aligned} \mathbf{R}_1^{(e)} &= \int m_e \mathbf{v}^{(e)} C_{ei}^{(0)}(f_e^{(1)})\, d\mathbf{v}^{(e)}, \\ W_{kl}^{(i)} &= \frac{\partial V_k}{\partial x_l} + \frac{\partial V_l}{\partial x_k} - \frac{2}{3} \delta_{kl} \operatorname{div} \mathbf{V}. \end{aligned}\right\} \tag{6.48}$$

In the electron equation (6.47), the term W_{kl} is omitted; this corresponds to neglect of the electron viscosity.

Equations (6.47) are solved in the paper of Braginskii, who also calculates the desired quantities $\mathbf{R}^{(e)} = -\mathbf{R}^{(i)}$, Q_α, $\mathbf{q}^{(\alpha)}$ and $\pi_{kl}^{(i)}$:

$$\left.\begin{array}{c}
\mathbf{R}^{(e)} = -\frac{m_e n_e}{\tau_e}(\mathbf{V}^{(e)} - \mathbf{V}^{(i)})0.51 - 0.71 n_e \nabla T_e, \\
\mathbf{q}^{(e)} = 0.71 n_e T_e (\mathbf{V}^{(e)} - \mathbf{V}^{(i)}) - \varkappa^{(e)} \nabla T_e, \\
q^{(i)} = -\varkappa^{(i)} \nabla T_i, \quad Q_e = -\mathbf{R}^{(e)}(\mathbf{V}^{(e)} - \mathbf{V}^{(i)}) - Q_i, \\
Q_i = 3\mu \frac{n_e}{\tau_e}(T_e - T_i), \quad \pi_{kl}^{(i)} = -\eta^{(i)} W_{kl}^{(i)}, \\
\varkappa^{(e)} = 3.16 \frac{n_e T_e \tau_e}{m_e}, \quad \varkappa^{(i)} = 3.9 \frac{n_i T_i \tau_i}{m_i}, \\
\eta^{(i)} = 0.96 n_i T_i \tau_i.
\end{array}\right\} \qquad (6.49)$$

§ 6.3. Ion-Acoustic Instability of a Fully Ionized Plasma in an Electric Field

Suppose that there is a static electric field $\mathbf{E}_0$ in the plasma. In accordance with (6.49), it gives rise to a motion of the electrons relative to the ions with a velocity

$$\mathbf{V}_0^{(e)} = 0.51 \frac{e_e}{m_e} \tau_e \mathbf{E}_0. \qquad (6.50)$$

[This is a refinement of Eq. (6.17), which is derived under the assumption that f_e is a Maxwellian function shifted by $\mathbf{V}_0^{(e)}$.] As we have shown in § 3.4, a relative motion of the components in a collisionless plasma can lead to excitation of ion-acoustic and ion plasma oscillations. Let us now consider the stability of a collisional plasma. To be specific we assume

$$\mathbf{V}_0^{(i)} = 0, \quad \mathbf{E}_0 \parallel \mathbf{z}, \quad \mathbf{k} \parallel \mathbf{z}.$$

We shall restrict the treatment to oscillations with $kd \ll 1$. In Poisson's equation the term with div $\mathbf{E}$ can then be ignored and the equation reduces to the condition of electrical neutrality of the plasma:

$$n_e = n_i \equiv n. \qquad (6.51)$$

We now linearize Eqs. (6.49) and (6.51). With allowance for (6.51), the continuity equations for the ions and electrons yield

$$n' = \frac{kV'_i}{\omega} n_0 = \frac{V'_i - V'_e}{V_0^e} n_0. \qquad (6.52)$$

Taking this equation into account, we reduce the equations of motion to

$$\left.\begin{aligned} &\frac{n'}{n_0}+e_e\psi+\frac{T'_e}{T_0}\left(1.7+\mathrm{i}\,1.5\,\frac{m_eV_{0e}}{k^2T_0\tau_{0e}}\right)=0,\\ &\omega\,\frac{n'}{n_0}=k^2\,\frac{T_0}{m_i}\left(2\,\frac{n'}{n_0}+\frac{T'_e+T'_i}{T_0}\right)\left(\omega+\mathrm{i}\,\frac{4}{3}\cdot 0.96\,\frac{k^2T_0\tau_i}{m_i}\right). \end{aligned}\right\} \tag{6.53}$$

In the equation of motion of the electrons we have neglected the electron inertia, which is small, ~ $(\omega/kv_{T_e})^2$. We have also assumed $T_{0e} = T_{0i} \equiv T_0$.

Using (6.52), we reduce the linearized equations of heat balance to

$$\left.\begin{aligned} &\frac{T'_e}{T_0}\left[\frac{3}{2}\,\omega-\left(\frac{3}{2}+0.7\right)k_zV_{0e}+\mathrm{i}\,3.16\,\frac{k^2T_0\tau_{0e}}{m_e}+\right.\\ &\left.+\,\mathrm{i}\,\frac{\mu}{\tau_{0e}}+\mathrm{i}\,\frac{3}{2}\,0.5\,\frac{m_eV_{0e}^2}{\tau_{0e}T_0}\right]-\mathrm{i}\,3\,\frac{\mu}{\tau_{0e}}\cdot\frac{T'_i}{T_0}-(\omega-k_zV_{0e})\,\frac{n'}{n_0}=0,\\ &\left(\frac{3}{2}\,\omega+\mathrm{i}\,3.9\,\frac{T_0\tau_{0i}}{m_i}+\mathrm{i}\,3\,\frac{\mu}{\tau_{0e}}\right)\frac{T'_i}{T_0}-\omega\,\frac{n'}{n_0}-\mathrm{i}\,3\,\frac{\mu}{\tau_{0e}}\cdot\frac{T'_e}{T_0}=0. \end{aligned}\right\} \tag{6.54}$$

Here we have omitted the subscript zero in $\tau_{0\alpha}$. Let us consider perturbations with a frequency in the range

$$\tau_e\,\frac{k^2T_0}{m_e}\gg\omega\gg\tau_i\,\frac{k^2T_0}{m_i},$$

which correspond to a high electron and a low ion thermal conductivity. In this case, Eqs. (6.54) simplify:

$$\left.\begin{aligned} &\frac{T'_e}{T_0}=-\mathrm{i}\,\frac{\omega-kV_{0e}}{3.2\tau_{0e}\,\dfrac{k^2T_e}{m_e}}\cdot\frac{n'}{n_0},\\ &\frac{T'_i}{T_0}=\frac{2}{3}\cdot\frac{n'}{n_0}\left(1-\mathrm{i}\,\frac{2\mu}{\omega\tau_e}-\mathrm{i}\,\frac{2}{3}\cdot 3.9\,\frac{k^2T_0\tau_i}{m_i\omega}\right). \end{aligned}\right\} \tag{6.55}$$

Equating the determinant of the system of equations (6.52), (6.55) to zero, we obtain the dispersion equation. The real and imaginary parts of the frequency are

$$\left.\begin{aligned} &\operatorname{Re}\omega=k\left(\frac{8}{3}\cdot\frac{T_e}{m_i}\right)^{1/2},\\ &\gamma=-\frac{\mu}{\tau_e}\left(\frac{3}{2}-\frac{kV_{0e}}{\operatorname{Re}\omega}\right)-\frac{2}{3}\left(0.96+\frac{3}{8}\cdot 3.9\right)\frac{k^2T_0}{m_i}\,\tau_i. \end{aligned}\right\} \tag{6.56}$$

The growth rate is positive if

$$V_{0e} > \left(\frac{8}{3}\cdot\frac{T_0}{m_i}\right)^{1/2}\left[\frac{3}{2}+\frac{2}{3}\left(0.96+\frac{3}{8}\,3.9\right)\tau_e\tau_i\frac{k^2T_0}{m_e}\right]. \qquad (6.57)$$

It can be seen that ion-acoustic oscillations are excited if the electron velocity is several times larger than the thermal velocity of the ions. In contrast to a collisionless plasma, instability is possible even if $T_{0e} = T_{0i}$. As the instability develops, the energy of the directed motion is transformed into oscillatory energy because the plasma conductivity is negative, $\omega \operatorname{Im} \varepsilon < 0$. This negative conductivity is due to the perturbation of the temperature of the electron component of the plasma.

§ 6.4. Ion-Acoustic Instability of a Weakly Ionized Plasma in an Electric Field

In contrast to § 6.3, we shall now assume that the collisions between the charged particles and the neutrals are more important than the collisions between the charged particles. We take the frictional force $\mathbf{R}^{(\alpha)}$ in the form (6.23). We shall neglect the perturbations of the electron and ion temperatures. The Boltzmann equation then yields the following system of hydrodynamic equations:

$$\left.\begin{aligned}&\frac{\partial n_\alpha}{\partial t}+\operatorname{div}(n_\alpha\mathbf{V}^{(\alpha)})=0,\\&\frac{\partial\mathbf{V}^{(\alpha)}}{\partial t}+(\mathbf{V}^{(\alpha)}\nabla)\,\mathbf{V}^{(\alpha)}=\frac{e_\alpha}{m_\alpha}\,\mathbf{E}-\nu_\alpha\mathbf{V}^{(\alpha)}-\frac{T^{(\alpha)}\nabla n_\alpha}{m_\alpha n_\alpha}.\end{aligned}\right\} \qquad (6.58)$$

If electron inertia is ignored, we obtain

$$\mathbf{V}^{(e)}=\frac{e_eE}{m_e\nu_e}-\frac{T_e\nabla n_e}{m_e n_e\nu_e}. \qquad (6.59)$$

Substituting this result into the electron continuity equation and taking into account the condition of equilibrium

$$\frac{e_e}{m_e}\,\mathbf{E}_0=\nu_e\mathbf{V}_0^{(e)}, \qquad (6.60)$$

we obtain

$$\left(\omega-\mathbf{k}\mathbf{V}_0+i\frac{k^2T_e}{m_e\nu_e}\right)n'_e-\frac{\mathbf{k}\mathbf{E}'e_e}{m_e\nu_e}\,n_0=0. \qquad (6.61)$$

For simplicity, we shall assume that the ions are cold ($T_i = 0$). We shall neglect the frictional force between the ions and the neutrals. It then follows from (6.58) that

$$n'_i = i\frac{e_i n_0}{m_i \omega^2} kE'. \tag{6.62}$$

With allowance for the electrical neutrality of the perturbations, Eqs. (6.61) and (6.62) yield the dispersion equation

$$\omega^2 + i\mu\nu_e(\omega - kV_0) - \frac{k^2 T_e}{m_i} = 0. \tag{6.63}$$

The perturbations grow if

$$V_0 > V_{cr} = (T_e/m_i)^{1/2}. \tag{6.64}$$

For

$$\mu\nu_e < k(T_e/m_i)^{1/2}$$

we have

$$\left.\begin{aligned} \mathrm{Re}\,\omega &= k(T_e/m_i)^{1/2}, \\ \gamma &\simeq \mu\nu_e. \end{aligned}\right\} \tag{6.65}$$

For perturbations with a longer wavelength, $\mu\nu_e > k(T_e/m_i)^{1/2}$,

$$\left.\begin{aligned} \mathrm{Re}\,\omega &= kV_0, \\ \gamma &= (kV_0)^2\frac{m_i}{m_e\nu}. \end{aligned}\right\} \tag{6.66}$$

Problem. Taking the collision integral in the form*

$$C_\alpha = -\nu_\alpha(f_\alpha - f_{0\alpha}) + \nu_\alpha\frac{f_\alpha^{(0)}}{n_{0\alpha}}\int(f_\alpha - f_{0\alpha})\,d\mathbf{v}, \tag{6.67}$$

where f_α is the total and

$$f_{0\alpha} = (m_\alpha/2\pi T_\alpha)^{3/2}\exp[-m_\alpha(\mathbf{v} - \mathbf{V}_{0\alpha})^2/2T_\alpha]$$

*A collision integral of the form (6.67) is a special case of the Bhatnagar-Gross-Krook model integral.

is the unperturbed distribution function, investigate the ion-acoustic instability of a weakly ionized plasma in an electric field for arbitrary ν_e/kv_{T_e}.

Solution. Using (6.67) and the transport equation (I), we find the perturbed distribution function

$$f'_\alpha = -\frac{i}{\omega + i\nu_\alpha - \mathbf{kv}}\left(\frac{e}{m}\mathbf{E}'\frac{\partial f_0}{\partial \mathbf{v}} - \nu_\alpha \frac{f_{0\alpha}}{n_{0\alpha}} n'_\alpha\right). \tag{6.68}$$

We then calculate the density of each species of charge, substituting the result into Poisson's equation. The dispersion equation obtained is

$$1 + \sum_{\alpha=i,\,e} \frac{1}{(kd_\alpha)^2}\cdot\frac{1 + i\sqrt{\pi}xW(x)}{1 - \dfrac{\sqrt{\pi}\,\nu_\alpha}{|k|v_{T\alpha}}W(x)} = 0;$$

$$x = \frac{\omega + i\nu_\alpha - \mathbf{kV}_{0\alpha}}{|k|v_{T\alpha}}. \tag{6.69}$$

In the limit $|\omega + i\nu_e - \mathbf{kV}_{0e}| \ll |kv_{T_e}, |\omega| > (kv_{T_i}, \nu_i)$, this yields the result of the collisionless approximation (§ 3.3). For $\nu_e > |k|v_{T_e}, |\omega| > (kv_{T_i}, \nu_i)$ we obtain the results given in the present section.

We shall now assume $\nu_e \simeq kv_{T_e} \gg \omega - \mathbf{kV}_{0e}$. It then follows from (6.69) that

$$\left.\begin{aligned} \mathrm{Re}\,\omega &= k\left(\frac{T_e}{m_i}\right)^{1/2}, \\ \gamma &= \left(\frac{\pi}{8}\mu\right)^{1/2}\left(\frac{V_{0e}}{\left(\frac{T_e}{m_i}\right)^{1/2}} - 1\right)\zeta\left(\frac{\nu_e}{|k|v_{T_e}}\right), \end{aligned}\right\} \tag{6.70}$$

where

$$\zeta(x) = e^{x^2}\left(1 - \frac{2}{\sqrt{\pi}}\int_0^x e^{-t^2}dt\right)\left[1 - \sqrt{\pi}xe^{x^2}\left(1 - \frac{2}{\sqrt{\pi}}\int_0^x e^{-t^2}dt\right)\right]^{-1}.$$

Bibliography

1. L. D. Landau, Zh. Eksp. Teor. Fiz., 7:203 (1937). An expression is obtained for the collision integral in a fully ionized plasma [Eq. (6.2)].
2. S. I. Braginskii, in: Reviews of Plasma Physics, Vol. 1, Consultants Bureau, New York (1965), p. 205. S. I. Braginskii, Zh. Eksp. Teor. Fiz., 33:459 (1957). The system of hydrodynamic equations discussed in § 6.2 is derived in these papers.
3. H. Dreicer, Phys. Rev., 117:329 (1960). The quasistationary state of a plasma in an electric field (§ 6.1) is discussed.

4. A. F. Kuckes, Phys. Fluids, 7:511 (1964). The collisional ion-acoustic instability is investigated (§ 6.3).
5. P. L. Bhatnagar, E. P. Gross, and M. Krook, Phys. Rev., 94:511 (1954). A model collision integral (see the problem to § 6.4) is derived.
6. B. B. Kadomtsev, Plasma Turbulence, Academic Press, London (1965). Some aspects of the stability theory of a collisional plasma are discussed.
7. A. V. Nedospasov, Usp. Fiz. Nauk, 94:439 (1968).
8. L. Pekarek, Usp. Fiz. Nauk, 94:463 (1968).
9. E. P. Velikhov and A. M. Dykhne, Proc. Sixth Intern. Conf. Phen. Ion. Gaz., Paris, 4:511 (1963). Some of the types of oscillation and instabilities of a weakly ionized plasma that are not considered in this chapter are discussed in [7-9].
10. B. A. Trubnikov, in: Reviews of Plasma Physics, Vol. 1, Consultants Bureau, New York (1965), p. 105.
11. D. V. Sivukhin, in: Reviews of Plasma Physics, Vol. 4, Consultants Bureau, New York (1966), p. 93. The structure of the collision integral of the transport equation for a fully ionized plasma is discussed in detail in the reviews [10, 11].
12. G. V. Gordeev, Zh. Eksp. Teor. Fiz., 22:230 (1952). The ion-acoustic instability of a weakly ionized plasma (§ 6.4) is discussed.

Part II

PLASMAS IN A MAGNETIC FIELD

Chapter 7

Permittivity of a Plasma in a Magnetic Field

§ 7.1. Permittivity of a Cold Plasma

Suppose the temperature of a group of particles is equal to zero, T = 0. The behavior of these particles in the perturbed electrostatic field and the constant magnetic field is described by the hydrodynamic equations

$$\left.\begin{aligned} &-i(\omega - k_z V_0)\, n' + i\mathbf{k}\mathbf{V}' n_0 = 0, \\ &-i(\omega - k_z V_0)\, \mathbf{V}' = -\frac{ie}{m}\, \mathbf{k}\psi + [\mathbf{V}' \boldsymbol{\omega}_B]. \end{aligned}\right\} \tag{7.1}$$

These equations follow from (1.10). It is assumed that the equilibrium velocity $\mathbf{V}_0$ of the particles is directed along $\mathbf{B}_0$. The vector $\boldsymbol{\omega}_B$ is $\omega_B \mathbf{e}_z$, where $\omega_B = eB_0/mc$ is the cyclotron frequency of the given species of particles and ψ is the potential of the perturbed field determined by the equation $\mathbf{E}' = -\nabla\psi$. It is assumed that $\mathbf{B}_0 \| z$.

Solving the system (7.1), we find the perturbed charge density $\rho = en'$ and then, using a formula of the type (1.36), the contribution that the group of particles in which we are interested makes to the permittivity:

$$\varepsilon_0^{(\alpha)} = -\frac{\omega_p^2 \cos^2\theta}{(\omega - k_z V_0)^2} - \frac{\omega_p^2 \sin^2\theta}{(\omega - k_z V_0)^2 - \omega_B^2}. \tag{7.2}$$

Here, θ is the angle between $\mathbf{k}$ and $\mathbf{B}_0$. The terms with $\cos^2\theta$ and $\sin^2\theta$ in $\varepsilon_0^{(\alpha)}$ are due to the perturbed motion of the particles along and at right angles to $\mathbf{B}_0$, respectively.

In the special case of a plasma at rest,

$$\varepsilon_0^{(\alpha)} = -\left(\frac{\omega_p}{\omega}\right)^2 \cos^2\theta - \frac{\omega_p^2 \sin^2\theta}{\omega^2 - \omega_B^2}. \tag{7.3}$$

Comparison of (7.2) and (1.36') shows that a magnetic field has little influence on the permittivity if $|\omega - k_z V_0| \gg \omega_B$ or $\sin^2\theta \ll 1$. If neither of these inequalities is satisfied, the magnetic field plays an important role.

§ 7.2. Permittivity of a Hot Plasma

1. Solution of the Boltzmann Equation. To obtain $\varepsilon_0^{(\alpha)}$ for the case when the thermal motion of the particles is important, it is first of all necessary to solve the Boltzmann equation

$$\frac{\partial f}{\partial t} + \mathbf{v}\nabla f + [\mathbf{v}\boldsymbol{\omega}_B]\frac{\partial f}{\partial \mathbf{v}} = \frac{e}{m}\nabla\psi\frac{\partial f_0}{\partial \mathbf{v}}. \tag{7.4}$$

This equation is obtained from (I) by assuming that the collisions between the particles are unimportant and that the perturbed field $\mathbf{E}'$ is derived from a potential, i.e., is electrostatic.

The solution of (7.4) can be represented in a form similar to (5.7):

$$f = \frac{e}{m}\int_{-\infty}^{t} \nabla\psi \frac{\partial f_0}{\partial \mathbf{v}}\, dt', \tag{7.5}$$

where the integration is along the particle trajectories (cf. § 5.1). In practice, this integration is performed as follows.

The unperturbed distribution function f_0 can be represented in the form $f_0(\mathbf{v}) = F(\varepsilon_\perp, v_z)$, where $\varepsilon_\perp = v_\perp^2/2$, and $v_\perp$ and v_z are the transverse and longitudinal velocities of the particles. Then

$$\frac{\partial f_0}{\partial \mathbf{v}} = \mathbf{v}_\perp \frac{\partial F}{\partial \varepsilon_\perp} + \mathbf{e}_z \frac{\partial F}{\partial v_z}. \tag{7.6}$$

Substituting (7.6) into (7.5) and taking the constants of the motion $\partial F/\partial\varepsilon_\perp$ and $\partial F/\partial v_z$ in front of the integral over t', we obtain

$$f = \frac{e\psi}{m}\left\{\frac{\partial F}{\partial \varepsilon_\perp} - \left[(\omega - k_z v_z)\frac{\partial F}{\partial \varepsilon_\perp} + k_z \frac{\partial F}{\partial v_z}\right] I\right\}, \tag{7.7}$$

where

$$I = -i \int_0^\infty \exp\left[i\omega t' - i\mathbf{k} \int_{\tau - t'}^{\tau} \mathbf{v}(\mathbf{r}_0, \mathbf{v}_0, \tau')\, d\tau'\right] dt'. \tag{7.8}$$

Here, we have used $\psi \sim e^{-i\omega t + i\mathbf{k}\mathbf{r}}$ and have integrated by parts using the formula

$$\mathbf{v}_\perp \nabla \psi = \frac{d\psi}{dt} + i(\omega - k_z v_z)\psi. \tag{7.9}$$

The remaining details of the derivation of (7.7) can be found by consulting § 5.1.

We shall now calculate the integral I. Remembering that the velocity of a particle that moves in a magnetic field depends on the time in accordance with the law

$$\left.\begin{aligned} v_x(t) &= v_\perp \cos[\alpha_0 - \omega_B(t - t_0)], \\ v_y(t) &= v_\perp \sin[\alpha_0 - \omega_B(t - t_0)], \\ v_z(t) &= v_z, \quad \alpha_0 = \tan^{-1}\frac{v_y(t_0)}{v_x(t_0)}, \end{aligned}\right\} \tag{7.10}$$

we carry out an integration in the argument of the exponential function:

$$\mathbf{k} \int_{\tau - t'}^{\tau} \mathbf{v}(\mathbf{r}_0, \mathbf{v}_0, \tau')\, d\tau' = k_z v_z t' - \xi\{\sin(\alpha_0 - \omega_B \tau - \Psi) - \\ - \sin[\alpha_0 - \omega_B(\tau - t') - \Psi]\}. \tag{7.11}$$

Here

$$\xi = k_\perp v_\perp / \omega_B \equiv k_\perp \rho, \qquad \Psi = \tan^{-1} k_y / k_x.$$

We represent the part of the argument of the exponential that contains $\sin[\alpha_0 - \omega_B(\tau - t')]$ as a series in Bessel functions:

$$\exp\{-i\xi \sin[\alpha_0 - \omega_B(\tau - t') - \Psi]\} = \\ = \sum_{n=-\infty}^{\infty} J_n(\xi) \exp[-in[\alpha_0 - \omega_B(\tau - t') - \Psi]]. \tag{7.12}$$

We then integrate over t', obtaining

$$I = \exp\left[i\xi \sin(\alpha - \Psi)\right] \sum_{n=-\infty}^{\infty} \zeta_n J_n e^{-in(\alpha-\Psi)}. \qquad (7.13)$$

Here $\zeta_n = (\omega - n\omega_B - k_z v_z)^{-1}$; $\alpha = \tan^{-1} v_x/v_y$.

2. Permittivity. Using (7.7) and (7.13), we find the perturbed charged density $\rho = en'$, and then, as in § 7.1, we use Eq. (1.36) to calculate

$$\varepsilon_0^{(\alpha)} = -\frac{4\pi e^2}{mk^2}\left\langle \frac{\partial F}{\partial \varepsilon_\perp} - \left[(\omega - k_z v_z)\frac{\partial F}{\partial \varepsilon_\perp} + k_z \frac{\partial F}{\partial v_z}\right] \sum_{n=-\infty}^{\infty} \frac{J_n^2(\xi)}{\omega - n\omega_B - k_z v_z}\right\rangle. \qquad (7.14)$$

Here, $\langle \ldots \rangle$ denotes the integral $\int (\ldots) v_\perp dv_\perp dv_z$. In (7.14) we have integrated over the angle α by means of the formula

$$\int_0^{2\pi} \frac{d\alpha}{2\pi} \exp(i\xi \sin\alpha - in\alpha) = J_n(\xi). \qquad (7.15)$$

The normalization of F is chosen such that $\langle F \rangle = n_0$.

Using the relation $\Sigma J_n^2(\xi) = 1$, we can write the right-hand side of (7.14) in the different form

$$\varepsilon_0^{(\alpha)} = \frac{4\pi e^2}{mk^2} \sum_{n=-\infty}^{\infty} \left\langle \frac{J_n^2}{\omega - n\omega_B - k_z v_z}\left(n\omega_B \frac{\partial F}{\partial \varepsilon_\perp} + k_z \frac{\partial F}{\partial v_z}\right)\right\rangle. \qquad (7.16)$$

Let us consider some special cases of formula (7.14).

A. Plasma at Rest with a Maxwellian Velocity Distribution. Let $F = \pi^{-1/2}(m/T)^{3/2} \times \exp(-mv^2/2T)$. The integration over v_z in (7.14) is performed as in § 2.4. The functions $W((\omega - n\omega_B)/k_z v_T)$ then appear on the right-hand side of (7.14). [The function W (x) is defined by (2.45).] The integrals over the transverse velocities can be calculated by means of the formula

$$\int_0^{\infty} e^{-\sigma^2 x^2} J_n(\alpha x) J_n(\beta x)\, x dx = \frac{1}{2\sigma^2} \exp\left(-\frac{\alpha^2 + \beta^2}{4\sigma^2}\right) I_n\left(\frac{\alpha\beta}{2\sigma^2}\right), \qquad (7.17)$$

where I_n is a Bessel function of imaginary argument. The result is [cf. (2.44)]

$$\varepsilon_0^{(\alpha)} = \frac{1}{k^2 d^2}\left[1 + i\sqrt{\pi}\,\frac{\omega}{|k_z| v_T}\sum_n W\left(\frac{\omega - n\omega_B}{|k_z| v_T}\right) e^{-z} I_n(z)\right],$$
$$z = k_\perp^2 T/m\omega_B^2 . \tag{7.18}$$

B. Moving Plasma with a Maxwellian Velocity Distribution, $F \propto \exp[-m(\mathbf{v}-\mathbf{V})^2/2T]$, $\mathbf{V}\,\|\,\mathbf{B}_0\,\|\,z$. The procedure for calculating the integrals over v_z and $v_\perp$ remains the same. The outcome is an expression for $\varepsilon_0^{(\alpha)}$ that differs from (7.18) only by the substitution $\omega \to \omega - k_z V$:

$$\varepsilon_0^{(\alpha)} = \frac{1}{(kd)^2}\left[1 + i\sqrt{\pi}\,\frac{\omega - k_z V}{|k_z| v_T}\sum_n W\left(\frac{\omega - k_z V - n\omega_B}{|k_z| v_T}\right) e^{-z} I_n(z)\right]. \tag{7.19}$$

C. Plasma with $T_\perp \neq T_\parallel$, $F \propto \exp(-mv_\perp^2/2T_\perp - mv_z^2/2T_\parallel)$. In this case, (7.18) is replaced by

$$\varepsilon_0^{(\alpha)} = \frac{1}{(kd_\parallel)^2}\sum_{n=-\infty}^{\infty} e^{-z_\perp} I_n(z_\perp)\left\{1 + \right.$$
$$\left. + i\sqrt{\pi}\,\frac{\omega}{|k_z| v_{T\parallel}}\left[1 - \frac{n\omega_B}{\omega}\left(1 - \frac{T_\parallel}{T_\perp}\right)\right] W\left(\frac{\omega - n\omega_B}{|k_z| v_{T\parallel}}\right)\right\}, \tag{7.20}$$

where $d_\parallel^2 = T_\parallel/4\pi e^2 n_0$; $z_\perp = k_\perp^2 T_\perp/m\omega_B^2$

For $T_\parallel = 0$ Eq. (7.20) yields

$$\varepsilon_0^{(\alpha)} = -\omega_p^2 \sum_{n=-\infty}^{\infty}\left[\frac{I_n e^{-z_\perp}\cos^2\theta}{(\omega - n\omega_B)^2} + \frac{nI_n e^{-z_\perp}\sin^2\theta}{z_\perp \omega_B(\omega - n\omega_B)}\right] \tag{7.21}$$

The expression for $\varepsilon_0^{(\alpha)}$ can also be simplified in the limit of low- and high-frequency perturbations [see §§ 7.3 and 7.4].

§ 7.3. Permittivity in the Low-Frequency Approximation

We shall say that oscillations are of low frequency (for each particular species of charges) if $|\omega - k_z v_z| \ll \omega_B$. For oscillations of this kind, only the term with n = 0 in the sum (7.14) is important; therefore, in this case

$$\varepsilon_0^{(\alpha)} = -\frac{4\pi e^2}{mk^2}\left\langle \frac{\partial F}{\partial \varepsilon_\perp}(1 - J_0^2) - \frac{k_z J_0^2}{\omega - k_z v_z} \cdot \frac{\partial F}{\partial v_z} \right\rangle. \tag{7.22}$$

If the particles have a Maxwellian velocity distribution, this means that

$$\varepsilon_0^{(\alpha)} = \frac{1}{(kd)^2}\left[1 + i\sqrt{\pi}\frac{\omega}{|k_z| v_T} W\left(\frac{\omega}{|k_z| v_T}\right) I_0(z) e^{-z}\right]. \tag{7.23}$$

Since we have assumed $k_z v_z \ll \omega_B$ in deriving (7.22) and (7.23), it is only necessary to retain all powers of the parameter $(k_\perp \rho)^2$ in these expressions if $(k_z/k_\perp) \ll (v_z/v_\perp)$. In particular, if the particle velocity distribution is isotropic and Maxwellian, this corresponds to angles $\cos\theta \ll 1$.

In the limit of long wavelengths, $k_\perp \rho \ll 1$, and arbitrary $\cos\theta$, the expression for $\varepsilon_0^{(\alpha)}$ that follows from (7.22) has the form

$$\varepsilon_0^{(\alpha)} = \left(\frac{\omega_p}{\omega_B}\right)^2 \sin^2\theta + \frac{4\pi e^2 k_z}{mk^2}\left\langle \zeta_0 \frac{\partial F}{\partial v_z} \right\rangle. \tag{7.24}$$

Similarly, if F is Maxwellian, it follows from (7.23) that

$$\varepsilon_0^{(\alpha)} = \frac{\omega_p^2}{\omega_B^2}\sin^2\theta + \frac{1}{(kd)^2}\left[1 + \frac{i\sqrt{\pi}\omega}{|k_z| v_T} W\left(\frac{\omega}{|k_z| v}\right)\right]. \tag{7.25}$$

It should be noted that the first terms on the right-hand sides of (7.24) and (7.25) must be taken into account only if $\omega \gg k_z v_z$, since they are otherwise small, $\sim (k\rho)^2$.

§ 7.4. Permittivity in the High-Frequency Approximation

Suppose ω and **k** are such that $|\omega - k_z v_z| \gg \omega_B$, $k\rho \gg 1$. In this case a large number of terms in the sum over n make a contribution to (7.14) (if $k_\perp \rho$ is not too small). If quantities of order $\omega_B/|\omega - k_z v_z|$ are neglected, these terms can be summed and (7.14) is replaced by

$$\varepsilon_0^{(\alpha)} = -\frac{4\pi e^2}{mk^2}\left\langle \frac{\partial F}{\partial \varepsilon_\perp} - \left[(\omega - k_z v_z)\frac{\partial F}{\partial \varepsilon_\perp} + k_z \frac{\partial F}{\partial v_z}\right] \frac{1}{[(\omega - k_z v_z)^2 - (k_\perp v_\perp)^2]^{1/2}}\right\rangle. \tag{7.26}$$

The procedure for making the transition from (7.14) to (7.26) is described in the paper by Lominadze and Stepanov.

The expression (7.26) can also be obtained directly by means of Eqs. (7.7) and (7.8). To do this, we must assume that the velocity $\mathbf{v}(\mathbf{r}_0, \mathbf{v}_0, \tau'')$ in (7.8) is constant. Then the integral I is equal to $I = (\omega - \mathbf{kv})^{-1}$. Substituting this I into (7.7), we perform the integration over the angle in the space of transverse velocities and then arrive in the standard manner at (7.26).

The expression (7.26) simplifies even further for $\omega \gg k_z v_z$:

$$\varepsilon_0^{(\alpha)} = -\frac{4\pi e^2}{mk^2}\left\langle \frac{\partial F}{\partial \varepsilon_\perp}\left[1 - \frac{\omega}{(\omega^2 - k_\perp^2 v_\perp^2)^{1/2}}\right]\right\rangle. \tag{7.27}$$

In the special case of a δ-function distribution over the transverse velocities, $F \sim \delta(\varepsilon_\perp - v_0^2/2)$, this yields

$$\varepsilon_0^{(\alpha)} = -\frac{\omega_p^2 \omega}{(\omega^2 - k_\perp^2 v_0^2)^{3/2}}. \tag{7.28}$$

If F is Maxwellian, this expression is replaced by

$$\varepsilon_0^{(\alpha)} = \frac{1}{k_\perp^2 d^2}\left[1 + \frac{i\sqrt{\pi}\,\omega}{k_\perp v_T} W\left(\frac{\omega}{k_\perp v_T}\right)\right]. \tag{7.29}$$

One can show that in the case $k_z \lessgtr k_\perp$ and $k_z v_z < \omega_B$ the limiting transition from (7.14) to (7.26) is correct provided $\mathrm{Im}\,\omega > \omega_B$. This condition has a simple physical meaning: a trajectory can be assumed to be rectilinear only if the particle does not succeed in making a single revolution about its Larmor center during the time in which the amplitude of the perturbed field changes appreciably. If $k_z v_z > \omega_B$, the thermal motion is more effective than the Larmor gyration and the condition $\mathrm{Im}\,\omega > \omega_B$ need not be satisfied.

Bibliography

1. E. P. Gross, Phys. Rev., 82:232 (1951). The Boltzmann equation is solved for perturbations with $k_z = 0$, $E_z = 0$. The dispersion relation is obtained for electrostatic perturbations with $k_z = 0$.
2. G. V. Gordeev, Zh. Eksp. Teor. Fiz. 23:660 (1952). The dispersion relation is obtained for electrostatic perturbations of a Maxwellian plasma with arbitrary k_z.
3. A. G. Sitenko and K. N. Stepanov, Zh. Eksp. Teor. Fiz., 31:642 (1956). Derivation of $\varepsilon_{\alpha\beta}$ for a Maxwellian plasma.
4. V. D. Shafranov, in: Plasma Physics and the Problem of Controlled Thermonuclear Reactions, Vol. 4, Pergamon Press, Oxford (1960), p. 489.
5. M. N. Rosenbluth and N. Rostoker, Phys. Fluids, 2:23 (1959).
6. J. E. Drummond, Phys. Rev., 110:293 (1958). The trajectory integral method used in § 7.2 is developed in [4-6].
7. I. B. Bernstein, Phys. Rev., 109:10 (1958). A general dispersion equation is given for a Maxwellian plasma in a magnetic field.
8. R. Z. Sagdeev and V. D. Shafranov, Proceedings of the Second United Nations International Conference on the Peaceful Uses of Atomic Energy, Geneva, 1958, publ. U. N. Geneva (1958), Vol. 31, p. 118.
9. R. Z. Sagdeev and V. D. Shafranov, Zh. Eksp. Teor. Fiz., 39:181 (1960) [Soviet Phys. —JETP, 12:130 (1961)]. In [8, 9], a general expression for the permittivity tensor of a plasma with an anisotropic particle velocity distribution is derived.
10. A. B. Kitsenko and K. N. Stepanov, Zh. Eksp. Teor. Fiz., 38:1840 (1960) [Soviet Phys. —JETP, 11:1323 (1960)]. An expression is given for the components $\varepsilon_{\alpha\beta}$ of an anisotropic plasma in the approximation $\omega \ll \omega_{Bi}$, $k\rho_i \ll 1$.
11. K. N. Stepanov, Zh. Eksp. Teor. Fiz., 34:1292 (1958) [Soviet Phys. —JETP, 7:892 (1958). The tensor $\varepsilon_{\alpha\beta}$ for a Maxwellian plasma is expressed in an integral form suitable for the investigation of perturbations with $\omega \gg \omega_{Bi}$.
12. K. N. Stepanov, Zh. Eksp. Teor. Fiz., 35:1155 (1958) [Soviet Phys. —JETP, 8:808 (1959)]. An expression is given for the permittivity of a Maxwellian plasma.
13. E. G. Harris, Phys. Rev. Lett., 2:34 (1959). An expression is given for ε_0 under the assumption of a δ-function particle velocity distribution.
14. K. N. Stepanov and A. B. Kitsenko, Zh. Tekh. Fiz., 31:167 (1961) [Soviet Phys. — Tech. Phys., 6:120 (1961)]. Expressions are given for $\varepsilon_{ik}^{(\alpha)}$ in the case of a cold stream and in the case of a stream with a Maxwellian distribution of the particles over the random velocities; this paper also contains expressions for the permittivity of cold and hot streams.
15. A. B. Kitsenko and K. N. Stepanov, Zh. Tekh. Fiz., 31:176 (1961) [Soviet Phys. — Tech. Phys., 6, 127 (1961)].. Expressions are given for $\varepsilon_{\alpha\beta}$ in the case of a δ-function particle velocity distribution and for f_0 in the form $f_0 \sim \delta(v_\perp - v_0)\exp(-v_z^2/v_T^2)$.
16. A. B. Mikhailovskii, Nucl. Fusion, 2:162 (1962). Expressions are given for $\varepsilon_{\alpha\beta}$ for $\omega \ll \omega_B$ and arbitrary $k_\perp\rho$ for a Maxwellian plasma.

17. A. B. Mikhailovskii, Zh. Eksp. Teor. Fiz., 43:230 (1962) [Soviet Phys. — JETP, 16:364 (1963)]. Expressions are given for $\varepsilon_{\alpha\beta}$ for $\omega \ll \omega_{B_i}$ and arbitrary $k_\perp \rho$ in the case of an anisotropic particle velocity distribution.
18. V. D. Shafranov, in: Reviews of Plasma Physics, Vol. 3, Consultants Bureau, New York (1967), p. 1. The tensor $\varepsilon_{\alpha\beta}$ for a hot plasma is derived from the equations of two-fluid hydrodynamics with an effective adiabatic exponent (the problem of § 3 in [18]).
19. M. N. Rosenbluth and R. F. Post. Phys. Fluids, 8:547 (1965).
20. V. B. Krasovitskii and K. N. Stepanov, Zh. Tekh. Fiz., 34:1013 (1964) [Soviet Phys. — Tech. Phys., 9:786 (1964)].
21. D. G. Lominadze and K. N. Stepanov, Zh. Tekh. Fiz., 35:441 (1965) [Soviet Phys. — Tech. Phys., 10:347 (1965)]. The papers [19-21] contain exprssions for the scalar permittivity and permittivity tensor in the approximation $\omega \gg \omega_{B_i}$.
22. A. I. Akhiezer et al., Collective Oscillations in a Plasma, Oxford (1967).
23. T. H. Stix, The Theory of Plasma Waves, McGraw-Hill, New York (1962).
24. A. A. Rukhadze and V. P. Silin, Usp. Fiz. Nauk, 74:223 (1961) [Soviet Phys. — Uspekhi, 4:459 (1961)]. The reviews [22-24] and also [18] contain general expressions and various limiting expressions for the permittivity tensor of a plasma in a magnetic field.
25. I. S. Gradshtein and I. M. Ryzhik, Table of Integrals, Series, and Products, Academic Press, New York and London (1965). This book contains a number of helpful relations, including integrals of Bessel functions, their asymptotic behavior, etc.

Chapter 8

Branches of Electrostatic Oscillations of a Plasma in a Magnetic Field

§ 8.1. Oscillations of a Cold Plasma

The aim of this chapter is to obtain a coherent picture of electrostatic oscillations of a Maxwellian plasma in a magnetic field.

The different types of oscillation of a plasma can be classified in different ways. The choice will be dictated by the group of phenomena under consideration. One of the fields of application of the theory of plasma oscillations is the problem of the propagation of waves excited by external sources. In this case the frequency of the source is assumed given and one seeks the refractive index $N = ck/\omega$. When the problem is posed in this manner, the dispersion equation $D(k, \omega) = 0$ must be solved for the function $N = N_\alpha(\omega)$, $\alpha = 1, 2, 3, \ldots$. In the problem of the propagation of waves it is therefore natural to classify the types of oscillations in accordance with the branches of the function $N_\alpha(\omega)$.

In what follows we shall be interested in the problem of plasma stability and we shall therefore consider the excitation of oscillations of the plasma with one or another value of the wave vector $\mathbf{k}$. In this case the dispersion equation is solved for $\omega = \omega_\alpha(\mathbf{k})$, $\alpha = 1, 2, 3, \ldots$. When the problem is posed in this manner, it is more convenient to adopt a classification of the type of oscillation in accordance with the branches of the function $\omega_\alpha(\mathbf{k})$. We shall do this below.

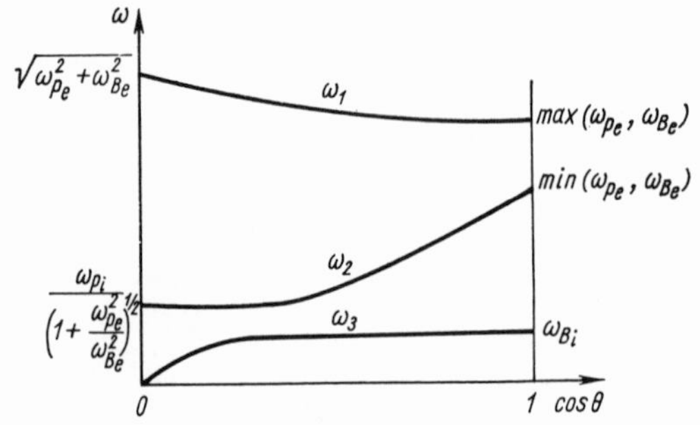

Fig. 8.1. Branches of oscillations of a cold plasma in a magnetic field when $\omega_{p_e} \gg \omega_{B_i}$.

Not all types of oscillation are of equal importance in stability problems; the most important are those that possess a phase velocity that is small compared with the velocity of light (electrostatic oscillations). This is because, by and large, the particles can interact in a resonant manner only with such oscillations.

In this section we shall consider the oscillations of a cold plasma and then take into account the temperature of the particles.

1. Electron Oscillations. In accordance with (7.3) and (1.35), the electron electrostatic oscillations of a cold plasma at rest are described by the dispersion equation

$$1-\frac{\omega_{p_e}^2}{\omega^2}\cos^2\theta-\frac{\omega_{p_e}^2}{\omega^2-\omega_{B_e}^2}\sin^2\theta=0. \tag{8.1}$$

The solution of (8.1) has the form

$$\omega_{1,2}^2=\frac{1}{2}\{\omega_{p_e}^2+\omega_{B_e}^2\pm[(\omega_{p_e}^2+\omega_{B_e}^2)^2-4\omega_{p_e}^2\omega_{B_e}^2\cos^2\theta]^{1/2}\}. \tag{8.2}$$

The dependence of $\omega_{1,2}$ on cos θ is shown in Fig. 8.1. (See § 8.1.3 for a discussion of the frequency ω_3.) The frequencies ω_1 and ω_2 lie in the ranges

$$\left.\begin{aligned}&\max(\omega_{p_e}^2, \omega_{B_e}^2)\leqslant\omega_1^2\leqslant(\omega_{p_e}^2+\omega_{B_e}^2),\\&\min(\omega_{p_e}^2, \omega_{B_e}^2)\geqslant\omega_2^2\geqslant\frac{\omega_{p_e}^2\omega_{B_e}^2}{\omega_{p_e}^2+\omega_{B_e}^2}(\cos^2\theta)_{\min}\end{aligned}\right\} \tag{8.3}$$

The values on the left correspond to $\cos\theta \to 1$ and those on the right to $\cos\theta \to 0$. As $\cos\theta$ decreases, the frequency ω_2 tends to the formal limit zero:

$$\omega_2^2 = \frac{\omega_{p_e}^2 \cos^2\theta}{1 + \frac{\omega_{p_e}^2}{\omega_{B_e}^2}}. \tag{8.4}$$

However, if $\cos\theta$ has values that are too small, the ions become important in the oscillations of the ω_2 branch, and this factor is not taken into account in Eq. (8.1). In the second equation of (8.3), $(\cos\theta)_{\min}$ stands for the minimal value of $\cos\theta$ for which Eq. (8.1) is still valid. An expression for $(\cos\theta)_{\min}$ will be given in § 8.1.2.

In the limiting cases of large and small $\omega_{p_e}^2/\omega_{B_e}^2$ the frequencies ω_1 and ω_2 have the form

$$|\omega_{1,2}| = \left(\omega_{p_e} + \frac{\omega_{B_e}^2}{2\omega_{p_e}} \sin^2\theta;\ \omega_{B_e} \cos\theta\right) \tag{8.5}$$

for $\omega_{p_e}^2 \gg \omega_{B_e}^2$, and

$$|\omega_{1,2}| = \left(\omega_{B_e} + \frac{\omega_{p_e}^2}{2\omega_{B_e}} \sin^2\theta;\ \omega_{p_e} \cos\theta\right) \tag{8.6}$$

for $\omega_{p_e}^2 \ll \omega_{B_e}^2$

2. Influence of the Motion of the Ions on the ω_2 Branch. To find the limit of ω_2 as $\cos\theta \to 0$ in Eq. (8.1), one must take into account the contribution of the ions. Then, instead of (8.1), Eqs. (7.3) and (1.35) yield

$$1 - \frac{\omega_{p_e}^2}{\omega^2}\cos^2\theta + \frac{\omega_{p_e}^2}{\omega_{B_e}^2} - \frac{\omega_{p_i}^2}{\omega^2} = 0. \tag{8.7}$$

Here, we have assumed $\omega_{B_i} \ll \omega \ll \omega_{B_e}$ and $\cos^2\theta \ll 1$. In accordance with (8.7), Eq. (8.2) for ω_2 ceases to hold when

$$\cos^2\theta \lesssim \mu \quad \left(\mu \equiv \frac{m_e}{m_i}\right). \tag{8.8}$$

The frequency of the oscillations of this branch when the condition (8.8) holds is given by the expression

$$\omega_2^2 = \frac{\omega_{p_i}^2}{1 + \dfrac{\omega_{p_e}^2}{\omega_{B_e}^2}}. \tag{8.9}$$

In a plasma with $\omega_{p_e} \gg \omega_{B_e}$ this frequency is equal to the hybrid frequency – the geometric mean of the electron and ion cyclotron frequencies:

$$|\omega_2| = |\omega_{B_i} \omega_{B_e}|^{1/2}. \tag{8.10}$$

In a less dense plasma, $\omega_{B_i} \ll \omega_{p_e} \ll \omega_{B_e}$, the oscillations of the branch ω_2 have a frequency equal to the ion plasma frequency, $\omega_2 = \omega_{p_i}$. These are purely ion oscillations. In a plasma without a magnetic field, oscillations with $\omega \approx \omega_{p_i}$ are possible only if $T_e \gg T_i$ (see § 2.5), whereas the latter condition is not essential if a magnetic field is present.

3. Oscillations with $\omega \lesssim \omega_{B_i}$. In this case the dispersion equation for the oscillations of a cold plasma is

$$1 - \frac{\omega_{p_e}^2}{\omega^2} \cos^2 \theta + \frac{\omega_{p_e}^2}{\omega_{B_e}^2} \sin^2 \theta - \frac{\omega_{p_i}^2}{\omega^2 - \omega_{B_i}^2} \sin^2 \theta = 0. \tag{8.11}$$

This shows that the branches ω_1 and ω_2 are augmented by a further branch of oscillations, which we shall denote by ω_3 (see also Fig. 8.1).

If $\omega_{p_e} \gg \omega_{B_i} \omega_3$ and $\cos\theta$ is not too small, ω_3 is very near the ion cyclotron frequency:

$$\left| \frac{\omega_3}{\omega_{B_i}} - 1 \right| \lesssim \mu^{1/2}. \tag{8.12}$$

If $\cos\theta \ll 1$ and $\omega_{p_i} \gg \omega_{B_i}$, it follows from (8.11) that

$$\omega_3^2 = \frac{\omega_{B_i}^2}{1 + \mu/\cos^2\theta}. \tag{8.13}$$

It can be seen that ω_3 can differ appreciably from ω_{B_i} only if $\cos\theta \lessapprox \mu^{1/2}$. If $\cos^2\theta < \mu$, then Eq. (8.13) shows that the oscillations of the ω_3 branch are of low frequency for both the electrons and the ions ($\omega_3 \ll \omega_{B_i}$):

$$\omega_3 = \omega_{B_i} \cos\theta \left(\frac{m_i}{m_e}\right)^{1/2}. \tag{8.14}$$

4. Oscillations of a Plasma of Very Low Density $\omega_{p_i} \lesssim \omega_{B_i}$. In this case the frequencies of the branches ω_2 and ω_3 are comparable for $\cos^2\theta \ll 1$. If the small term $(\omega_{p_e}^2/\omega_{B_e}^2)\sin^2\theta$ in Eq. (8.11) is neglected, we find

$$\omega_{2,3}^2 = \frac{1}{2}\{\omega_{B_i}^2 + \omega_{p_i}^2 \sin^2\theta + \omega_{p_e}^2 \cos^2\theta \pm [(\omega_{B_i}^2 + \omega_{p_i}^2 \sin^2\theta + \omega_{p_e}^2 \cos^2\theta)^2 - 4\omega_{p_e}^2 \omega_{B_i}^2 \cos^2\theta]^{1/2}\}. \tag{8.15}$$

The frequencies ω_2 and ω_3 determined by this equation are shown in Fig. 8.2. It can be seen that for both high and low densities

$$\left.\begin{array}{l} \omega_{B_i} \leqslant |\omega_2| \leqslant \omega_{B_e}, \\ |\omega_3| \leqslant \omega_{B_i}. \end{array}\right\} \tag{8.16}$$

If the plasma density is so low that $\omega_{p_e} < \omega_{B_i}$, the frequency of the branch ω_3 for all $\cos\theta$ is low compared with the ion-cyclotron frequency:

$$|\omega_3| = \omega_{p_e} \cos\theta < \omega_{B_i}. \tag{8.17}$$

Formally, the dispersion law of this branch is the same as for the branch ω_2 in a plasma with $\mu \ll \omega_{p_e}^2/\omega_{B_e}^2 \ll 1$ and for not too small $\cos\theta$.

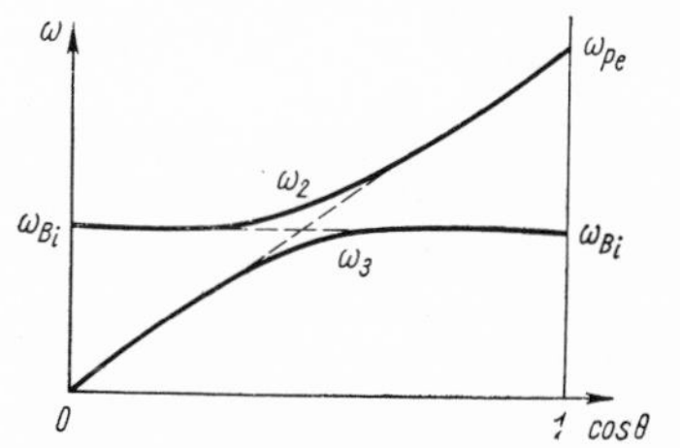

Fig. 8.2. Oscillation branches ω_2 and ω_3 in a low-density plasma ($\omega_{p_i} < \omega_{B_i}$).

8.2. Oscillations with Frequencies Near Electron Cyclotron Harmonics

If the ratio of the electron Larmor radius to the transverse wavelength is finite – that is, if $k_\perp \rho_e \neq 0$ – the branches obtained in the foregoing section are augmented by an infinite number of branches of oscillations whose frequencies lie near the harmonics of the electron cyclotron frequency (or between the harmonics). To see this, we set $k_z = 0$ in Eq. (7.18) and neglect the ion contribution in the dispersion equation $\varepsilon_0 = 0$. Then the latter becomes

$$1-\sum_{n=1}^{\infty}\frac{\omega_{p_e}^2}{\omega^2-(n\omega_{B_e})^2}\cdot\frac{2n^2 I_n(z_e)\,e^{-z_e}}{z_e}=0. \tag{8.18}$$

In the case $\omega_{p_e} \lesssim \omega_{B_e}$, Eq. (8.18) has the approximate solution

$$\omega^2=\omega_e^2(n,\,k_\perp)\equiv n^2\left(\omega_{B_e}^2+\omega_{p_e}^2\,\frac{2I_n(z_e)\,e^{-z_e}}{z_e}\right),\quad n=1,\,2,\ldots. \tag{8.19}$$

If $n = 1$ and $z_e < 1$, this yields an expression for the frequency of the upper branch of oscillations of a cold plasma [cf. with (8.2) for $\cos\theta = 0$]:

$$\omega_e^2(1,\,k)=\omega_1^2=\omega_{p_e}^2+\omega_{B_e}^2\,. \tag{8.20}$$

In accordance with Eq. (8.19) and the asymptotic formula $i_n e^{-z} = (2\pi z)^{-1/2}$ the frequency of short-wavelength oscillations, $z_e \gg n$, is

$$\omega_e(n,\,k)=n\omega_{B_e}\left(1+\frac{1}{k^2 d_e^2}\cdot\frac{1}{\sqrt{2\pi z_e}}\right). \tag{8.21}$$

The oscillation branches with the frequencies ω_e (n, k) determined by Eq. (8.19) are shown in Fig. 8.3.

A similar, but slightly more complicated picture is obtained for the oscillation branches when $\omega_{p_e} > \omega_{B_e}$ [see, for example, Crawford's review].

Let us consider the conditions under which the electron cyclotron oscillations are weakly damped. The condition that the decay rate of the oscillations should be exponentially small is

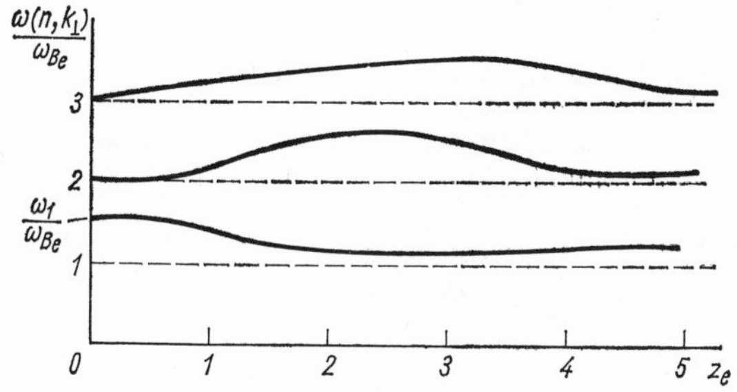

Fig. 8.3. Electron-cyclotron harmonics in a low-density plasma.

$|\omega - n\omega_{B_e}| \gg k_z v_{T_e}$. For the characteristic value $k_\perp \simeq 1/\rho_i$, we therefore obtain

$$\cos\theta \ll 1, \tag{8.22}$$

i.e., the wave vector must be almost (or strictly) perpendicular to the direction of the magnetic field.

§ 8.3. Ion-Acoustic and Ion Plasma Oscillations of a Plasma in a Magnetic Field

In the approximation $\omega \ll k_z v_{T_e}$ in a plasma with $B_0 \neq 0$ there can exist two new branches of oscillations, which we shall denote by ω_4 (k) and ω_5 (k). Using the expression (7.18) for $\varepsilon_0^{(e)}$ and (7.3) for $\varepsilon_0^{(i)}$, we obtain the dispersion equation for $\omega \ll k_z v_{T_e}$ and $T_i \to 0$. [In the expression (7.18) we expand $W(\omega/k_z v_T)$ in a series in the small argument.] The imaginary terms of the dispersion equation are of the order $\omega/k_z v_{T_e} \ll 1$. The equation for the real part of the frequency is

$$\mathrm{Re}\,\varepsilon_0(\omega, k) = 1 + \frac{1}{k^2 d_e^2} - \frac{\omega_{p_i}^2 \cos^2\theta}{\omega^2} - \frac{\omega_{B_i}^2 \sin^2\theta}{\omega^2 - \omega_{B_i}^2} = 0. \tag{8.23}$$

Therefore, for perturbations with $k^2 d_e^2 \ll 1$, we find

$$\omega_{4,5}^2 = \frac{1}{2}\left[\omega_{B_i}^2 + k^2 V_s^2 \pm \sqrt{(\omega_{B_i}^2 + k^2 V_s^2)^2 - 4k_z^2 V_s^2 \omega_{B_i}^2}\right]. \tag{8.24}$$

Here, $V_s = (T_e/m_i)^{1/2}$ is the velocity of propagation of the ion-acoustic oscillations in the absence of a magnetic field [cf. (2.65)].

Let us consider the meaning of the solutions (8.24) and the limits of applicability of these expressions.

Suppose that, in addition to the condition that the plasma be nonisothermal, $T_e \gg T_i$, and the condition $k_\perp \rho_i \ll 1$, the inequality

$$k^2 V_s^2 \gg \omega_{B_i}^2 \tag{8.25}$$

is also satisfied; then the solution $\omega = \omega_4(k)$ corresponds to high-frequency oscillations ($\omega \gg \omega_{B_i}$) and is identical with (2.65). In this limiting case the frequency of the ω_4 branch is quite independent of the magnetic field. However, the magnetic field does affect the damping of the oscillations. In accordance with Eq. (2.17), the damping of the oscillations is determined by the imaginary part of the expression for ε_0 – in the given case by

$$\operatorname{Im} \varepsilon_0 = \frac{\sqrt{\pi}\,\omega}{|k_z|\, v_{T_e}} \cdot \frac{1}{(kd_e)^2} \sum_{n=-\infty}^{\infty} I_n(z_e) \exp\left[-z_e - \left(\frac{\omega - n\omega_{B_e}}{k_z v_{T_e}}\right)^2\right]. \tag{8.26}$$

This shows that, in contrast to the ion-acoustic oscillations for $\mathbf{B}_0 = 0$, the decay rate of the branch ω_4 depends strongly on the magnetic field. This dependence is weak only if $z_e \gg 1$, $k_z v_{T_e} \gg \omega_{B_e}$. In this limiting case the series (8.26) can be summed and the final result is independent of B_0.

We shall say that the oscillations of the branch ω_4 when the condition (8.25) is satisfied are **high-frequency ion-acoustic oscillations**. At even larger k ($k^2 d_e^2 \gtrsim 1$), the branch ω_4 becomes the ion plasma oscillations discussed in § 2.5.

As the wave number decreases to values satisfying

$$k^2 V_s^2 \lesssim \omega_{B_i}^2 \tag{8.27}$$

the frequency of the branch ω_4 decreases and is then of the order of the ion-cyclotron frequency (see Fig. 8.4).

If the strong inequality (8.27) is satisfied, the frequency of the branch ω_4 tends to the ion-cyclotron frequency. In this case,

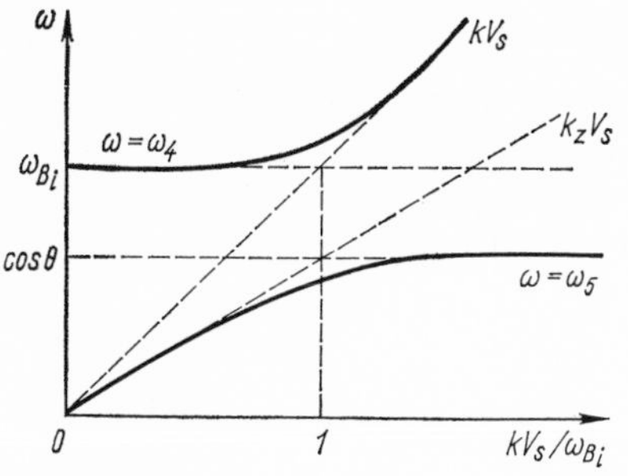

Fig. 8.4. High-frequency, $\omega = \omega_4$, and low-frequency, $\omega = \omega_5$, ion-acoustic oscillations.

the frequency of the branch ω_5 is determined by the equation

$$\omega_5^2 = k_z^2 V_s^2 . \tag{8.28}$$

In the opposite limiting case of large k [condition (8.25)] the frequency of the branch ω_5 is approximately equal to

$$\omega_5 = \omega_{B_i} \cos\theta. \tag{8.29}$$

In contrast to (2.49), we shall call the oscillations of the type (8.28) low-frequency ion-acoustic oscillations.

In the approximation of low-frequency oscillations ($\omega \ll \omega_{B_i}$), Eqs. (8.23) and (8.26) reduce to

$$1 + \frac{1}{(kd_e)^2}\left(1 + i\sqrt{\pi}\,\frac{\omega}{|k_z|\, v_{T_e}}\right) + \frac{\omega_{p_i}^2 \sin^2\theta}{\omega_{B_i}^2} - \frac{\omega_{p_i}^2 \cos^2\theta}{\omega^2} = 0. \tag{8.30}$$

From this we obtain the corrections to (8.28) due to the fact that the parameters $(kd_e)^2$, $(k_\perp V_s)^2$ and $\omega/k_z v_{T_e}$ are finite:

$$\omega^2 = (k_z V_s)^2 \left[1 + k^2 d_e^2 + \frac{k_\perp^2}{\omega_{B_i}^2} V_s^2\right]^{-1} \tag{8.31}$$

$$\gamma = -\sqrt{\frac{\pi}{8}}\,\mu^{1/2}\,\frac{\omega^2}{|k_z|\, V_s}\left[1 + k^2 d_e^2 + \frac{k_\perp^2}{\omega_{B_i}^2} V_s^2\right]^{-1}. \tag{8.32}$$

Relationship between the Branches ω_2 and ω_4 in the Case $\cos^2\theta \le \mu$. As $\cos\theta$ decreases, the ratio $\omega/k_z v_{T_e}$ for the branch ω_4 [see Eq. (2.60)] increases and at $\cos\theta \approx \mu^{1/2}$ is of order unity. Under these conditions, the oscillations are described by the dispersion equation

$$1+\frac{1}{(kd_e)^2}\left[1+i\sqrt{\pi}xI_0(z_e)e^{-z_e}W(x)\right]-\frac{\omega_{p_i}^2}{\omega^2}=0,$$
$$x\equiv\frac{\omega}{|kz|\,v_{T_e}}. \tag{8.33}$$

In the limiting case $\omega \gg k_z v_{T_e}$, Eq. (8.33) yields

$$1+\frac{\omega_{p_e}^2}{\omega_{B_e}^2}\cdot\frac{1-I_0e^{-z_e}}{z_e}-\frac{\omega_{p_e}^2}{\omega^2}I_0e^{-z_e}\cos^2\theta-\frac{\omega_{p_i}^2}{\omega^2}=0. \tag{8.34}$$

If we assume $z_e \ll 1$ in (8.34), it reduces to Eq. (8.7) for the branch ω_2. The oscillations described by Eq. (8.34) can therefore be regarded as a continuation of the branch ω_2 to the region of shorter wavelengths, $z_e \gtrless 1$. For arbitrary z_e, the solution of (8.34) has the form

$$\omega_2^2=\frac{\omega_{p_i}^2+\omega_{p_e}^2 I_0e^{-z_e}\cos^2\theta}{1+\frac{\omega_{p_e}^2}{\omega_{B_e}^2}\cdot\frac{1-I_0e^{-z_e}}{z_e}}. \tag{8.35}$$

Figure 8.5 shows the dependence $\omega_2 = \omega_2(k)$ for $\cos\theta = 0$ in a plasma with $\omega_{p_e} \gg \omega_{B_e}$. Under these conditions and for $z_e \ll 1$, the frequency of the oscillations is equal to the hybrid frequency (8.10). At large wave numbers in the interval

$$\frac{1}{\rho_e}<k<\frac{1}{d_e}, \tag{8.36}$$

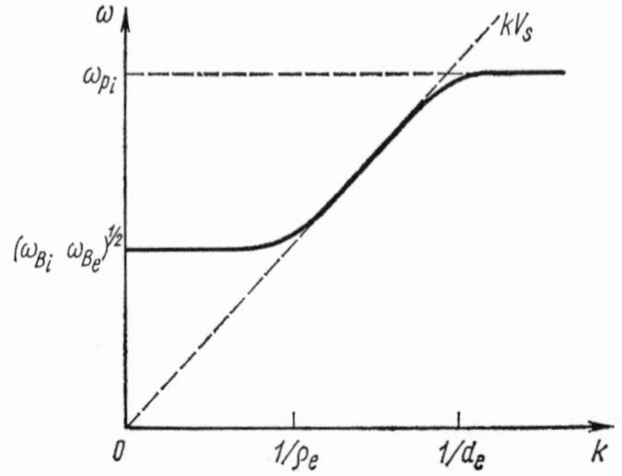

Fig. 8.5. Correspondence between the branches ω_2 and ω_4 for small $\cos\theta$.

the frequency of the branch ω_2 is, in accordance with (8.35), the same as for the high-frequency ion-acoustic branch.

8.4. Ion-Cyclotron and Electron-Acoustic Oscillations

1. Ion-Cyclotron Oscillations for $k_z = 0$. If $\omega \ll \omega_{B_e}$, $k_\perp \rho_e \ll 1$, $k_z = 0$, and $k_\perp \rho_i \neq 0$ the dispersion equation (1.35) with ε_0 of the form (7.18) reduces to

$$1+\left(\frac{\omega_{p_e}}{\omega_{B_e}}\right)^2+\frac{1}{(kd_i)^2}\left(1-\sum_{n=-\infty}^{\infty}\frac{\omega I_n(z_i)\,e^{-z_i}}{\omega-n\omega_{B_i}}\right)=0. \tag{8.37}$$

Except for the notation, this equation is identical with that for the electron-cyclotron oscillations (8.18).

The graphical representation of the branches of the ion-cyclotron oscillations in the limiting case $\omega_{p_i} \ll \omega_{B_i}$ is as in Fig. 8.3 if ω_1 is replaced by ω_2, where ω_2 is defined by Eq. (8.15).

2. Low-Frequency Electron-Acoustic Oscillations. In the limit of cold ions, the oscillations of the low-frequency part of the branch ω_3 (see Fig. 8.1) are described by Eq. (8.14). If allowance is made for the fact that the ion Larmor radius is finite ($z_i \neq 0$), Eq. (8.11) is replaced by

$$\varepsilon_0=1-\frac{\omega_{p_e}^2}{\omega^2}\cos^2\theta+\frac{\omega_{p_i}^2}{\omega_{B_i}^2}\cdot\frac{1-I_0(z_i)\,e^{-z_i}}{z_i}=0. \tag{8.38}$$

This equation follows from (7.24) under the assumptions $z_e \ll 1$, $(k_z v_{T_e},\ k_z v_{T_i}) \ll \omega \ll \omega_{B_i}$. From (8.38)

$$\omega^2=k_z^2\,\frac{T_i}{m_e}\cdot\frac{1}{1+k^2d_i^2-I_0(z_i)\,e^{-z_i}}. \tag{8.39}$$

In the limit $z_i \gg 1$ and $\omega_{p_i}^2 \gg \omega_{B_i}^2$, the frequency of these oscillations is

$$\omega=\omega_6\equiv k_z\,(T_i/m_e)^{1/2}. \tag{8.40}$$

This result is identical with that for the frequency of the low-frequency ion-acoustic oscillations (8.28) if the ion and electron subscripts are exchanged. In this connection we shall say that oscillations of the type (8.40) are low-frequency electron-acoustic oscillations. The restriction imposed by the low-frequency condition $\omega \ll \omega_{B_i}$ on the interval of angles between **k** and $\mathbf{B}_0$ has the following form for these oscillations:

$$\cos\theta < (\mu/z_i)^{1/2}. \qquad (8.41)$$

Like the ion-acoustic oscillations, the branch (8.40) can exist only if the plasma is nonisothermal. However, the relationship between the ion and electron temperatures must be the opposite of that needed in the case of ion-acoustic oscillations – the ions must be hotter:

$$T_i \gg T_e. \qquad (8.42)$$

If this is not so, Eq. (8.40) shows that the condition $\omega \gg k_z v_{T_e}$ will be violated.

In a low-density plasma, $\omega_{p_i} \ll \omega_{B_i}$, the dispersion law of short-wavelength oscillations, $z_i \gtrsim 1$, has, in accordance with (8.38), the same form as in the case $z_i \ll 1$ and is determined by the formula (8.17) obtained in § 8.1 for magnetized plasma oscillations. This is because the contribution of the ions to the permittivity at such low densities is negligibly small for all z_i.

3. Electron-Acoustic and Ion-Cyclotron Oscillations for $1 \gg \cos\theta \gg \sqrt{\mu/z_i}$. If $\cos\theta > \sqrt{\mu/z_i}$, the frequency of the electron-acoustic oscillations becomes comparable with ω_{B_i}, and this leads to an appreciable distortion of the simple dispersion law (8.40) by cyclotron effects. In accordance with (7.18), the dispersion equation has the form

$$\varepsilon_0 = 1 + \frac{\omega_{p_e}^2}{\omega_{B_e}^2} - \frac{\omega_{p_e}^2}{\omega^2}\cos^2\theta + \frac{1}{k^2 d_i^2}\left(1 - \omega \sum_{n=-\infty}^{\infty} \frac{I_n(z_i)\,e^{-z_i}}{\omega - n\omega_{B_i}}\right) = 0. \qquad (8.43)$$

Noting that the functions $I_n(z_i)e^{-z_i}$ are numerically small for subscripts $n \neq 0$, we find that the dispersion law of oscillations with frequencies not too near cyclotron harmonics is determined, as in the case $\omega \ll \omega_{B_i}$, by Eq. (8.39). On the other hand, if k_z is such that the right-hand side of Eq. (8.39) is not too near $(n\omega_{B_i})^2$, the frequency of the cyclotron oscillations is

$$\omega = n\omega_{B_i}\left\{1 + I_n(z_i)\,e^{-z_i}\left[1 - I_0 e^{-z_i} + k^2 d_i^2\left(1 + \frac{\omega_{p_e}^2}{\omega_{B_e}^2} - \frac{\omega_{p_e}^2 \cos^2\theta}{n^2\omega_{B_i}^2}\right)\right]^{-1}\right\}. \tag{8.44}$$

Equations (8.39) and (8.44) cease to be valid if the frequency of the electron-acoustic oscillations is approximately equal to $n\omega_{B_i}$. This is the case when

$$\cos\theta \simeq \cos\theta_0 \equiv n\sqrt{\mu/z_i}\left[1 - I_0(z_i)\,e^{-z_i} + k^2 d_i^2\left(1 + \frac{\omega_{p_e}^2}{\omega_{B_e}^2}\right)\right]^{1/2}. \tag{8.45}$$

An approximate solution of Eq. (8.43) for $\cos\theta \approx \cos\theta_0$ can be found in the same way as the solution of the dispersion equation (1.46) in the problem of the resonant two-stream instability (see § 1.5.2). The general form of the oscillation branches that is obtained is shown in Fig. 8.6. This figure shows that the branches corresponding to electron-acoustic oscillations go over into cyclotron oscillations when $\cos\theta$ is increased and, conversely, the branches that correspond to cyclotron oscillations for $\cos\theta <$ $\cos\theta_0$ become electron-acoustic oscillations.

4. Electron-Acoustic Oscillations for $\omega \gg \omega_{B_i}$

If $\omega \gg \omega_{B_i}$, the frequency intervals in which the dispersion law of the electron-acoustic oscillations changes become very narrow. If these intervals are ignored, the behavior of the ions in the field of the oscillations is the same as for $B_0 = 0$. This corresponds to the high-frequency approximation discussed in § 7.4. In this ap-

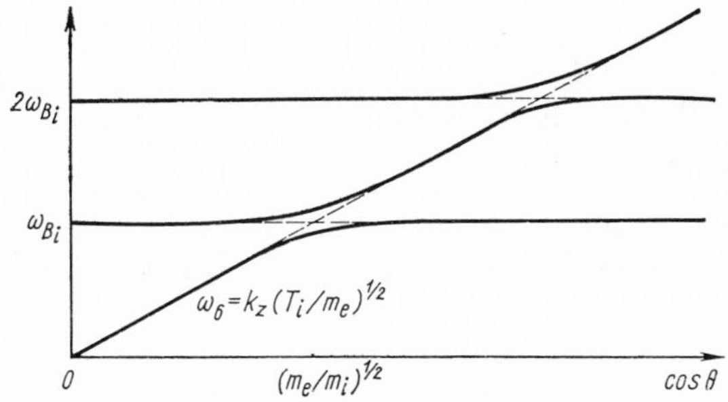

Fig. 8.6. Electron-acoustic and ion-cyclotron oscillations.

proximation, the permittivity of the ions is determined by Eq. (7.29), in which one must set $\omega \ll kv_{T_i}$. In this high-frequency limit, the electron-acoustic oscillations are described by the dispersion equation

$$\varepsilon_0 = 1 + \frac{\omega_{p_e}^2}{\omega_{B_e}^2} - \frac{\omega_{p_e}^2}{\omega^2}\cos^2\theta + \frac{1}{k^2 d_i^2} = 0. \tag{8.46}$$

The frequency of the oscillations is

$$\omega = \omega_6 \equiv k_z (T_i/m_e)^{1/2} (1 + k^2 d_i^2 + k^2 \rho_i^2 \mu)^{-1/2}. \tag{8.47}$$

For finite $k_\perp \rho_e$, Eq. (8.47) is replaced by an equation obtained by means of (7.23) and (7.29):

$$\omega^2 = \frac{k_z^2 T_e}{m_e}\left[3 + \frac{I_0(z_e)\,e^{-z_e}}{1 - I_0(z_e)\,e^{-z_e} + \tau^{-1} + k^2 d_e^2}\right],$$
$$\tau = T_i/T_e. \tag{8.48}$$

This equation also takes into account small terms of order $(k_z v_{T_e}/\omega)^2$.

This result satisfies the original assumption $\omega \gg k_z v_{T_e}$ provided z_e is small:

$$z_e \ll 1. \tag{8.49}$$

At larger values of z_e, the electron-acoustic oscillations are strongly damped, $\operatorname{Im}\omega \approx \operatorname{Re}\omega$.

5. Ion-Cyclotron Oscillations in a Plasma with a Finite Electron Temperature. If the parameter $k_z v_{T_e}/\omega$ has a finite value, the dispersion equation (8.18) is replaced by (for $z_e \ll 1$)

$$\varepsilon_0 = 1 + \frac{\omega_{p_e}^2}{\omega_{B_e}^2} + \frac{1}{k^2 d_e^2}\left[1 + i\sqrt{\pi}\,xW(x)\right] + \frac{1}{k^2 d_i^2}\left[1 - \omega\sum \frac{I_n(z_i)\,e^{-z_i}}{\omega - n\omega_{B_i}}\right] = 0, \tag{8.50}$$

where $x = \omega/k_z v_{T_e}$. It follows that in a plasma with $T_e \gtrsim T_i$ oscillations of the electron-acoustic type are strongly damped. For the ion-cyclotron branches of oscillations, Eq. (8.50) yields

$$\omega = n\omega_{B_i}\Big\{1 + I_n(z_i)\,e^{-z_i}\Big[1 - I_0 e^{-z_i} + k^2 d_i^2\Big(1 + \frac{\omega_{p_e}^2}{\omega_{B_e}^2}\Big) + \tau\Big(1 + i\sqrt{\pi}x_n W(x_n)\Big)\Big]^{-1}\Big\}, \tag{8.51}$$

$$x_n = n\omega_{B_i}\,|k_z|\,v_{T_e}.$$

In the limiting case $k_z v_{T_e} \gg n\omega_{B_i}$, the frequency of the oscillations (8.51) is almost real, the real part satisfying

$$\mathrm{Re}\,\omega = n\omega_{B_i}\Big\{1 + \frac{I_n(z_i)\,e^{-z_i}}{1 + \tau + k^2 d_i^2 - I_0(z_i)\,e^{-z_i}}\Big\}. \tag{8.52}$$

For $z_i \ll 1$ and $n = 1$, these oscillations correspond to the left-hand section of the branch ω_4 in Fig. 8.4.

The condition that the ion-cyclotron damping of oscillations of the type (8.52) be small leads to a restriction on the interval of angles between **k** and $\mathbf{B}_0$ for which these oscillations exist:

$$\cos\theta < \frac{n}{\sqrt{z_i}} \cdot \frac{I_n(z_i)\,e^{-z_i}}{1 + \tau + k^2 d_i^2 - I_0(z_i)\,e^{-z_i}}. \tag{8.53}$$

In calculating the decay rate of the oscillations in Eq. (8.50), it must be remembered that the parameter $k_z v_{T_i}/(\omega - n\omega_{B_i})$ is finite. Taking into account these terms and assuming that $\omega \simeq n\omega_{B_i}$, $\omega \ll k_z v_{T_e}$ and $\omega_{p_i}^2 \gg \omega_{B_i}^2$, we can represent the permittivity in the form

$$\mathrm{Re}\,\varepsilon_0 = \frac{1}{(kd_i)^2}\Big[1 + \tau - I_0 e^{-z_i} - \frac{n\omega_{B_i}}{\Delta} I_n e^{-z_i}\Big(1 + \frac{k_z^2 T_i}{m_i \Delta^2}\Big)\Big]; \tag{8.54}$$

$$\mathrm{Im}\,\varepsilon_0 = \frac{\sqrt{\pi}\,\omega}{k^2 d_e^2 |k_z| v_{T_e}}\Big[1 + \mu^{-1/2}\tau^{-3/2} I_n e^{-z} e^{-\Delta^2/k_z^2 v_{T_i}^2}\Big], \tag{8.55}$$

where $\Delta = \omega - n\omega_B$. Using (8.54), (8.55), and (2.17), we find

$$\gamma_n = -\frac{\Delta_0^2\sqrt{\pi}}{|k_z| v_{T_e}}\,\tau\Big[1 + \mu^{-1/2}\tau^{-3/2} I_n e^{-z} e^{-\left(1 + \Delta^2/k_z^2 v_{T_i}^2\right)}\Big], \tag{8.56}$$

where Δ_0 is Δ calculated without allowance for the thermal corrections in (8.54) [i.e., $\Delta_0 = \mathrm{Re}\,\omega - n\omega_B$ for $\mathrm{Re}\,\omega$ equal to (8.52)].

Bibliography

1. A. I. Akhiezer and L. É. Pargamanik, Trudy Fiz. Mat. Fak. Khar'k. Univ., 27:75 (1948). The electron oscillations of a cold plasma (§ 8.1) are considered.
2. E. P. Gross, Phys. Rev., 82:232 (1951). A study is made of electron perturbations with $k_z = 0$ and $k_\perp \rho_e \ll 1$, allowance being made for terms of order $(k_\perp \rho_e)^2$ (§ 8.2).
3. G. V. Gordeev, Zh. Eksp. Teor. Fiz., 23:660 (1952). Electron perturbations are investigated with allowance for thermal motion (§§ 8.1 and 8.2).
4. A. G. Sitenko and K. N. Stepanov, Zh. Eksp. Teor. Fiz., 31:642 (1956). A study is made of the influence of thermal motion on electron oscillations.
5. S. I. Braginskii, Dokl. Akad. Nauk SSSR, 115:475 (1957). The two-fluid hydrodynamic model with an effective adiabatic exponent is used to obtain a dispersion equation from which the dependence $\omega = \omega_\alpha(k)$, $\alpha = 1, 2, \ldots$, is found for a number of limiting cases. The types of oscillation of a plasma in a magnetic field are classified on the basis of the correspondence between the types of oscillation and the set of dispersion branches $\omega = \omega_\alpha(k)$. (To a certain extent, this classification is similar to the classification adopted in the present chapter.)
6. I. B. Bernstein, Phys. Rev., 109:10 (1958). A direct calculation shows that electrostatic perturbations of a Maxwellian plasma cannot grow in time. A study is made of the damping of the electrostatic perturbations of the branches ω_1 and ω_2. A dispersion equation for low-frequency ion-acoustic oscillations (§ 8.3) is derived and investigated.
7. K. N. Stepanov, Zh. Eksp. Teor. Fiz., 35:1155 (1958) [Soviet Phys. — JETP, 8:1155 (1958)]. A dispersion equation is derived for the electrostatic oscillations of the branch ω_2 for $\cos\theta < (m_e/m_i)^{1/2}$ (§ 8.1). A study is made of the kinetic damping of electrostatic electron-ion oscillations (§§ 8.1, 8.3, 8.4).
8. Yu. N. Dnestrovskii and D. P. Kostomarov, Zh. Eksp. Teor. Fiz., 41:1527 (1961) [Soviet Phys. — JETP, 14:1089 (1962)]. Electron perturbations with $k_z = 0$ (§ 8.2) are studied.
9. W. E. Drummond and M. N. Rosenbluth, Phys. Fluids, 5:1507 (1962).
10. D. G. Lominadze and K. N. Stepanov, Zh. Tekh. Fiz., 34:1823 (1964) [Soviet Phys. — Tech. Phys., 9:1408 (1965)]. The papers [9, 10] contain investigations of ion-cyclotron oscillations. Allowance is made for the finite values of T_e and T_i (§§ 8.3 and 8.4). In [10] ion-acoustic branches of oscillations (§ 8.3) are also investigated.
11. A. B. Mikhailivskii, Zh. Tekh. Fiz., 37:1365 (1967) [Soviet Phys. — Tech. Phys., 12:993 (1968)]. Electron-acoustic oscillations are discussed.
12. F. W. Crawford, Nucl. Fusion, 5:73 (1965). This paper reviews oscillations with frequencies near electron-cyclotron harmonics.
13. V. D. Shafranov, in: Reviews of Plasma Physics, Vol. 3, Consultants Bureau, New York (1967), p. 1.
14. A. I. Akhiezer et al., Collective Oscillations in a Plasma, Oxford (1967).

15. T. H. Stix, The Theory of Plasma Waves, McGraw-Hill, New York (1962).
16. A. A. Rukhadze and B. P. Silin, Usp. Fiz. Nauk, 74:223 (1961) [Soviet Phys. — Uspekhi, 4:459 (1961)]. Aspects of the theory of plasma oscillations in a magnetic field are discussed in [13-16] in more detail than in the present chapter.

Chapter 9

Plasma with Longitudinal Electron Beams

9.1. Cold Plasmas and a Beam for $\omega_{p_e} > \omega_{B_e}$

In the approximation $B_0 = 0$, we have shown (§ 1.5) that the hydrodynamic two-stream instability can arise in a plasma consisting of cold beams. The results obtained in this approximation remain in force when $\mathbf{B}_0 \neq 0$ if $k_z \gg k_\perp$ (it is assumed that $V \parallel B_0$), or if $\gamma > \omega_B$. If neither of these conditions is fulfilled, the influence of the magnetic field on the motion of the particles must be taken into account.

The dispersion equation for a system of cold beams in a magnetic field can be obtained by using the expression for the scalar permittivity of a beam (7.2) and Eq. (1.35):

$$1 - \Sigma\omega_p^2 \left[\frac{\cos^2\theta}{(\omega - k_z V)^2} + \frac{\sin^2\theta}{(\omega - k_z V)^2 - \omega_B^2} \right] = 0, \tag{9.1}$$

where the summation is extended over the streams.

Let us consider the effect that a magnetic field has on the two-stream instability that arises when a low-density beam passes through a plasma at rest; we shall assume that the magnetic field is not too strong, $\omega_{p_e} \gtrsim \omega_{B_e}$. In this case, Eq. (9.1) can be conveniently represented in the form

$$\varepsilon_0^{(0)} + \varepsilon_0^{(1)} = 0. \tag{9.2}$$

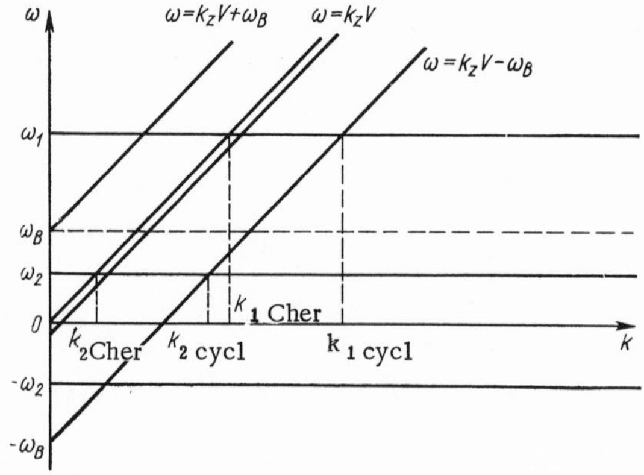

Fig. 9.1. Branches of oscillations of plasma and beam in the approximation $\alpha \to 0$.

Here

$$
\left.
\begin{aligned}
\varepsilon_0^{(0)} &= 1 - \frac{\omega_p^2}{\omega^2}\cos^2\theta - \frac{\omega_p^2 \sin^2\theta}{\omega^2 - \omega_B^2},\\
\varepsilon_0^{(1)} &= -\frac{\alpha\omega_p^2 \cos^2\theta}{(\omega - k_z V)^2} - \frac{\alpha\omega_p^2 \sin^2\theta}{(\omega - k_z V)^2 - \omega_B^2},
\end{aligned}
\right\} \qquad (9.3)
$$

$\alpha = n_1/n_0$; V is the velocity of the beam; ω_p is the plasma frequency of the plasma at rest.

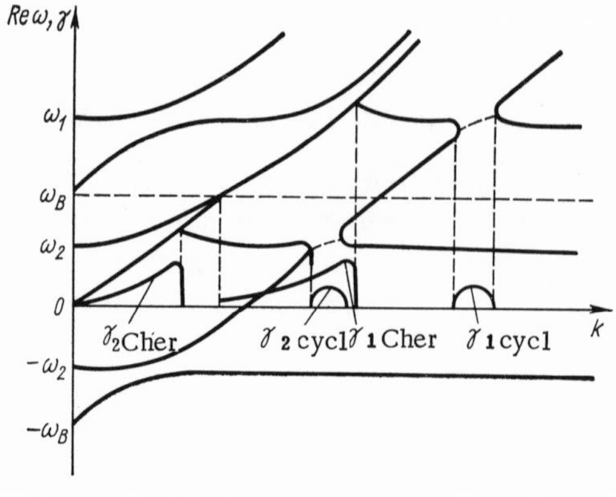

Fig. 9.2. Branches of oscillations of plasma and beam for finite α.

In the zeroth approximation in α, Eq. (9.2) describes the oscillations of the plasma at rest with the frequencies (8.2) and beam oscillations with

$$\omega = k_z V + n\omega_B, \qquad n = 0, \pm 1. \tag{9.4}$$

The roots of (9.2) obtained in this approximation are shown in Fig. 9.1. The double line $\omega = k_z V$ in this figure corresponds to two equal roots. We shall give the name Cherenkov branches to those of the type $\omega = k_z V$ and cyclotron branches to those of the type $\omega = k_z V \pm \omega_B$.

For finite (i.e., nonvanishing) values of α, Fig. 9.1 is replaced by Fig. 9.2. The latter is based on the treatment that follows. As in the case $B_0 = 0$ (see §1.5.2), the frequencies of the oscillations are in general complex if allowance is made for the finite value of α. In Fig. 9.2 we have plotted the real and imaginary parts of ω (cf. Fig. 1.1).

1. Instability of Cherenkov Beam Oscillations. The frequencies of the beam branches (9.4) with $n = 0$ and nonvanishing α are [cf. (1.47)]

$$\omega = k_z V \pm \frac{\alpha^{1/2}\omega_p\,|\cos\theta\,|}{\left(\sqrt{\varepsilon_0^{(0)}(k,\,\omega)}\right)_{\omega=k_z V}}. \tag{9.5}$$

In accordance with (9.3), $\varepsilon_0^{(0)}(\mathbf{k}, \omega)$ is negative if ω lies in one of the intervals

$$\left.\begin{array}{l} 0 < \omega < \omega_2, \\ \omega_B < \omega < \omega_1, \end{array}\right\} \tag{9.6}$$

where ω_1 and ω_2 are defined by Eq. (8.2). From this it follows that the solutions (9.5) are complex for k_z that satisfy one of the following conditions (see Fig. 9.2):

$$\left.\begin{array}{l} \omega_B/V < k_z < \omega_1/V, \\ 0 < k_z < \omega_2/V. \end{array}\right\} \tag{9.7}$$

For k_z in these intervals the growth rate of the oscillations (9.5) is

$$\gamma = \frac{\alpha^{1/2}\omega_p\,|\cos\theta\,|}{\left(\sqrt{-\varepsilon_0^{(0)}}\right)_{\omega=k_z V}}. \tag{9.8}$$

The maximal growth rate is attained at the upper limits of the intervals (9.7) corresponding to the "intersection" of the beam branches $\omega = k_z V$ with the branches ω_1 and ω_2 of the plasma oscillations (see Fig. 9.1), i.e., for

$$k_z = k_{z1,2} \equiv \omega_{1,2}/V. \tag{9.9}$$

If k_z satisfies this condition, $\varepsilon_0^{(0)} = 0$, and, as a result, Eq. (9.5) becomes inapplicable. Using the method of § 1.5.2, we find that the maximal growth rate is

$$\gamma_{1,2\max} = \frac{\sqrt{3}}{2}\alpha^{1/3}\left(\frac{\omega_p^2 \cos^2\theta}{\partial\varepsilon_0^{(0)}/\partial\omega}\right)^{1/3}_{\omega=\omega_{1,2}}. \tag{9.10}$$

In the limiting case $\omega_p \gg \omega_B$, Eqs. (9.9) and (9.10) yield

$$k_{z1,2} = \left(\frac{\omega_p}{V}, \frac{\omega_B\cos\theta}{V}\right), \tag{9.11}$$

$$\gamma_{1,2\max} = \frac{\sqrt{3}}{2^{4/3}}\alpha^{1/3}\omega_p\left((\cos\theta)^{2/3}, \frac{\omega_B}{\omega_p}\cos\theta\,(\sin\theta)^{2/3}\right). \tag{9.12}$$

The real part of the expression for the frequency is approximately equal to (8.5) (without allowance for terms of order $\alpha^{1/3}$).

The value of $\gamma_{1\max}$ differs from that in the expression (1.52) obtained in the approximation $B_0 = 0$ by the factor $(\cos\theta)^{2/3}$. This is because in deriving (9.10) we ignored the contribution of the last term of the right-hand side of the first equation in (9.3). When $\sin\theta \approx \cos\theta$, this approximation is justified if $\gamma_{1\max} \ll \omega_B$, i.e., for

$$\omega_B/\omega_p > \alpha^{1/3}. \tag{9.13}$$

2. Instability at the "Intersection" of the Cyclotron Beam Branches and the Branches ω_1 and ω_2 of the Plasma Oscillations. The cyclotron beam branches [see Eq. (9.4) with $n \neq 0$] neither grow or decay, $\operatorname{Im}\omega = 0$, if their frequency is not near ω_1 or ω_2 (see Fig. 9.1). Let us now assume that

$$k_z V \pm \omega_B = \omega_{1,2}. \tag{9.14}$$

In this case, allowance for the finite value of α in Eq. (9.2) leads to the following equation for the correction $\delta\omega$ to the frequencies of the intersecting branches:

$$(\delta\omega_{1,2}^{(\pm)})^2 = \pm \frac{\alpha\omega_p^2 \sin^2\theta}{2\omega_B \left(\dfrac{\partial\varepsilon_0^{(0)}}{\partial\omega}\right)_{\omega=\omega_{1,2}}}. \tag{9.15}$$

The indices 1, 2, and ($\pm$) indicate the numbers of the intersecting branches. In what follows we shall assume $\omega_{1,2} > 0$ and $\omega_B > 0$.

Equation (9.15) shows that if the branches ω_1 or ω_2 intersect the branch $\omega = k_z V + \omega_B$ (normal Doppler effect) the correction is real (see Fig. 9.2). Instability occurs only when $\omega_{1,2} = k_z V - \omega_B$ (anomalous Doppler effect; see Fig. 9.2). The wave number and the growth rate of the oscillations that are excited are determined by the equations

$$k_{z1,2} = (\omega_B + \omega_{1,2})/V, \tag{9.16}$$

$$\gamma_{1,2} = \alpha^{1/2}\omega_p \sin\theta \left(\frac{1}{2\omega_B \partial\varepsilon_0^{(0)}/\partial\omega}\right)^{1/2}_{\omega=\omega_{1,2}}. \tag{9.17}$$

If $\omega_p^2 \gg \omega_B^2$, Eqs. (9.16) and (9.17) yield

$$k_{z1,2} \simeq \left(\frac{\omega_p}{V},\ \frac{\omega_B(1+|\cos\theta|)}{V}\right), \tag{9.18}$$

$$\gamma_{1,2} = \frac{\alpha^{1/2}}{2}\omega_p \sin\theta \left\{\left(\frac{\omega_p}{\omega_B}\right)^{1/2},\ \frac{\omega_B}{\omega_p}(\cos\theta)^{1/2}\sin\theta\right\}. \tag{9.19}$$

The expression (9.19) for γ_1 is valid if ω_B/ω_p is restricted below by the condition (9.13), so that $\gamma_1 < \gamma_{B_0=0\,\max}$.

Let us compare the instability conditions obtained above and the growth rates with those for $B_0 = 0$ (see § 1.5.2). In contrast to the case $B_0 = 0$, for which instability arises if $\omega \approx k_z V$, two types of resonance are possible when $B_0 \neq 0$: $\omega = k_z V$ (Cherenkov) and $\omega = k_z V - \omega_B$ (cyclotron). The difference in the case $B_0 \neq 0$ is that two, and not one, branches of plasma oscillations can be excited.

The Cherenkov and cyclotron growth rates do not exceed $\gamma_{B_0=0}$. For $B_0 \neq 0$, the Cherenkov growth rate is larger, being, like $\gamma_{B_0=0}$, proportional to $\alpha^{1/3}$.

§ 9.2. Cold Plasmas and Beam in a Strong Magnetic Field

We shall now investigate the instabilities of a plasma and a beam in a strong magnetic field, $\omega_B \gg \omega_p$. The presence of the small parameter ω_p/ω_B greatly simplifies this problem. As a result, it is a relatively simple problem to analyze not only the case of a low-density beam, $\alpha \equiv n_1/n_0 \ll 1$, but also the case $\alpha \approx 1$.

The perturbations of the plasma-beam system are described by Eqs. (9.2) and (9.3). These equations take into account the four types of beam-plasma interaction analyzed in § 9.1 for $\omega_p > \omega_B$. Let us now consider each of these elementary interactions in the case $\omega_B \gg \omega_p$.

1. Cherenkov – Cherenkov Interaction. If $\omega_B \gg \omega_p$, one can distinguish a class of low-frequency long wavelength perturbations, $(\omega, k_z V) \ll \omega_B$. For such perturbations, Eqs. (9.2) and (9.3) yield

$$1-\left(\frac{\omega_p}{\omega}\cos\theta\right)^2-\alpha\frac{(\omega_p\cos\theta)^2}{(\omega-k_zV)^2}=0. \tag{9.20}$$

Equation (9.20) differs formally from Eq. (1.46) by the replacement of ω_p by $\omega_p \cos\theta$. To analyze (9.20) we can therefore use the results of § 1.5, in which we investigated the two-stream instability for $B_0 = 0$.

We then find that the perturbations grow if

$$k\leqslant\frac{\omega_p}{V}(1+\alpha^{1/3})^{3/2}. \tag{9.21}$$

Here $k \equiv (k_\perp^2 + k_z^2)^{1/2}$ is the total wave number. The range of frequencies of the growing perturbations is bounded above by the condition

$$\omega\leqslant k_zV/(1+\alpha^{1/3}). \tag{9.22}$$

In the special cases $\alpha = 1$ and $\alpha \ll 1$, the growth rate of the perturbations can be readily calculated. If $\alpha = 1$, this can be done in the same way as in § 1.5.1, in which the relative velocity of the streams is assumed equal to 2V and the frame of reference was chosen such that the mean velocity of the particles vanishes. Therefore, to use the results of § 1.5.1, we must here make the substitutions $\omega_p \to \omega_p \cos\theta$, $\omega \to \omega + kV/2$, $V \to V/2$. Then

$$\left.\begin{aligned} \gamma_{max} &= \frac{1}{2\sqrt{3}} k_{z\,opt} V, \\ \mathrm{Re}\,\omega &= k_{z\,opt} \frac{V}{2}, \\ k_{z\,opt} &= \sqrt{3}\,\omega_p \cos\theta \frac{1}{V}. \end{aligned}\right\} \tag{9.23}$$

If $\alpha \ll 1$, we follow § 1.5.2, obtaining

$$\left.\begin{aligned} \gamma_{max} &= \frac{\sqrt{3}}{2^{4/3}} k_{z\,opt} V, \\ \mathrm{Re}\,\omega &\simeq k_{z\,opt} V, \\ k_{z\,opt} &= \omega_p \cos\theta / V. \end{aligned}\right\} \tag{9.24}$$

2. Cyclotron – Cyclotron Interaction. If a beam is not present, the frequency of the branch ω_1 for $\omega_B \gg \omega_p$ is near the cyclotron frequency, $\omega_1 \approx \omega_B$. The cyclotron interacaction of the beam with this branch occurs if $k_z V = \omega_B + \omega_1 \approx 2\omega_B$. With allowance for this fact, Eqs. (9.2) and (9.3) yield the approximate equation

$$1 - \frac{(\omega_p \sin\theta)^2}{2\omega_B(\omega - \omega_B)} + \frac{\alpha(\omega_p \sin\theta)^2}{2\omega_B(\omega - k_z V + \omega_B)} = 0. \tag{9.25}$$

The solutions of (9.25) are complex if k_z lies in the interval $k_{z\,min} < k_z < k_{z\,max}$, where

$$k_{z\,max,\,min} = \frac{2\omega_B}{V}\left[1 + \left(\frac{\omega_p \sin\theta}{2\omega_B}\right)^2 (1 + \alpha \pm \sqrt{\alpha})\right]. \tag{9.26}$$

The maximal growth rate is

$$\gamma_{max} = \frac{\sqrt{\alpha}}{2} \cdot \frac{(\omega_p \sin\theta)^2}{\omega_B}. \tag{9.27}$$

3. Cherenkov – Cyclotron Interaction. Let us now consider perturbations for which

$$\left.\begin{aligned} &|\omega| \ll \omega_B \\ &|\omega - k_z V + \omega_B| \ll \omega_B. \end{aligned}\right\} \tag{9.28}$$

In this case $k_z \approx \omega_B/V$. Then Eqs. (9.2) and (9.3) can be reduced to the equation

$$1-\left(\frac{\omega_p \cos\theta}{\omega}\right)^2+\frac{\alpha(\omega_p \sin\theta)^2}{2\omega_B(\omega-k_zV+\omega_B)}=0. \tag{9.29}$$

Since the parameter ω_p/ω_B is small, we find

$$\left.\begin{aligned}\gamma_{max}&=\left(\alpha\frac{\omega_p}{\omega_B}\cos\theta\right)^{1/2}\frac{\omega_p \sin\theta}{2},\\ \mathrm{Re}\,\omega&\simeq\omega_p\cos\theta.\end{aligned}\right\} \tag{9.30}$$

These results are valid for $\alpha \ll 1$ and $\alpha \approx 1$.

4. Cyclotron – Cherenkov Interaction. The branch of cyclotron oscillations of the plasma, $\omega_1 \approx \omega_B$, can be excited as a result of interaction with the Cherenkov branch of the beam. To study this effect, we assume

$$\left.\begin{aligned}|\omega-\omega_B|&\ll\omega_B,\\ |\omega-k_zV|&\ll\omega_B.\end{aligned}\right\} \tag{9.31}$$

In this case, the wave number k_z is approximately $k_z \approx \omega_B/V$. If the conditions (9.31) hold, Eqs. (9.2) and (9.3) reduce to

$$1-\frac{(\omega_p \sin\theta)^2}{2\omega_B(\omega-\omega_B)}-\frac{\alpha(\omega_p\cos\theta)^2}{(\omega-k_zV)^2}=0. \tag{9.32}$$

If $\alpha < (\omega_p/\omega_B)^2$,

$$\gamma_{max}=\frac{\sqrt{3}}{2^{4/3}}\omega_p\left[\frac{\omega_p}{\omega_B}\sin^2\theta\cos^2\theta\right]^{1/3}. \tag{9.33}$$

However, if $\alpha > (\omega_p/\omega_B)^2$, this is replaced by

$$\gamma_{max}=\frac{\omega_p\sin\theta}{2}\left(\alpha\frac{\omega_p}{\omega_B}\cos\theta\right)^{1/2}. \tag{9.34}$$

§ 9.3. Instability of a "Hot" Beam in a Dense Cold Plasma

The approximation of a cold beam adopted in § 9.1 is valid if $\gamma \gg k_z v_{T1}$, $k_\perp \rho_1 \ll 1$. If α is sufficiently small, the first of these inequalities is violated before the other when the thermal spread of the beam is increased. It follows that the kinetic instabilities caused by a beam with a finite but small v_{T1}/V can be investigated in the approximation $k_\perp \rho_i \ll 1$. In accordance with Eq. (7.19), the only terms that then remain in the expression for $\varepsilon_0^{(1)}$ are those that correspond to the Cherenkov and first cyclotron resonances, i.e.,

$$\varepsilon_0^{(1)} = \alpha \frac{\omega_{p0}^2 m}{k^2 T_1} \left\{ 1 + i\sqrt{\pi}\, x_0 \left[W(x_0) + \frac{k_\perp^2 T_1}{m\omega_B^2} \left(W(x_1) + W(x_{-1}) \right) \right] \right\}. \quad (9.35)$$

Here $x_n = \dfrac{\omega - n\omega_B - k_z V}{|k_z| v_{T1}}$, $n = 0 \pm 1$, $\alpha = n_1/n_0$.

1. Cherenkov Interaction of the Beam with the Plasma Oscillations. The expression (9.10) for the growth rate of the hydrodynamic Cherenkov instability is valid if

$$\frac{v_{T1}}{V} < \frac{\gamma}{\omega} \simeq \frac{\alpha^{1/3} \omega_p \cos\theta}{\omega \left[\omega_p (\partial \varepsilon_0 / \partial \omega) \right]^{1/3}}. \quad (9.36)$$

If $\omega_p \gtrless \omega_B$ and $\cos\theta \simeq \sin\theta$, the condition (9.36) for both the branches ω_1 and ω_2 is the same in order of magnitude as (3.6). If $\omega_p \ll \omega_B$, the same is true for the branch ω_2. For the branch ω_1 the condition (9.36) in this case means

$$v_{T1}/V < \alpha^{1/3} (\omega_p/\omega_B)^{4/3}. \quad (9.37)$$

If the condition (9.36) is violated, the Cherenkov instability of a beam in a plasma becomes essentially a kinetic instability. We can find the kinetic growth rate from Eq. (2.17) by taking $\mathrm{Re}\,\varepsilon_0$ in this equation from the expression (9.3) for $\varepsilon_0^{(0)}$ and $\mathrm{Im}\,\varepsilon_0$ equal to the term of the imaginary part of (9.35) proportional to $\mathrm{Re}\,W(x_0)$. Thus, we obtain a result similar to (3.14):

$$\gamma_{1,2} = -\frac{2\sqrt{\pi}\,\alpha\omega_p^2}{v_{T1}^3 k^2 |k_z|} \left\{ \frac{\omega - k_z V}{\partial \varepsilon_0^0 / \partial \omega} e^{-\left(\frac{\omega - k_z V}{k_z v_{T1}} \right)^2} \right\}_{\omega = \omega_{1,2}}. \quad (9.38)$$

The maximum of the kinetic growth rate of a beam with $v_{T1} \ll V$ is attained at approximately the same values of k_z as in the case of a cold beam [the condition (9.9)], the value being

$$\gamma_{1,2\,\max} \simeq \alpha \left(\frac{V}{v_{T1}} \right)^2 \frac{(\omega_p \cos\theta)^2}{\left(\omega^2 \dfrac{\partial \varepsilon_0^{(0)}}{\partial \omega} \right)_{\omega = \omega_{1,2}}}. \quad (9.39)$$

The growth rate depends on the parameter V/v_{T1} in the same way as for $B_0 = 0$ [cf. (9.39) and (3.16) and Fig. 3.2].

In the limiting cases of weak and strong magnetic fields, Eq. (9.39) gives

$$\gamma_{1,2\max} \simeq \frac{\alpha}{2}\left(\frac{V}{v_{T1}}\right)^2 (\omega_p \cos^2\theta,\ \omega_B \cos\theta \sin\theta), \qquad \omega_p \gg \omega_B, \tag{9.40}$$

$$\gamma_{1,2\max} \simeq \frac{\alpha}{2}\left(\frac{V}{v_{T1}}\right)^2 \omega_p \left(\left(\frac{\omega_p}{\omega_B}\right)^3 \sin^2\theta\cos^2\theta,\ \cos\theta\right), \qquad \omega_p \ll \omega_B. \tag{9.41}$$

2. Cyclotron Excitation of Plasma Oscillations by a Beam. The thermal spread of the beam can be ignored in the excitation of oscillations with $\omega \approx k_z V \pm \omega_B$ if

$$\frac{v_{T1}}{V} < \frac{\gamma_{1,2}}{|\omega_{1,2}| + |\omega_B|} \simeq \frac{\alpha^{1/2}}{\omega_B} \cdot \frac{\omega_p \sin\theta}{|\omega_{1,2}| + |\omega_B|}\left(2\,\frac{\partial\varepsilon_0}{\partial\omega}\right)^{-1/2}_{\omega=\omega_{1,2}}. \tag{9.42}$$

Here, γ_1 and γ_2 are the growth rates of the hydrodynamic approximation (9.17). For weak and strong fields this condition amounts to

$$\frac{v_{T1}}{V} < \frac{\alpha^{1/2}}{2}\sin\theta\left\{\left(\frac{\omega_p}{\omega_B}\right)^{1/2};\ \frac{(\cos\theta)^{1/2}\sin\theta}{1+|\cos\theta|}\right\}, \qquad \omega_p \gg \omega_B, \tag{9.43}$$

$$\frac{v_{T1}}{V} < \frac{\alpha^{1/2}}{2}\sin\theta\left\{\frac{1}{2}\left(\frac{\omega_p}{\omega_B}\right)^2 \sin\theta,\ \left(\frac{\omega_p}{\omega_B}\right)^{3/2}(\cos\theta)^{1/2}\right\}, \qquad \omega_p \ll \omega_B. \tag{9.44}$$

These restrictions on the thermal spread of the beam are more stringent than in the case of Cherenkov excitation [cf. (9.36) and (9.37)].

If the opposite condition to (9.42) holds, the cyclotron interaction of the beam and the plasma oscillations leads to a kinetic instability with the growth rate

$$\gamma_{1,2} = -\frac{\sqrt{\pi}\,\alpha\omega_p^2 \sin^2\theta}{\omega_B^2 |k_z| v_{T1}}\left\{\frac{\omega - k_z V}{\partial\varepsilon_0^{(0)}/\partial\omega}\exp\left\{-\left(\frac{\omega - k_z V + \omega_B}{k_z v_{T1}}\right)^2\right\}\right\}_{\omega=\omega_{1,2}}. \tag{9.45}$$

Since the ratio v_{T1}/V is assumed to be small, the maximum of the kinetic growth rate is attained at values of k_z that approximately satisfy the relations (9.16). In order of magnitude

$$\gamma_{1,2} \simeq \alpha \frac{V}{v_{T1}} \cdot \frac{(\omega_p \sin\theta)^2}{\omega_B} \left\{ (\omega + \omega_B) \frac{\partial \varepsilon_0}{\partial \omega} \right\}^{-1}_{\omega=\omega_{1,2}}. \quad (9.46)$$

As v_{T1}/V increases, the cyclotron growth rate decreases more slowly than the Cherenkov growth rate (Fig. 9.3). If $\omega_p \approx \omega_B$ and $\cos\theta \approx \sin\theta$, the two growth rates are of the same order of magnitude for $v_{T1} \approx V$ (i.e., at the limit of applicability of the approximation $v_{T1} \ll V$ adopted here). At the same time $\gamma \approx \alpha\omega_B$.

In the limiting cases of large and small ω_p/ω_B, the relation (9.46) yields

$$\gamma_{1,2} \simeq \frac{\alpha}{2} \cdot \frac{V}{v_{T1}} \omega_B \sin^2\theta \left\{ \left(\frac{\omega_p}{\omega_B}\right)^2, \ \frac{\cos\theta \sin^2\theta}{1 + |\cos\theta|} \right\},$$

$$\alpha^{-1/3} > \frac{\omega_p}{\omega_B} \gg 1, \quad (9.47)$$

$$\gamma_{1,2} \simeq \frac{\alpha}{2} \cdot \frac{V}{v_{T1}} \left(\frac{\omega_p}{\omega_B}\right)^2 \omega_p \sin^2\theta \left\{ \frac{\omega_p}{\omega_B} \sin^2\theta, \ \cos\theta \right\},$$

$$\omega_p/\omega_B \ll 1. \quad (9.48)$$

A comparison of (9.48) and (9.41) shows that in a strong magnetic field ($\omega_B \gg \omega_p$) the cyclotron growth rates are small compared with the Cherenkov growth rates even if $v_{T1} \approx V$.

3. Instability of a Beam Whose Thermal Velocity is Comparable with the Directed Velocity. If $v_{T1} \gtrsim V$, it is necessary to take into account the finite Lar-

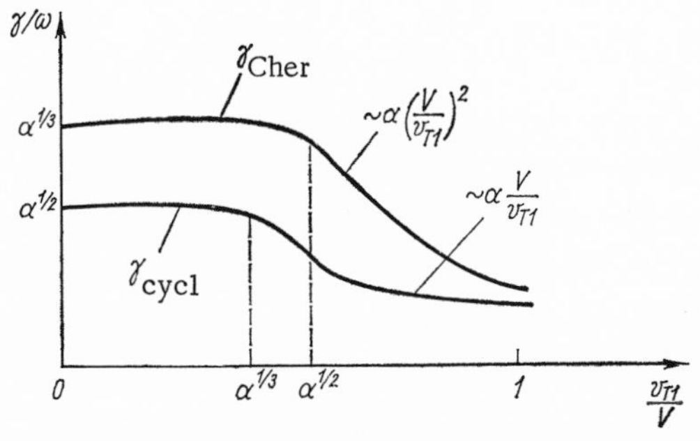

Fig. 9.3. Growth rates as functions of the thermal spread of the beam.

mor radius of the beam particles (if ω_p/ω_B or $\sin\theta$ are not too small).

For arbitrary $k_\perp\rho_1$ and a Maxwellian distribution of the particle velocities of the beam, the function $\varepsilon_0^{(1)}$ is determined by Eq. (7.19). Using this equation in the same way as in § 9.3.1, we find a general expression for the kinetic growth rate of the oscillations of the cold plasma

$$\gamma_{1,2} = -\frac{\sqrt{\pi}\,\alpha\omega_p^2}{k^2\,|k_z|\,v_{T1}^3}\left\{\frac{\omega - k_z V}{\partial\varepsilon_0^{(0)}/\partial\omega}\sum_{n=-\infty}^{\infty} I_n e^{-z}\exp\left\{-\left(\frac{\omega - n\omega_B - k_z V}{k_z v_{T1}}\right)^2\right\}\right\}_{\omega=\omega_{1,2}}. \tag{9.49}$$

If the thermal velocity does not exceed the directed velocity, $v_{T1} \lessgtr V$, the terms of the series (9.49) have a maximum at values of k_z near $k_{z\,\mathrm{opt}} \equiv (\omega_{1,2} + |n|\omega_B)/V$ $(n \leq 0)$ over an interval of width

$$\frac{\Delta k_z}{k_{z\,\mathrm{opt}}} \simeq \frac{v_{T1}}{V}. \tag{9.50}$$

The magnitude of the terms of the series is also strongly dependent on the transverse wave number $k_\perp$. The maximum of the term with $n = 0$ occurs at long wavelengths, $z < 1$, and that of the terms with $|n| \geq 1$ at $z \approx 1$. If $V \approx v_{T1}$ and $\omega_p \approx \omega_B$, the terms of the series with low numbers make the main contribution to the growth rate. In this case, the characteristic wavelengths of the excited oscillations are $k_z \approx k_\perp \approx 1/\rho_1$, and the growth rate of the oscillations is

$$\gamma \sim \alpha\omega_B. \tag{9.51}$$

§ 9.4. Instability of a Slow Beam in a Hot Plasma

If the plasma has a nonvanishing temperature, $\varepsilon_0^{(0)}$ in Eq. (9.2) must be taken equal to the expression that follows from (7.18):

$$\varepsilon_0^{(0)} = \frac{1}{(kd_0)^2}\left[1 + i\sqrt{\pi}\,\frac{\omega}{|k_z|\,v_{T0}}\sum_{n=-\infty}^{+\infty} e^{-z_0} I_n(z_0)\, W\left(\frac{\omega - n\omega_B}{|k_z|\,v_{T_0}}\right)\right]. \tag{9.52}$$

Here $z_0 = k_\perp^2 T_0/m\omega_B^2$, $d_0^2 = T_0/4\pi e^2 n_0$.

Below, we shall restrict the treatment to the limiting case of a strong magnetic field, $\omega_p \ll \omega_B$. Concerning the interaction of a beam with a hot plasma for $\omega_p \gtrless \omega_B$, we make the following observation. Using (9.52) and (9.2), we can show that for $\omega_p \gtrless \omega_B$ the plasma must be regarded as a "hot" plasma if $v_{T0} \gtrless V$; this is the same as in the approximation $B_0 = 0$ (see § 3.1). The nature of the instabilities is then similar to the case considered in § 3.3.

If $\omega_p \ll \omega_B$, the condition for the plasma to be "cold", $v_{T0} > T$, remains in force only for the case of Cherenkov–Cherenkov excitation ($\omega \approx \omega_p \cos\theta \approx k_z V$). In the case of Cherenkov–cyclotron excitation ($\omega \approx \omega_p \cos\theta$, $k_z \approx \omega_B/V$) the condition for a cold plasma is more stringent:

$$v_{T_0}/V < \omega_p/\omega_B. \tag{9.53}$$

An even more severe restriction on the thermal spread of the plasma follows for the cyclotron–Cherenkov ($\omega \approx \omega_B$, $k_z V \approx \omega$) and cyclotron–cyclotron ($\omega \approx \omega_B$, $k_z V \approx |\omega| + |\omega_B|$) interactions:

$$v_{T_0}/V < (\omega_p/\omega_B)^2. \tag{9.54}$$

A general picture of the instabilities of a plasma with $\omega_p \ll \omega_B$ and finite v_{T0}/V can be obtained by means of Fig. 9.4 and the relations that follow. If $v_{T0}/V > (\omega_p/\omega_B)^2$, the growth rates of the cyclotron–Cherenkov and cyclotron–cyclotron instabilities are determined by kinetic effects. Their maximal values are

$$\gamma_{\text{cycl-Cher}} = \frac{\sqrt{\pi}\,\alpha^{1/2}}{4} \cdot \frac{\omega_p^3}{\omega_B^2} \cdot \frac{V}{v_{T_0}} \cos\theta \sin^2\theta, \tag{9.55}$$

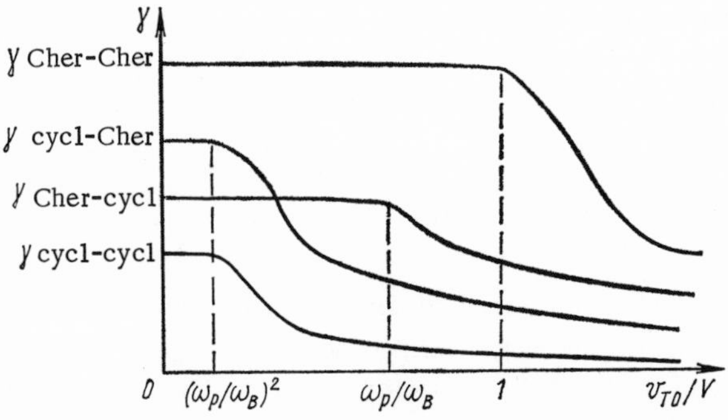

Fig. 9.4. Growth rates as a function of the thermal spread of the plasma for $\omega_p \ll \omega_B$.

$$\gamma_{\text{cycl-cycl}} = \frac{\sqrt{\pi}\,\alpha}{8}\cdot\frac{\omega_p^4}{\omega_B^3}\cdot\frac{V}{v_{T_0}}\sin^4\theta. \tag{9.56}$$

If $v_{T0}/V > \omega_p/\omega_B$, the Cherenkov–cyclotron instability also becomes kinetic in natute. Its maximal growth rate is

$$\gamma_{\text{Cher-cycl}} = \frac{\sqrt{\pi}\,\alpha e^{-1/2}\omega_p^4}{\omega_B^3}\cdot\frac{V^2}{v_{T_0}^2}\sin^2\theta\cos^2\theta. \tag{9.57}$$

For a fast beam, $v_{T0}/V < \omega_p/\omega_B$, the growth rates (9.55)-(9.57) are small compared with the growth rate of the Cherenkov–Cherenkov instability. If this condition is satisfied, the latter is determined by the hydrodynamic expression (9.24).

In the limit of a slow beam, $V/v_{T0} \ll 1$, the Cherenkov–Cherenkov instability is kinetic in nature and is described by relations similar to those given in § 3.3. To this instability there corresponds the minimal threshold velocity of a beam capable of exciting oscillations. At a sufficiently low beam velocity, the Cherenkov–Cherenkov instability is therefore decisive.

§ 9.5. Instability of a One-Dimensional Distribution with One Maximum

We have shown in § § 2.7 and 2.8 that a plasma with $B_0 = 0$ is stable against electrostatic perturbations if its one-dimensional distribution function has only a single maximum.

We shall now show that a plasma with a particle distribution velocity of this kind may be unstable if $B_0 \neq 0$.

Let us consider a distribution of the form $f_0 = f_0^{(0)} + f_0^{(1)}$, where $f_0^{(0)} = n_0\delta(\mathbf{v})$, $f_0^{(1)} = n_1\delta(\mathbf{v}_\perp)F(v_z)$, and

$$F(v_z) = \begin{cases} 1/v_1 & \text{for } 0 < v_z < v_1, \\ 0 & \text{for } v_z < 0,\ v_z > v_1. \end{cases} \tag{9.58}$$

The one-dimensional distribution over the velocities v_z (along $\mathbf{B}_0$) corresponding to the function f_0 is shown in Fig. 9.5. The one-dimensional distribution function has the same form for other directions of the velocity, so that this function has only a single maximum in all cases.

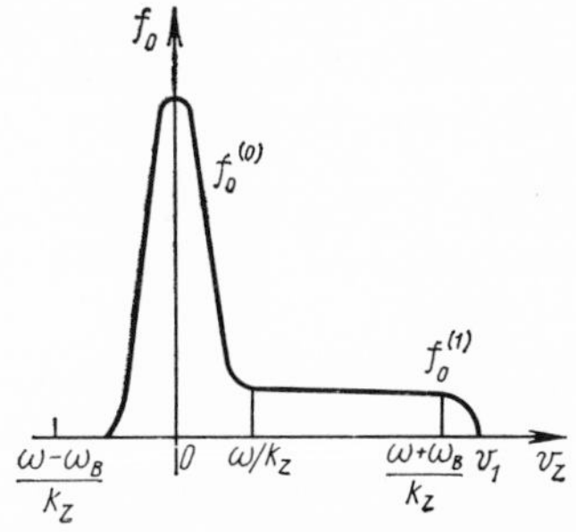

Fig. 9.5. Example of an unstable one-dimensional distribution with a single maximum.

Let us now turn to an investigation of the perturbations. We shall assume that $\alpha \equiv n_1/n_0 \ll 1$. In the zeroth approximation in this parameter, the perturbations have real frequencies ω_1 and ω_2 given by Eqs. (8.2). We can find the imaginary correction to the frequency i.e., the growth rate, by means of Eq. (2.17) by substituting into the latter $\mathrm{Re}\,\varepsilon_0 = \varepsilon_0^{(0)}$ defined by Eq. (9.3) and $\mathrm{Im}\,\varepsilon_0$ of the form

$$\mathrm{Im}\,\varepsilon_0 = -\pi\alpha \frac{(\omega_p \sin\theta)^2}{2\omega_B |k_z|}\left[F\left(\frac{\omega-\omega_B}{k_z}\right) - F\left(\frac{\omega+\omega_B}{k_z}\right)\right]. \qquad (9.59)$$

This expression can be obtained by means of Eq. (7.16), in which one must assume $k_\perp\rho \ll 1$. In deriving (9.59) we have used the fact that there is no contribution from the Cherenkov resonance $v_z = \omega/k_z$ since $\partial F/\partial v_z = 0$. A contribution to (9.59) arises only from the cyclotron resonances $v_z = (\omega - \omega_B)/k_z$ (normal Doppler effect) and $v_z = (\omega + \omega_B)/k_z$ (anomalous Doppler effect).

The sign of the growth rate is determined by the sign of the product $\omega\,\mathrm{Im}\,\varepsilon_0$. Using (9.59), we find that the growth rate is positive – instability arises – if

$$\frac{\omega}{\omega_B}\left[F\left(\frac{\omega+\omega_B}{k_z}\right) - F\left(\frac{\omega-\omega_B}{k_z}\right)\right] > 0. \qquad (9.60)$$

This condition is satisfied if each of the following inequalities is satisfied:

$$\left.\begin{aligned} &|\omega| < |\omega_B|, \\ &|\omega| + |\omega_B| < |k_z|\, v_1. \end{aligned}\right\} \qquad (9.61)$$

It follows from the first inequality in (9.61) that only the branch ω_2 can be excited. The second inequality gives a lower limit on the interval of wave numbers of growing perturbations:

$$k_z \gtrsim \omega_B/v_1. \tag{9.62}$$

However, k_z must not be too large:

$$k_z \ll \omega_2/v_{T_0}, \tag{9.63}$$

since otherwise the approximation of a cold plasma cannot be used. From (9.62) and (9.63) we obtain a restriction on the plasma temperature (one of the instability conditions):

$$v_{T_0}/v_1 \ll \omega_2/\omega_B. \tag{9.64}$$

In the case of a strong magnetic field, $\omega_B > \omega_p$, the frequency ω_2 does not exceed the plasma frequency, $\omega_2 \leq \omega_p$. In this case, (9.64) entails

$$v_{T_0}/v_1 \ll \omega_p/\omega_B. \tag{9.65}$$

In order of magnitude the growth rate satisfies

$$\gamma \simeq \alpha\,(\omega_p/\omega_B)^2\,\omega_2. \tag{9.66}$$

In practice, an instability of this kind can appear in a plasma with runaway electrons.

§ 9.6. Excitation of Plasma Oscillations by a Radially Inhomogeneous Electron Beam

If a low-density beam passes through a plasma, one can distinguish a class of perturbations whose frequency is essentially determined by the properties of the plasma. The effect of the beam responsible for the excitation of these oscillations can be found by the method of successive approximations. Above we have considered some examples of the excitation by a beam of oscillations of a plasma under the assumption that the beam and plasma are of infinite extent and spatially homogeneous. We shall now investigate the excitation of oscillations of a homogeneous bounded plasma of cylindrical symmetry through which a beam with an arbitrary radial distribution function of the particle density passes.

The investigation of the excitation of oscillations of a bounded plasma (by solving the problem with boundary conditions) is necessary not only when the transverse wavelength of the perturbations is comparable with the radius R of the plasma but also when $k_\perp R \gg 1$ if the condition (5.30) of the local approximation is not satisfied. If oscillations are excited by a beam of small radius, $a_\perp \ll R$, the beam radius $a_\perp$ plays the role of the length L in Eq. (5.30) and V_{gr} is the radial component of the group velocity $\partial\omega/\partial K_r$.

An investigation of the oscillations of a bounded plasma is not necessary if $\gamma_{loc} \gg \frac{\partial\omega}{\partial K_r}\frac{1}{a_\perp}$, but is necessary if this condition is not satisfied.

<u>Cold Beam.</u> Suppose a cold, cylindrically symmetrical beam whose particles have a velocity $\mathbf{V} = (0, 0, V)$ that does not depend on the radius, $V = \text{const}$, $n_1 = n_1(r)$, passes through a cold, cylindrically symmetric plasma. Let us consider the electrostatic perturbations of such a system, assuming that they are axisymmetric, $\partial/\partial\varphi = 0$ (φ is a polar angle). Under these assumptions, Poisson's equation becomes

$$(\hat{L}_0 + \hat{L}_1)\psi = 0. \tag{9.67}$$

Here

$$\hat{L}_\alpha\psi = \frac{1}{r}\cdot\frac{\partial}{\partial r}\left(r\varepsilon_\perp^{(\alpha)}\frac{\partial\psi}{\partial r}\right) - k_z^2\varepsilon_\parallel^{(\alpha)}\psi, \quad \alpha = 0, 1, \tag{9.68}$$

$$\left.\begin{aligned} &\varepsilon_\perp^{(0)} = 1 - \frac{\omega_{p0}^2}{\omega^2 - \omega_B^2}, \quad \varepsilon_\parallel^{(0)} = 1 - \left(\frac{\omega_{p0}}{\omega}\right)^2; \\ &\varepsilon_\perp^{(1)} = -\omega_{p1}^2/[(\omega - k_zV)^2 - \omega_B^2], \\ &\qquad \varepsilon_\parallel^{(1)} = -\omega_{p1}^2/(\omega - k_zV)^2. \end{aligned}\right\} \tag{9.69}$$

Equation (9.67) can be derived by proceeding from the equation

$$\operatorname{div}\mathbf{D} = 0 \tag{9.70}$$

and then using the equations $D_i = \varepsilon_{ik}E_k$, $\mathbf{E} = -\nabla\psi$, and appropriate expressions for the tensor ε_{ik}.

To be specific, we shall assume that at r = R the plasma is bounded by a perfectly conducting metallic vessel whose potential is fixed. Then the boundary condition for the potential of the perturbation at r = R has the simple form $\psi(R) = 0$.

Assuming that the terms with $\hat{L}_1$ in Eq. (9.67) are small, we solve this equation by the method of successive approximations. We represent the potential and the frequency of the oscillations in the form

$$\left.\begin{aligned} \psi &= \psi^{(0)} + \psi^{(1)}, \\ \omega &= \omega^{(0)} + \omega^{(1)}. \end{aligned}\right\} \tag{9.71}$$

Here, $\psi^{(0)}$ and $\omega^{(0)}$ are the eigenfunctions and eigenvalues of the equation

$$\hat{L}_0(\omega^{(0)})\,\psi^{(0)} = 0 \tag{9.72}$$

with the boundary conditions $\psi^{(0)}(R) = 0$ and $\psi^{(0)}(0)$ equal to a finite quantity. The corrections $\psi^{(1)}$ and $\omega^{(1)}$ are determined by the equation

$$\hat{L}_0(\omega^{(0)})\,\psi^{(1)} = -\left\{\omega^{(1)}\frac{\partial \hat{L}_0(\omega^{(0)})}{\partial \omega^{(0)}}\psi^{(0)} + \hat{L}_1(\omega^{(0)} + \omega^{(1)})\,\psi^{(0)}\right\}. \tag{9.73}$$

We find $\omega^{(1)}$ from the condition that the right-hand side of (9.73) is orthogonal to the solution of the zeroth approximation. This condition has the form

$$\omega^{(1)}\int \psi^{(0)*}\frac{\partial \hat{L}_0(\omega^{(0)})}{\partial \omega^{(0)}}\psi^{(0)}\,d\mathbf{r} + \int \psi^{(0)*}\hat{L}_1(\omega^{(0)} + \omega^{(1)})\,\psi^{(0)}\,d\mathbf{r} = 0. \tag{9.74}$$

If $\hat{L}_1$ does not have a singularity at $\omega = \omega^{(0)}$, then $\omega^{(1)}$ in the expression for $\hat{L}_1$ can be neglected compared with $\omega^{(0)}$. In this case, the correction to the eigenvalue of the frequency due to the beam has the usual form of perturbation theory:

$$\omega^{(1)} = -\frac{\int \psi^{(0)*}\hat{L}_1(\omega^{(0)})\,\psi^{(0)}\,d\mathbf{r}}{\int \psi^{(0)*}\frac{\partial \hat{L}_0(\omega^{(0)})}{\partial \omega^{(0)}}\psi^{(0)}\,d\mathbf{r}}. \tag{9.75}$$

However, we are interested in the case when the correction $\omega^{(1)}$ to the operator $\hat{L}_1$ is important (cf. the similar situation in

§ 9.1). Substituting the actual expressions for $\hat{L}_0$ and $\hat{L}_1$ [see Eqs. (9.68) and (9.69)] into (9.74), we reduce the latter to the form

$$\omega^{(1)} \frac{\partial \varepsilon_0^{(0)}(\omega^{(0)})}{\partial \omega^{(0)}} + \varepsilon_{\text{eff}}^{(1)} = 0, \tag{9.76}$$

where

$$\varepsilon_{\text{eff}}^{(1)} = -\frac{\alpha_{\parallel}(\omega_{p0}\cos\theta)^2}{(\omega - k_z V)^2} - \frac{\alpha_{\perp}(\omega_{p0}\sin\theta)^2}{(\omega - k_z V)^2 - \omega_B^2} = 0. \tag{9.77}$$

Here

$$\left.\begin{aligned} &\theta = \tan^{-1}\frac{k_\perp}{k_z}, \qquad \varepsilon_0^{(0)} = \varepsilon_\perp^{(0)}\sin^2\theta + \varepsilon_\parallel^{(0)}\cos^2\theta,\\ &\alpha_\parallel = \frac{\int n_1(r)\,|\psi^{(0)}|^2\,d\mathbf{r}}{n_0\int|\psi^{(0)}|^2\,d\mathbf{r}}, \qquad \alpha_\perp = \frac{\int n_1(r)\,|\nabla_\perp\psi^{(0)}|^2\,d\mathbf{r}}{k_\perp^2 n_0\int|\psi^{(0)}|^2\,d\mathbf{r}},\\ &\psi^{(0)} = CJ_0(k_\perp r). \end{aligned}\right\} \tag{9.78}$$

The frequency $\omega^{(0)}$ of the zeroth approximation satisfies the condition $\varepsilon_0^{(0)}(\omega^{(0)}) = 0$, and the "transverse wave number" $k_\perp$ satisfies the condition

$$J_0(k_\perp R) = 0. \tag{9.79}$$

Equation (9.76) describes the hydrodynamic instabilities discussed in § § 9.1 and 9.2. The only difference is that the ratio α of the densities in the dispersion equation is replaced by its average value.

As an example, let us consider a beam with a Gaussian density distribution:

$$n_1(r) = Ne^{-r^2/a^2}. \tag{9.80}$$

In this case, (9.78) yields

$$\left.\begin{aligned} \alpha_\parallel &= \alpha_0\frac{a^2}{R^2}\cdot\frac{I_0(\nu)\,e^{-\nu}}{J_1^2(k_\perp R)},\\ \alpha_\perp &= \alpha_0\frac{a^2}{R^2}\cdot\frac{I_1(\nu)\,e^{-\nu}}{J_1^2(k_\perp R)}. \end{aligned}\right\} \tag{9.81}$$

Here, $\alpha_0 = N/n_0$ is the ratio of the beam to the plasma density at $r = 0$ and $\nu = (k_\perp a)^2/2$.

Using (9.81), we find that for $k_\perp a \gg 1$ the growth rate of Cherenkov excitation of oscillations by a Gaussian beam is determined by the relation

$$\gamma/\gamma_0 \simeq (a/R)^{1/3}, \tag{9.82}$$

where γ_0 is the growth rate corresponding to a spatially homogeneous beam. The expression for γ_0 is given by Eq. (9.10) and, in the limiting cases of large and small $\omega_{p_e}/\omega_{B_e}$, by Eqs. (9.12), (9.24), and (9.34).

Similarly, for cyclotron excitation ($\omega = k_z V - \omega_B$)

$$\gamma/\gamma_0 \simeq (a/R)^{1/2}, \tag{9.83}$$

where γ_0 is given by Eq. (9.17) or (9.19), (9.27), and (9.30).

Inhomogeneous Beam with Longitudinal Thermal Spread. The expression (9.77) can be generalized to the case of a beam with a longitudinal thermal spread by the method indicated in the problem in § 2.1. In this case

$$\varepsilon_{\mathrm{eff}}^{(1)} = -\alpha_{\parallel}\omega_p^2 \cos^2\theta \int \frac{F_1 dv_z}{(\omega - k_z v_z)^2} - \alpha_{\perp}\omega_p^2 \sin^2\theta \int \frac{F_1 dv_z}{(\omega - k_z v_z)^2 - \omega_B^2}. \tag{9.84}$$

Here, $F_1 \equiv F(r, v_z)/n_1(r)$ is the beam distribution function, normalized to unity (it is assumed that F_1 does not depend on r); the ratios of the effective densities $\alpha_\parallel$ and $\alpha_\perp$ are determined by Eqs. (9.78).

Using (9.84) and (9.76), we find that the hydrodynamic instabilities with the growth rates (9.82) and (9.83), respectively, are possible only if the thermal spread of the beam satisfies the conditions

$$v_{T1}/V < (\alpha_0 a/R)^{1/3}, \tag{9.85}$$

$$v_{T1}/V < (\alpha_0 a/R)^{1/2}. \tag{9.85}$$

If this is not the case, the instability of a beam in a plasma is kinetic in nature. In order of magnitude, the corresponding growth rate is given by

$$\gamma/\gamma_0 \simeq a/R, \tag{9.87}$$

where γ_0 is the growth rate in the case of a spatially homogeneous beam of the same density (obtained in § 9.3).

§ 9.7. Thresholds of Two-Stream or Beam Instabilities in a Bounded Plasma in a Strong Magnetic Field

We have seen in § 9.2 that, in the case of a strong magnetic field, $\omega_B \gg \omega_p$, the two-stream or beam instabilities can be divided into long-wavelength, $k \leq \omega_p/V$, and short-wavelength, $k \simeq \omega_B/V$, instabilities. The long-wavelength instabilities have the larger growth rate. It is these instabilities that are least sensitive to the thermal spread of the particles. It follows that these are the most important instabilities. However, the long-wavelength instabilities cannot develop if the plasma density is too low since the instability condition $k \leq \omega_p/V$ is not satisfied in the case of low ω_p because of the spatial confinement of the plasma. The instability condition leads to the following restriction on the plasma density:

$$n_0 > \frac{V^2}{4\pi e^2}(k_\perp^2 + k_z^2). \tag{9.88}$$

In the most important case of a plasma with $L \gg a$, this means

$$n_0 > \frac{V^2}{4\pi e^2} k_\perp^2. \tag{9.89}$$

Let us now consider what must be substituted in place of $k_\perp^2$ in the case of a cylindrically symmetric plasma.

Plasma Bounded by a Conducting Vessel. Let R be the radius of the plasma. Then $k_\perp$ is determined by the condition $J_0(k_\perp R) = 0$. We are interested in the smallest of the possible values of $k_\perp$. This corresponds to the first zero of the Bessel function. Thus,

$$k_{\perp\,\min} = 2.4/R, \tag{9.90}$$

so that the limiting density (9.89) of an unstable plasma is in this case

$$n_{0\lim} = \left(\frac{2.4}{R}\right)^2 \frac{V^2}{4\pi e^2}. \tag{9.91}$$

Plasma Bounded by Vacuum at $r = a$; Metallic Wall at $r = R \gg a$; Plasma Length Greater than the Radius of the Wall ($L \gg R$). We obtain an expression for $k_\perp$ by fitting at $r = a$ the solution for the potential of the perturbation in the plasma

$$\psi_1 = C_1 J_0 (k_\perp r) \tag{9.92}$$

to the solution outside the plasma

$$\psi_2 = C_2 K_0 (k_z r) + C_3 I_0 (k_z r). \tag{9.93}$$

In making the fitting, we use the boundary conditions

$$\left.\begin{aligned} \psi_1 (a) = \psi_2 (a), \quad \psi'_1 (a) = \psi'_2 (a), \\ \psi_2 (R) = 0. \end{aligned}\right\} \tag{9.94}$$

As a result we find that $k_\perp^2$ in (9.89) must be taken in the form

$$k_\perp^2 = \frac{2}{a^2 \ln \frac{R}{a}}. \tag{9.95}$$

The above results refer to the case of a low-density beam, $n_1 \gg n_0$. If $n_1 \gtrless n_0$, the substitution

$$n_0 \to (n_0^{1/3} + n_1^{1/3})^3 \tag{9.96}$$

must be made in (9.89).

Bibliography

1. J. R. Pierce, J. Appl. Phys., 19:231 (1948). The dispersion equation is obtained for a system of cold streams in a magnetic field [Eq. (9.1)].
2. V. V. Zheleznyakov, Izv. Vyssh. Ucheb. Zaved., Radiofiz., 2:14 (1959).
3. M. S. Kovner, Izv. Vyssh. Ucheb. Zaved., Radiofiz., 3:631 (1960).
4. M. S. Kovner, ibid, p. 746.
5. V. O. Rappoport, ibid, p. 767.
6. K. N. Stepanov and A. B. Kitsenko, Zh. Tekh. Fiz., 31:167 (1961) [Soviet Phys. — Tech. Phys., 6:120 (1967)].
7. I. B. Bernstein and K. Trehan, Nucl. Fusion, 1:3 (1960).
8. J. Neufeld and P. H. Doyle, Phys. Rev., 121:654 (1961).

9. J. Neufeld and H. Wright, Phys. Rev., 129:1489 (1963).
10. A. I. Akhiezer et al., Collective Oscillations in a Plasma, Oxford (1967). In [2-10] investigations are made of the instabilities of a beam (either cold or with a Maxwellian distribution about a mean velocity) moving along a magnetic field. The electrostatic instabilities of a cold beam in a cold plasma (§§ 9.1 and 9.2) are discussed in [5, 6, 10] and those of a hot beam (§ 9.3) in [3, 4, 6, 10]. The interaction between a cold beam and a hot plasma (§ 9.4) is investigated in [10]. In each of the investigations [2-10] allowance is made for the nonelectrostatic nature of the perturbations. In [1, 7, 9] it is assumed that $k_\perp = 0$ and perturbations of an electromagnetic type ($E_z = 0$) are investigated. In [10] the stability of a one-dimensional plasma distribution in a magnetic field is also discussed (§ 9.5).
11. C. Étiévant and M. Perulli, C. R. Acad. Sci., 255:2739 (1962).
12. D. G. Lominadze and K. N. Stepanov, Zh. Tekh. Fiz., 35:205 (1965) [Soviet Phys. — Tech. Phys., 10:169 (1965)]. The excitation of oscillations when plasmas of equal density collide (§ 9.2) is investigated in [11-12].
13. M. F. Gorbatenko, Zh. Tekh. Fiz., 33:1070 (1963). [Soviet Phys. — Tech. Phys., 8:798 (1964)].
14. A. B. Mikhailovskii and K. Jungwirth, Zh. Tekh. Fiz., 36:777 (1966) [Soviet Phys. — Tech. Phys., 11:581 (1966)].
15. E. G. Harris, Phys. Fluids, 7:1572 (1964).
16. V. Ya. Kislov and E. V. Bogdanov, Radiotekh. Elektron. 5:1974 (1960).
17. D. L. Book, Phys. Fluids, 10:198 (1967).
18. J. E. Simpson and D. A. Dunn, J. Appl. Phys., 37:4201 (1966). The instabilities of a beam with finite transverse dimensions (§§ 9.6 and 9.7) are investigated in [13-18].

Chapter 10

Plasmas with Non-Maxwellian Electrons

§ 10.1. Introductory Remarks on the Properties of a Strongly Anisotropic Plasma

In this chapter we shall investigate the instabilities of a plasma with a non-Maxwellian particle velocity distribution in the absence of directed particle streams along the magnetic field. If the velocity distribution function is anisotropic, a new mechanism of growth of oscillations peculiar to a plasma in a magnetic field may arise.

1. Elementary Derivation of the Dispersion Equation of the Oscillations of a Strongly Anisotropic Plasma. Suppose the electron velocity distribution is strongly anisotropic:

$$\frac{\partial f_0}{\partial v_z} \gg \frac{\partial f_0}{\partial v_\perp}. \tag{10.1}$$

Let us consider the electrostatic perturbations of such a plasma. The assumption that the electrostatic treatment is valid, which is justified in the limit $\beta_\perp \to 0$, enables us to neglect the perturbed magnetic field in the transport equation and the condition (10.1) enables us to neglect the perturbed electric field at right angles to $\mathbf{B}_0$. The transport equation simplified in this manner has the following form in Lagrangian variables:

$$\frac{df}{dt} = -\frac{e}{m} E_z \frac{\partial f_0}{\partial v_z}. \tag{10.2}$$

On the right-hand side of this equation, only $E_z \equiv E_z[\mathbf{r}, (t), t]$ depends on the time; therefore

$$f = -\frac{e}{m}\cdot\frac{\partial f_0}{\partial v_z}\int^{t} E_z[\mathbf{r}(t'), t']\,dt'. \qquad (10.3)$$

We take the dependence of the field on the coordinates and the time in the form $E = E_0 \exp\{-i\omega t + ik_z z + ik_x x\}$. The coordinates x and z in the argument of the exponential function must be taken at the point at which the particle is situated at the time t' (on the particle trajectory). The particle moves along the magnetic field freely and therefore $z(t') = z_0 + v_z t'$. The transverse trajectory of the particle is its Larmor circle. At the points of this circle

$$x(t') = x_0 - \frac{v_\perp}{\omega_B}[\cos(\alpha_0 - \omega_B t') - \cos\alpha_0].$$

Therefore, in the integrand in (10.3) we have an exponential function of the form

$$\exp\left[-i(\omega - k_z v_z)t' - \frac{ik_x v_\perp}{\omega_B}\cos(\alpha_0 - \omega_B t')\right]. \qquad (10.4)$$

Let us establish the meaning of this exponential function. Suppose $v_\perp = 0$, so that the particle is displaced only along the magnetic field. This displacement occurs at a constant velocity. In this case a field with the displaced frequency $\omega' = \omega - k_z v_z$ acts on the particle. This effect corresponds to the term with ω' in the exponential function (10.4).

The motion of the particle at right angles to the magnetic field is neither rectilinear nor uniform. As a result, it cannot be reduced to a simple shift of the effective frequency of the field acting on the particle. However, every periodic process can be represented as a superposition of sinusoidal processes. In particular, the exponential function (10.4) corresponds to the following superposition of simple exponential functions:

$$\sum_{n=-\infty}^{\infty} J_n\left(\frac{k_\perp v_\perp}{\omega_B}\right)\exp\{-i(\omega - k_z v_z - n\omega_B)t' - in\alpha_0\}. \qquad (10.5)$$

It follows from this that the equilibrium motion of the particle at right angles to the magnetic field leads to a splitting of the single effective frequency ω' into an infinite set of effective frequencies $\omega'' = \omega - k_z v_z - n\omega_B$ (splitting into cyclotron harmonics). The relative importance of each harmonic is characterized by $J_n\,(k_x v_\perp/\omega_B)$.

In accordance with (10.3)-(10.5), the perturbed distribution function can be expressed as a superposition of quantities corresponding to each of the effective frequencies:

$$f = \sum_n f_n, \tag{10.6}$$

$$f_n = -\frac{ie}{m}\cdot\frac{\partial f_0}{\partial v_z}\,E_z J_n\left(\frac{k_\perp v_\perp}{\omega_B}\right)\frac{1}{\omega - k_z v_z - n\omega_B}\times$$

$$\times\exp\left\{-i(\omega - k_z v_z - n\omega_B)t + i\alpha_0 n + \right.$$

$$\left. + i k_z z_0 + i k_x\left(x_0 + \frac{v_\perp \cos\alpha_0}{\omega_B}\right)\right\}. \tag{10.7}$$

We now express the function of the Lagrangian variables $\mathbf{r}_0$ and $\mathbf{v}_0$ on the right-hand side of (10.7) in terms of the Eulerian variables $\mathbf{r}$ and $\mathbf{v}$. This transition is effected as follows:

$$\left.\begin{aligned} &\alpha_0 \to \alpha + \omega_B t, \\ &x_0 + \frac{v_\perp}{\omega_B}\cos\alpha_0 \to x + \frac{v_\perp}{\omega_B}\cos\alpha, \\ &z_0 \to z + v_z t. \end{aligned}\right\} \tag{10.8}$$

As a result, Eq. (10.7) reduces to

$$f_n = -\frac{ie}{m}E_{0z}\exp\{-i\omega t + i k_z z + i k_x x\}\frac{\partial f_0}{\partial v_z}\times$$

$$\times\frac{J_n\left(\frac{k_\perp v_\perp}{\omega_B}\right)\exp\left\{i\frac{k_\perp v_\perp}{\omega_B}\cos\alpha - i n\alpha\right\}}{\omega - n\omega_B - k_z v_z}. \tag{10.9}$$

Note that with this transformation of the variables the frequency of each of the harmonics f_n is equal to the frequency ω of the field.

Integrating (10.9) with respect to the velocities, we find the perturbed charge density. Summed over all the harmonics, it is equal to

$$\rho = \int ef d\mathbf{v} = -\frac{ie^2}{m} E_z \sum_n \int J_n^2 \left(\frac{k_\perp v_\perp}{\omega_B}\right) \frac{\partial f_0/\partial v_z}{\omega - k_z v_z - n\omega_B} d\mathbf{v}. \quad (10.10)$$

Substituting this result into Poisson's equation, we obtain the desired dispersion equation of the oscillations of a strongly anisotropic plasma

$$1 + \frac{4\pi e^2 k_z}{mk^2} \sum_{n=-\infty}^{\infty} \int J_n^2 \frac{\partial f_0/\partial v_z}{\omega - n\omega_B - k_z v_z} d\mathbf{v} = 0. \quad (10.11)$$

A more general equation valid for any degree of anisotropy can be obtained by means of (7.20).

2. On the Anisotropic Instability Mechanism. Suppose that the equilibrium function f_0 is equal to the product of longitudinal and transverse parts:

$$f_0 = n_0 f_\perp (v_\perp) f_\parallel (v_z). \quad (10.12)$$

Then the n-th harmonic of the perturbed distribution function, f_n, integrated over the transverse velocities, has, in accordance with (10.9), the form

$$f_n(v_z) \equiv \int f_n d\mathbf{v}_\perp = -\frac{ieE_z \langle J_n^2 \rangle \frac{\partial f_\parallel}{\partial v_z}}{m(\omega - n\omega_B - k_z v_z)}. \quad (10.13)$$

Here, $\langle J_n^2 \rangle = \int_0^\infty v_\perp dv_\perp J_n^2 f_\perp$. Let us compare this result with the perturbed distribution function of a beam moving along the magnetic field:

$$f(v'_z) = -\frac{ieE_z}{m} \cdot \frac{\partial f_\parallel / \partial v'_z}{\omega - k_z V - k_z v'_z} \quad (10.14)$$

(v_z' is the spread of the beam over the velocities). It can be seen that the n-th harmonic has the same effect as a beam with an effective velocity V:

$$V_{\text{eff}}^{(n)} = \frac{n\omega_B}{k_z}. \tag{10.15}$$

It follows from this analogy that a plasma with an anisotropic particle velocity distribution can sustain instabilities of the same kind as a plasma with streams. Depending on the relationship between the parameters in (10.11), these instabilities may appear as "hydrodynamic" or as "kinetic" two-stream instabilities (considered in Chapters 1 and 3, respectively).

The physical difference between a strongly anisotropic plasma and a collection of ordinary streams is as follows. In their system of rest, the particles of a stream are subjected to a field with a sinusoidal time dependence, i.e., they are subjected to a field with a single frequency. In this case the existence of perturbed motions with different effective frequencies is due to the relative motion of the particles. A particle that moves in a magnetic field with a nonvanishing transverse velocity is subjected in its system of rest to a nonsinusoidal field, i.e., to a collection of many fields with different effective frequencies. The "multistream" effect arises because one and the same particle participates simultaneously in different types of perturbed motion.

§ 10.2. Instability of Hydrodynamic Type of an Anisotropic Plasma

Suppose the velocity distribution of the plasma electrons is strongly anisotropic:

$$\bar{v}_{\perp e} \gg \bar{v}_{\parallel e}. \tag{10.16}$$

To investigate the stability of such a plasma we shall use the approximate dispersion equation (10.11), which was discussed in general terms in § 10.1.1. We shall neglect the contribution of the ions to the permittivity, $\varepsilon_0^{(i)} \to 0$. Let us consider perturbations for which

$$|\omega - n\omega_{B_e}| \gg k_z \bar{v}_{\parallel e}. \tag{10.17}$$

We neglect the effects of resonant particles, which are small in this approximation. It then follows from (10.3) that

$$1-\omega_{p_e}^2\cos^2\theta\sum_{n=-\infty}^{\infty}\frac{<J_n^2>}{(\omega-n\omega_{B_e})^2}=0. \tag{10.18}$$

1. Long-Wavelength Perturbations. In the simplest case of long-wavelength perturbations, $k_\perp\rho_e\ll 1$, this equation corresponds to a "system of three beams" (see § 10.1.2):

$$\left.\begin{aligned}&1-\cos^2\theta\,\frac{\omega_{p_e}^2}{\omega^2}-k_\perp^2\rho_e^2\omega_{p_e}^2\frac{\cos^2\theta}{2}\left(\frac{1}{(\omega-\omega_{B_e})^2}+\right.\\&\qquad\left.+\frac{1}{(\omega+\omega_{B_e})^2}\right)=0,\\&\qquad\rho_e^2=\frac{1}{\omega_{B_e}^2}\left\langle\frac{v_\perp^2}{2}\right\rangle.\end{aligned}\right\} \tag{10.19}$$

The effective density of the component "at rest" exceeds the effective density of the "moving streams" by a factor of $(k_\perp\rho_e)^{-2}$ so that in the approximation of long-wavelength perturbations the problem of the anisotropic instability is completely analogous to the problem considered in § 1.5.2 of the interaction of a low-density beam with a plasma. As is shown in § 1.5, such a beam can lead to the excitation of two types of oscillation: oscillations of the plasma at rest (resonant instability) and beam oscillations (nonresonant instability). Similar effects obtain in the case of anisotropic plasmas. The resonant instability is due to the excitation of oscillations whose frequencies are approximately equal to

$$\omega\approx\omega_{p_e}\cos\theta\approx\omega_{B_e}. \tag{10.20}$$

The growth rate of these oscillations is

$$\gamma\simeq\omega_{B_e}(k_\perp\rho_e)^{2/3}. \tag{10.21}$$

The interval of frequencies of the excited oscillations is of the same order.

The frequency of the oscillations corresponding to the nonresonant instability satisfies the condition

$$\omega\approx\omega_{B_e}<\omega_{p_e}\cos\theta. \tag{10.22}$$

The growth rate for the nonresonant instability is slightly less than γ_{res}:

$$\gamma \simeq \omega_{B_e} k_\perp \rho_e \equiv k_\perp \bar{v}_{\perp e}. \tag{10.23}$$

Equation (10.19) and the results that can be deduced from it are valid only if $k_\perp \rho_e$ and $\cos\theta$ are not too small. This is because the original equations (10.11) and (10.18) are themselves approximate in the sense thay they ignore the perturbed motion of the particles at right angles to the magnetic field.

If this effect, which is small for $k_\perp \rho_e \geqslant 1$ and $\bar{v}_{\perp e} \gg \bar{v}_{\parallel e}$, is taken into account, Eq. (7.20) does not lead to Eq. (10.18) but to

$$1 - \omega_{p_e}^2 \cos^2\theta \sum_{n=-\infty}^{\infty} \frac{\langle J_n^2 \rangle}{(\omega - n\omega_{B_e})^2} - \frac{2\omega_{p_e}^2 \sin^2\theta}{\omega_{B_e}} \sum_{n=-\infty}^{\infty} \frac{n\left\langle \frac{J_n J'_n}{\xi} \right\rangle}{\omega - n\omega_{B_e}} = 0. \tag{10.24}$$

For small $k_\perp \rho_e$ and $\omega \approx \omega_{B_e}$ we obtain

$$1 - \frac{\omega_{p_e}^2}{\omega^2}\cos^2\theta - \frac{\omega_{p_e}^2 \cos^2\theta}{(\omega - \omega_{B_e})^2} \cdot \frac{(k_\perp \rho_e)^2}{2} - \frac{\omega_{p_e}^2 \sin^2\theta}{2\omega_{B_e}(\omega - \omega_{B_e})} = 0. \tag{10.25}$$

Using (10.25), we find that the above results for resonant and nonresonant instabilities are valid for $k_\perp \rho_e$ and θ that satisfy respectively, the conditions

$$(\tan\theta)^{3/2} \ll k_\perp \rho_e \ll 1, \tag{10.26}$$

$$(\tan\theta)^2 \ll k_\perp \rho_e \ll 1. \tag{10.27}$$

Clearly these conditions can be satisfied only if

$$\sin\theta \ll 1. \tag{10.28}$$

If θ does not satisfy (10.26) and (10.27), the oscillations described by Eq. (10.25) are stable.

The solutions of Eq. (10.25) are shown in Fig. 10.1 (resonant instability, $\omega_{p_e} \approx \omega_{B_e}$) and Fig. 10.2 (nonresonant instability, $\omega_{p_e} > \omega_{B_e}$).

The figures show that, besides the branches ω_1 and ω_2 which also exist when there is a Maxwellian electron distribution (cf.

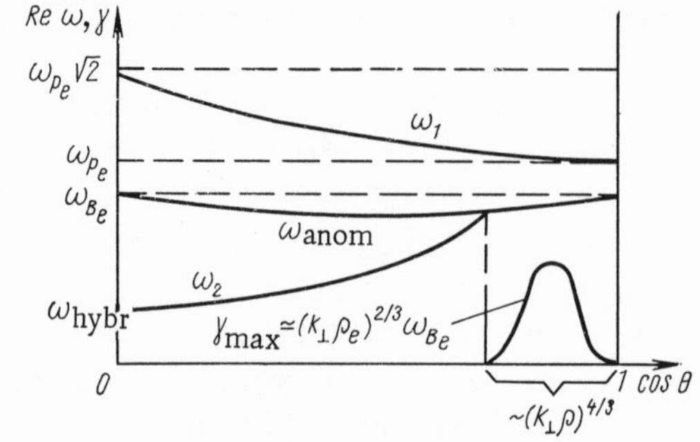

Fig. 10.1. Branches of oscillations for the resonant anisotropic instability ($\omega_{p_e} \approx \omega_{B_e}$).

Fig. 8.1), an additional (anomalous) branch of oscillations, $\omega = \omega_{anom} = \omega_{B_e}$, can arise in an anisotropic plasma. This branch is similar to the branches of beam oscillations in a plasma with streams (cf. Fig. 1.1). The resonant anisotropic instability — like the resonant two-stream instability — arises if the frequencies of three branches of oscillations (ω_1, ω_2, and ω_{anom}) are near each other. For this it is necessary that $\omega_{p_e} \approx \omega_{B_e}$. In the case of nonresonant instability, the frequencies of only two branches, ω_2 and ω_{anom}, are near each other.

2. Instability of Hydrodynamic Type for $k_\perp \rho_e \gtrsim 1$. The growth rates of the oscillations considered in § 10.2.1 increase with increasing $k_\perp \rho_e$. However, the actual expressions

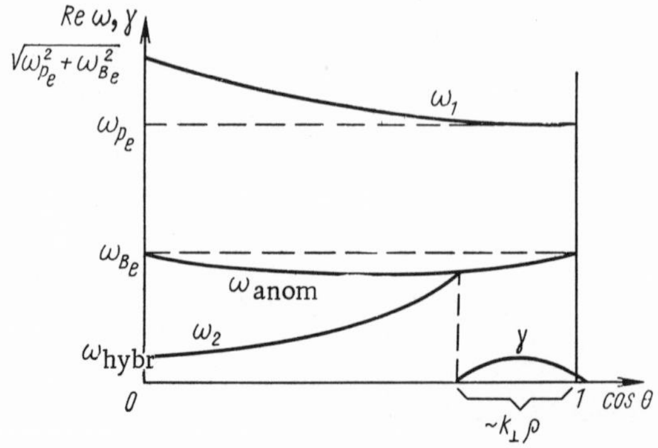

Fig. 10.2. Branches of oscillations for the nonresonant anisotropic instability ($\omega_{p_e} > \omega_{B_e}$).

for the growth rates (10.21) and (10.23) are valid only if $k_\perp \rho_e \ll 1$. To consider the instability of perturbations with $k_\perp \rho_e \gtrsim 1$, it is necessary to investigate Eq. (10.24), which is valid both when $k_\perp \rho_e \ll 1$ and $k_\perp \rho_e \gtrsim 1$. For simplicity, we shall assume that the distribution of the transverse velocities of the electrons is Maxwellian with temperature $T_\perp$. Then (10.24) can be written in the form

$$1+\frac{1-I_0 e^{-z}}{k^2 d_{\perp e}^2}-\frac{\omega_{p_e}^2 \cos^2\theta}{\omega^2} I_0 e^{-z}-\omega_{p_e}^2 \cos^2\theta \sum_{n\neq 0}\frac{I_n e^{-z}}{(\omega-n\omega_{B_e})^2}-$$
$$-\frac{\omega}{k^2 d_{\perp e}^2}\sum_{n\neq 0}\frac{I_n e^{-z}}{\omega-n\omega_{B_e}}=0, \qquad z\equiv k_\perp^2 T_{\perp e}/m_e\omega_{B_e}^2 \tag{10.29}$$

This equation can also be deduced from (7.21). The terms in (10.29) are grouped in such a way that if the frequency of the oscillations is near a definite harmonic, $\omega \approx n'\omega_{B_e}$, all the terms of the infinite sums except n = n' can be neglected. If the oscillation frequency is not near one of the cyclotron harmonics, each of the sums as a whole is small. We can therefore replace Eq. (10.29) by the simpler approximate equation

$$1+\frac{1-I_0 e^{-z}}{k^2 d_{\perp e}^2}-\frac{\omega_{p_e}^2 \cos^2\theta}{\omega^2} I_0 e^{-z}-$$
$$-\frac{\omega_{p_e}^2 \cos^2\theta\, I_n e^{-z}}{(\omega-n\omega_{B_e})^2}-\frac{\omega I_n e^{-z}}{k^2 d_{\perp e}^2(\omega-n\omega_{B_e})}=0. \tag{10.30}$$

If $n \neq 0$, the product $i_n(z)e^{-z}$ is numerically small – its maximum for all n and z does not exceed 0.2 (the maximum being attained at n = 1, z ≈ 1.5) – whereas $I_0 e^{-z}$ is small only if $z \gg 1$. For $z \approx 1$, Eq. (10.30) can therefore be analyzed by the approximate method used in § 10.2.1. Neglecting terms of order $I_n e^{-z}$ ($n \neq 0$), we find the frequency of the oscillations of the "cold component of the plasma":

$$\omega_0^2 = k_z^2 \frac{T_{\perp e}}{m_e}\cdot\frac{I_0 e^{-z}}{1-I_0 e^{-z}+k^2 d_{\perp e}^2}. \tag{10.31}$$

A resonant instability of these oscillations occurs at $\omega_0 \approx n\omega_{B_e}$. The growth rate of the instability is

$$\gamma=\frac{\sqrt{3}}{2}\left(\frac{1}{2}\cdot\frac{I_n}{I_0}\right)^{1/3}\omega_{B_e}. \tag{10.32}$$

In the limit of small $k_\perp\rho_e$, this result is identical with (10.21) [for simplicity, we have omitted the numerical coefficient in (10.21)]. As $k_\perp\rho_e$ increases, the growth rate tends to a maximum that is slightly less than ω_{B_e}. For large $k_\perp\rho_e$, – for which $I_n \approx I_0$ – the expression (10.32) is valid only in order of magnitude. The maximal growth rate can be estimated as

$$\gamma_{max} \simeq \frac{\omega_{B_e}}{2}. \qquad (10.33)$$

This resonant instability of hydrodynamic type is impossible if the plasma density is too low, since the condition $\omega_0 \simeq n\omega_{B_e}$ is then not satisfied. From (10.31) we find the relationship between the relative density of the plasma and the wave vector of unstable perturbations:

$$\frac{\omega_{p_e}^2}{\omega_{B_e}^2} = \left(\frac{\omega_{p_e}^2}{\omega_{B_e}^2}\right)_{crit} \equiv \left(\cos^2\theta I_0 e^{-z} - \sin^2\theta \frac{1 - I_0 e^{-z}}{z}\right)^{-1}. \qquad (10.34)$$

This yields a necessary condition for instability:

$$\frac{\omega_{p_e}^2}{\omega_{B_e}^2} > 1. \qquad (10.35)$$

If $\omega_{p_e}^2/\omega_{B_e}^2$ does not differ much from unity, the instability is associated with perturbations for which $\cos\theta \approx 1$ and $z \ll 1$. This accords with (10.20). As $\omega_{p_e}^2/\omega_{B_e}^2$ increases, the intervals of $\cos\theta$ and z are extended. The dependence of the critical density on z is shown in Fig. 10.3.

A nonresonant instability is possible only if the plasma density exceeds the critical density. In the case of not very large

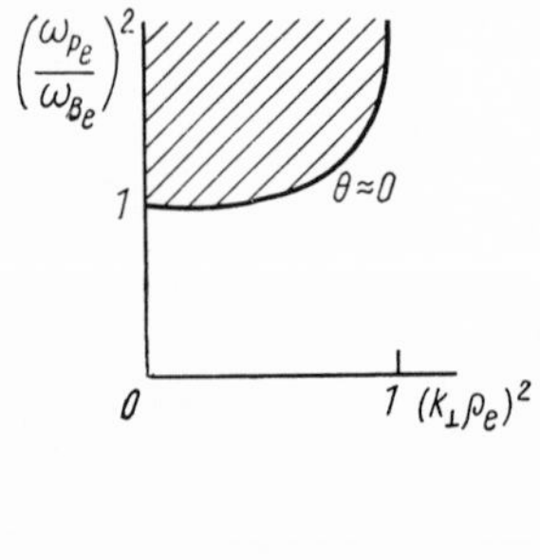

Fig. 10.3. Critical density of the anisotropic instability of hydrodynamic type (the instability region is shaded).

wave numbers — that is, such that $I_0 > i_n$ and $\omega_{p_e} \gg \omega_{p_e \text{crit}}$ — the frequency and the growth rate of the perturbations corresponding to the nonresonant instability have the form

$$\left.\begin{aligned} \operatorname{Re}\omega &= n\omega_{B_e}, \\ \gamma &\simeq \left(\frac{I_n}{I_0}\right)^{1/2} \omega_{B_e}. \end{aligned}\right\} \tag{10.36}$$

If $z \gtrsim 1$, the difference between the resonant and nonresonant instabilities is no longer significant, the more so since the very analogy with the problem of a low-density beam ceases to obtain.

3. Degree of Anisotropy Necessary for the Development of Instabilities of Hydrodynamic Type. The results of the two foregoing subsections have been obtained under the assumption that the effects of the resonant particles are small. If the distribution of the longitudinal velocities of the electrons is Maxwellian (with "temperature" $T_{\parallel}$) this assumption means

$$\gamma \gg k_z v_{T_{\parallel}}.$$

(We have omitted the subscript e in $T_{\parallel}$, $T_{\perp}$, and $v_{T_{\parallel}}$). It follows from this inequality, in particular, that a necessary condition for the long-wavelength resonant instability with the growth rate (10.21) is a small value of the ratio $T_{\parallel}/T_{\perp}$:

$$T_{\parallel}/T_{\perp} < (k_{\perp}\rho_e)^{2/3}. \tag{10.38}$$

In deriving (10.38) we have also taken into account (10.26). Using the relations (10.23), (10.27), and (10.37), we can obtain a similar condition for the long-wavelength nonresonant instability:

$$T_{\parallel}/T_{\perp} < k_{\perp}\rho_e. \tag{10.39}$$

The conditions (10.38) and (10.39) are also qualitatively correct if $k_{\perp}\rho_e \approx 1$. It should however be borne in mind that, in accordance with the general result (10.49) obtained below, the ratio $T_{\parallel}/T_{\perp}$ must not exceed 1/2 if an instability is to arise.

If $k_{\perp}\rho_e \gg 1$, a condition on the anisotropy can be obtained from the condition that the phase velocity of the oscillations of the type

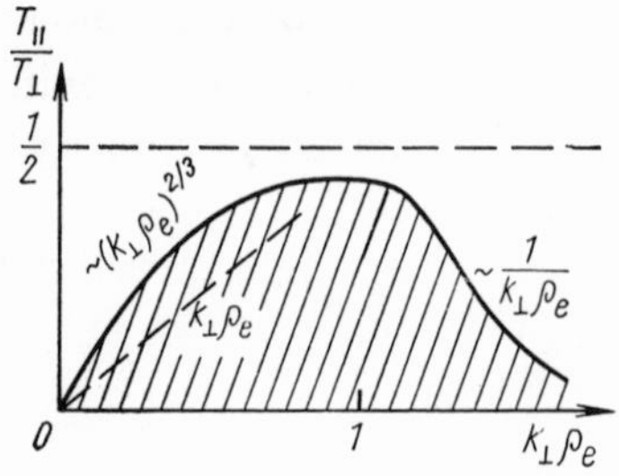

Fig. 10.4. Schematic representation of the boundary of the anisotropic instability of hydrodynamic type. The dashed line gives the boundary of the nonresonant instability (the instability region is shaded).

(10.31) be large compared with the longitudinal velocity of the electrons, $\omega_0/k_z > v_{T_{\|}}$. As a result, we have

$$\frac{T_{\|}}{T_{\perp}} < \frac{1}{k_{\perp}\rho_e\sqrt{2\pi}}. \tag{10.40}$$

Using (10.38)-(10.40), we can obtain a qualitative picture of the boundary of the anisotropic instability of hydrodynamic type for all $k_{\perp}\rho_e$. This is shown schematically in Fig. 10.4.

§ 10.3. Kinetic Instability of an Anisotropic Plasma

1. Necessary Condition for the Instability of an Anisotropic Plasma. As the value of the ratio $\bar{v}_{\|}^2/\bar{\varepsilon}_{\perp}$ increases to unity or greater, the structure of the dispersion equation (10.11) changes radically and the similarity noted in § 10.1 between the anisotropic and beam effects disappears. If $\bar{\varepsilon}_{\perp} \leqslant \bar{v}_{\|}^2$, and the particles have an equilibrium distribution of their transverse velocities $(\partial f_{\perp}/\partial\varepsilon_{\perp} \leqslant 0)$, the plasma is stable. We shall demonstrate this for a distribution of the bi-Maxwellian type $(T_{\perp} \neq T_{\|})$.

We shall first obtain a general condition of stability. We represent the general expression for the scalar permittivity (7.16) in the form

$$\varepsilon_0 = 1 - \frac{1}{k^2}\int_{-\infty}^{\infty} \frac{du P(u)}{u - w}, \tag{10.41}$$

where $w = \omega/k_z$;

$$P(u, k_z, k_{\perp}) = \frac{1}{k_z}\int_0^{\infty} 2\pi v_{\perp}\, dv_{\perp} \sum_{n=-\infty}^{\infty} \frac{4\pi e^2}{m} J_n^2(\xi) \times$$
$$\times \left[k_z \frac{\partial F(v_z, v_{\perp})}{\partial v_z} + n\omega_B \frac{\partial F(v_z, v_{\perp})}{\partial \varepsilon_{\perp}} \right]_{v_z = \frac{uk_z - n\omega_B}{k_z}}. \tag{10.42}$$

To be specific, we shall assume $k_z > 0$ so that $\operatorname{Im} w > 0$. The function $P(u)$ is similar to the derivative of the distribution function in the problems on plasma oscillations in the absence of a magnetic field [cf. (10.41) and (2.10)].

By analogy with § 2.7, we can conclude that a necessary and sufficient condition for stability is that for all $u = u_0$ for which $P(u_0) = 0$ and $P'(u_0) > 0$ the inequality

$$\mathcal{P}\int \frac{P(u)\,du}{u-u_0} \leqslant 0 \tag{10.43}$$

is satisfied.

It follows, in particular, that a plasma is stable if for all finite u the function $P(u)$ has not more than one zero.

Using these general results, we shall show that in the case of a bi-Maxwellian distribution the function P has only a single zero if $T_\perp \leqslant T_\parallel$. Substituting (2.98) into (10.42), we find that in this case

$$P(u) = -\sum_{n=-\infty}^{\infty} e^{-z} I_n(z) \left(\frac{m}{2\pi T_\parallel}\right)^{1/2} \exp\left[-\left(\frac{k_z u - n\omega_B}{k_z v_{T\parallel}}\right)^2\right] \times \\ \times \frac{m}{T_\parallel} \cdot \frac{4\pi e^2 n_0}{m}\left[u + \frac{n\omega_B}{k_z}\left(\frac{T_\parallel}{T_\perp} - 1\right)\right]. \tag{10.44}$$

In this sum, the only terms that need be retained—all the remainder being exponentially small—are those in which the sign of $n\omega_B$ is not opposite to that of u. It then follows from (10.44) that for $T_\parallel > T_\perp$ none of the retained terms vanishes anywhere except at $u = 0$. This corresponds to stability.

Using (10.44), we can estimate how small the ratio $T_\parallel / T_\perp$ must be for the anisotropic instability to arise. It follows from the above treatment that a necessary condition for instability is that for some n

$$1 - \frac{n\omega_B}{k_z u}\left(1 - \frac{T_\parallel}{T_\perp}\right) < 0. \tag{10.45}$$

On the other hand, $k_z u$ must not be appreciably different from $n\omega_B$, since otherwise the corresponding term would be exponentially small compared with the neighbors and would not change the sign of the complete sum. In addition to (10.45), we must therefore have the condition

$$\exp\left[-\frac{(k_z u - n\omega_B)^2}{(k_z v_{T\parallel})^2}\right] \geqslant \exp\left[-\left(\frac{k_z u-(n-1)\,\omega_B}{k_z v_{T\parallel}}\right)^2\right]. \qquad (10.46)$$

It follows from (10.45) and (10.46) that $k_z u$ must lie in the interval

$$\left(n-\frac{1}{2}\right)\omega_B < k_z u < n\omega_B\left(1-\frac{T_\parallel}{T_\perp}\right). \qquad (10.47)$$

We conclude that the n-th term of the sum (10.44) can change the sign of the complete sum only if

$$\frac{T_\parallel}{T_\perp} < \frac{1}{2n},\quad n = 1, 2, \ldots. \qquad (10.48)$$

A necessary condition for instability is therefore

$$T_\parallel < \frac{T_\perp}{2}. \qquad (10.49)$$

2. Kinetic Instability. It follows from § 10.2 that hydrodynamic instabilities are impossible if the plasma density is not sufficiently high (see Fig. 10.3) or if the ratio $T_\parallel/T_\perp$ is not too small (Fig. 10.4). Let us consider the role that kinetic effects – the interaction of resonant particles with oscillations – play in these conditions.

The absence of unstable perturbations of hydrodynamic type means that the solutions of Eq. (10.30) are real. In this case, the imaginary correction to the oscillation frequency can be found from the general relation (2.17), in which Re ε_0 must be replaced by the left-hand side of Eq. (10.30), ω_k by the solution of (10.30), and Im $\varepsilon_0(\omega_k)$ by the imaginary part of the permittivity determined by Eq. (7.20).

Suppose that the plasma density lies in the unshaded region of Fig. 10.3. Then the oscillation frequency (10.31) does not attain the electron cyclotron frequency and an instability of the hydrodynamic type is impossible. The growth rate of the kinetic instability calculated in the above manner is

$$\gamma = -\frac{\sqrt{\pi}\,\omega_k^2 \frac{T_\perp}{T_\parallel}}{|k_z| v_{T_\parallel} (1 - I_0 e^{-z} + k^2 d_\perp^2)} \times$$

$$\times \sum_{n=-\infty}^{\infty} I_n e^{-z} \left[1 - \frac{n\omega_B}{\omega_k}\left(1 - \frac{T_\parallel}{T_\perp}\right)\right] \exp\left[-\left(\frac{\omega_k - n\omega_B}{k_z v_{T_\parallel}}\right)^2\right]. \quad (10.50)$$

In this sum, only the terms n = 0 and n = 1 – corresponding to the Cherenkov and first cyclotron resonance – are important. In the given case, the Cherenkov resonance leads to damping of the oscillations. It follows that instability is possible if the contribution to γ of the first cyclotron resonance is not only positive but also exceeds the negative contribution of the Cherenkov resonance. Similar arguments have already been adduced in § 10.4.1. Proceeding in the same manner, we arrive at an instability condition of the type (10.47), in which we must set n = 1 and $k_z u = \omega_k$. Apart from the condition (10.49), it also follows from (10.47) that the oscillation frequency must not be too low:

$$\omega_k > \frac{\omega_B}{2}. \quad (10.51)$$

This means that the kinetic instability is not possible at an arbitrarily low plasma density but only at one such that

$$\omega_{p_e}^2 \geqslant \omega_{B_e}^2/4. \quad (10.52)$$

§ 10.4. Instability of a Plasma with a Non-Maxwellian Transverse Velocity Distribution of the Particles

In the approximation $B_0 = 0$, a plasma with a nonequilibrium distribution of the transverse particle velocities $(\partial f_0/\partial v_\perp > 0)$ with one maximum in stable (§ 2.8). If $B_0 \neq 0$, this conclusion is applicable only to perturbations in the limit $\omega/\omega_B \to \infty$. To consider the stability of such a plasma against perturbations with a lower frequency, i.e., $\omega \lessapprox \omega_B$, it is necessary to investigate the dispersion equation that takes into account the influence of a magnetic field on the perturbed motion of the particles. In what follows we shall do this for the very simple case of perturbations whose wave vector is at right angles to the magnetic field, $k_z = 0$.

In accordance with (7.16), we then have

$$\varepsilon_0 = 1 + \frac{4\pi e^2}{mk^2} \sum_{n=-\infty}^{\infty} \frac{n\omega_B}{\omega - n\omega_B} \left\langle J_n^2(\xi) \frac{\partial F}{\partial \varepsilon_\perp} \right\rangle. \tag{10.53}$$

If $\partial F/\partial \varepsilon_\perp \leqslant 0$ for all $v_\perp$, the corresponding oscillations do not grow. This can be seen by means of the Nyquist condition (§ 2.7); for it follows from § 2.7 that a necessary condition for instability is a negative value of $\text{Im}\,\varepsilon_0$ for at least one positive ω. In the given case

$$\text{Im}\,\varepsilon_0 = -\frac{4\pi e^2 \omega}{mk^2} \sum_{n=-\infty}^{\infty} \left\langle J_n^2(\xi) \frac{\partial F}{\partial \varepsilon_\perp} \right\rangle \delta(\omega - n\omega_B). \tag{10.54}$$

It can be seen that if $\partial F/\partial \varepsilon_\perp$ is everywhere less than or equal to zero and $\omega > 0$ then $\text{Im}\,\varepsilon_0$ is not negative.

We shall begin our investigation of the stability of nonequilibrium distributions with the case $F = n_0 \delta(\varepsilon_\perp - \varepsilon_{\perp 0})$. For a distribution of this kind, the expression (10.53) reduces to

$$\varepsilon_0 = 1 - \frac{\omega_p^2}{k^2 \varepsilon_{\perp 0}^2} \sum_{n=-\infty}^{\infty} \frac{n\omega_B \xi_0 J_n(\xi_0) J'_n(\xi_0)}{\omega - n\omega_B},$$

where

$$\xi_0 = kv_{\perp 0}/\omega_B, \qquad \varepsilon_{\perp 0} = v_{\perp 0}^2/2. \tag{10.55}$$

The dispersion equation $\varepsilon_0(\mathbf{k}, \omega) = 0$ has the solution $\omega = 0$ if

$$\frac{\omega_p^2}{\omega_B^2} = \frac{\xi_0}{(J_0^2)'}. \tag{10.56}$$

The corresponding values of ξ_0 lie in the intervals between the zeros and maxima of the function J_0^2, where $(J_0^2)' > 0$. The function $J_0^2(\xi)$ and the "critical" value of the parameter $(\omega_p/\omega_B)^2$

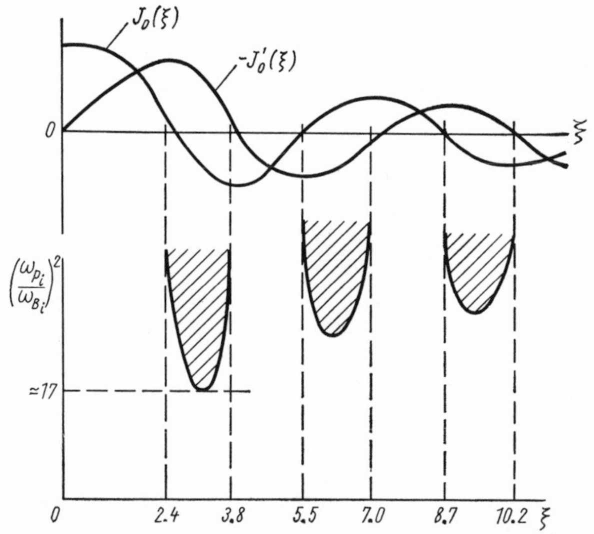

Fig. 10.5. Instability regions (shaded) of a plasma with a δ-function distribution of the transverse particle velocities.

that satisfies the condition (10.56) are shown in Fig. 10.5. The minimal value of the parameter $(\omega_p/\omega_B)^2$ and the corresponding optimal ξ_0 are

$$\left(\frac{\omega_p}{\omega_B}\right)^2_{\min} \simeq 17, \tag{10.57}$$

$$(\xi_0)_{\text{opt}} \simeq 3. \tag{10.58}$$

Let us now assume that the parameter $(\omega_p/\omega_B)^2$ differs slightly from (10.56). In this case the dispersion equation has a nontrivial solution with $|\omega| \ll \omega_B$. Expanding the denominator $1/(\omega - n\omega_B)$ in (10.55) with respect to the small parameter ω/ω_B, we find

$$\frac{\omega^2}{\omega_B^2} = \frac{(J_0^2)'}{\sum\limits_{n\neq 0}\left(\frac{1}{n^2}J_n^2\right)'}\left(\frac{\omega_p^2}{\omega_B^2}\cdot\frac{(J_0^2)'}{\xi_0} - 1\right). \tag{10.59}$$

The terms of the sum in the denominator (10.59) decrease rapidly with n. Retaining only the terms with n = ± 1, we have

$$\frac{\omega^2}{\omega_B^2} = \frac{J_0}{2J''_0}\left(\frac{\omega_p^2}{\omega_B^2}\cdot\frac{(J_0^2)'}{\xi_0} - 1\right). \tag{10.60}$$

For all ξ_0 in the critical intervals of Fig. 10.5 [i.e., those at which $(J_0^2)' > 0$], the ratio J_0/J_0'' is negative. This follows from the relationship between the Bessel function and its derivatives:

$$\frac{J''_0}{J_0} = -\left[1 + \frac{2(J'_0)^2}{\xi_0 (J_0^2)'}\right]. \tag{10.61}$$

Therefore, at densities exceeding the critical density:

$$\left(\frac{\omega_p}{\omega_B}\right)^2 > \frac{\xi_0}{(J_0^2)'}, \tag{10.62}$$

the plasma is unstable, $\omega^2 < 0$.

It follows from Fig. 10.5 that only perturbations with not too small ξ_0 can grow; this is because ξ_0 must at least exceed the first zero of the function $j_0(\xi_0)$:

$$\xi_0 \equiv \frac{kv_{\perp 0}}{\omega_B} \geqslant 2.4. \tag{10.63}$$

As the parameter $(\omega_p/\omega_B)^2$ increases from the values (10.56) and above, the growth rate of the unstable perturbations increases and becomes comparable with the cyclotron frequency:

$$\gamma \lesssim \omega_B. \tag{10.64}$$

For a plasma with an arbitrary particle velocity distribution, (10.60) is replaced by the equation

$$\left(\frac{\omega}{\omega_B}\right)^2 = \frac{1}{2} \cdot \frac{\left\langle \frac{1}{\xi} (J_0^2)' \right\rangle}{\left\langle \frac{1}{\xi} (J_1^2)' \right\rangle} \left[\left(\frac{\omega_p}{\omega_B}\right)^2 < \frac{(J_0^2)'}{\xi} > -1\right]. \tag{10.65}$$

Here, the symbol $<\ldots>$ denotes averaging over the distribution function normalized to unity. The necessary condition for instability is

$$\left\langle \frac{(J_0^2)'}{\xi} \right\rangle > 0. \tag{10.66}$$

If this condition is satisfied, the integral in the denominator (10.65) is always negative [cf. (10.61)]:

$$\left\langle \frac{1}{\xi} (J_1^2)' \right\rangle = -\left[2 \left\langle \frac{(J'_2)^2}{\xi^2} \right\rangle + \left\langle \frac{(J_0^2)'}{\xi} \right\rangle \right] < 0, \qquad (10.67)$$

and if the plasma density is not too low, instability arises. The necessary and sufficient condition for instability can therefore be represented in the form

$$\left(\frac{\omega_p}{\omega_B} \right)^2 > \frac{1}{\left\langle \frac{(J_0^2)'}{\xi} \right\rangle} > 0. \qquad (10.68)$$

Let us consider some examples. Suppose

$$F = \frac{m(T+T_1)}{2\pi T^2} \exp\left(-\frac{mv_\perp^2}{2T}\right)\left[1 - \exp\left(-\frac{mv_\perp^2}{2T_1}\right)\right]. \qquad (10.69)$$

If the distribution function has this form, then

$$\left\langle \frac{(J_0^2)'}{\xi} \right\rangle = -\left(1 + \frac{T_1}{T}\right) \frac{1}{z} [I_0(z_2)\, e^{-z_2} - I_0(z)\, e^{-z}], \qquad (10.70)$$

where $z = k^2T/m\omega_B^2$, $z_2/z = T_1/(T + T_1)$. This expression is negative for all z and T_1/T, so that the inequality (10.66) is not satisfied. This example shows that by no means every distribution with a single maximum is unstable.

Now suppose that the velocity distribution has the form

$$F = \frac{\alpha^{j+1}}{\pi j!} v_\perp^{2j} \exp(-\alpha v_\perp^2), \qquad j = 1, 2, 3. \qquad (10.71)$$

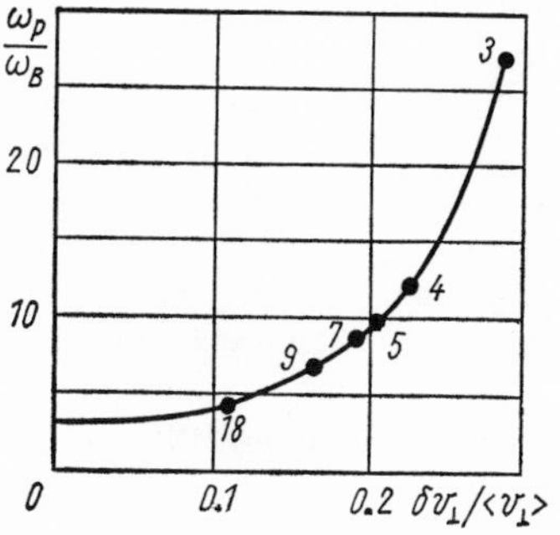

Fig. 10.6. Threshold value of ω_p/ω_B as a function of $\delta v_\perp/\langle v_\perp\rangle \equiv 1/2j$ for a distribution of the type (10.71).

The quantity $j^{1/2}\alpha$ plays the role of the mean velocity $\langle v_\perp \rangle$ and the velocity spread $\delta v_\perp$ is $\delta v_\perp = \alpha/(4j)^{1/2}$. In this case

$$\left\langle \frac{(J_0^2)'}{\xi} \right\rangle = \frac{2(-1)^j}{j!} \cdot \frac{1}{z} \alpha^j \left(\frac{\partial}{\partial \alpha}\right)^j (e^{-z} I_0(z)), \tag{10.72}$$

where $z = k^2/(2\alpha\omega_B^2)$. Substituting into this equation the actual values of the integral j, we find that this expression is negative for all z – there is stability – if

$$j < 3. \tag{10.73}$$

For $j \geq 3$ and certain intervals of z, the expression (10.72) is positive, this corresponding to instability if the density is sufficiently high. Determining the maximum of (10.72) for each $j \geq 3$ and substituting the result into (10.68), we can find the limiting density of an unstable plasma. These values are shown in Fig. 10.6, which is taken from the paper of Dory, Guest, and Harris. In the limit $j \to \infty$, the distribution (10.71) becomes a δ-function distribution, and the corresponding limiting value of the density is determined by the expression (10.57).

Bibliography

1. K. G. Malmfors, Arkiv Fys., 1:569 (1950). The question of the stability of the distribution $F \sim \delta(v_z)\delta(v_\perp - v_0)$ (§ 10.4) is posed.
2. H. K. Sen, Phys. Rev., 88:816 (1952). It is shown that a distribution of the Malmfors type is unstable against perturbations with $k_z = 0$.
3. E. G. Harris, Phys. Rev. Lett., 2:34 (1959). Application of the Nyquist method (see § 2.7) confirms Sen's conclusion that a distribution of the Malmfors type is unstable. Perturbations with $k_z \neq 0$ are also investigated.
4. R. Z. Sagdeev and B. D. Shafranov, Zh. Eksp. Teor. Fiz., 39:181 (1960) [Soviet Phys. – JETP, 12:130 (1961)]. It is shown that electromagnetic perturbations with $k_\perp = 0$ in a plasma with $T_\perp > T_\parallel$ are unstable.
5. V. V. Zheleznyakov, Izv. Vyssh. Ucheb. Zaved., Radiofiz., 3:57 (1960).
6. V. V. Zheleznyakov, Izv. Vyssh. Ucheb. Zaved., Radiofiz., 3:180 (1960). In [5, 6] a study is made of the electromagnetic instability of a plasma with $F \sim \delta(v_\perp - V_\perp)\delta(v_z - V_\parallel)$ for $k_\perp = 0$.
7. Y. Ozawa, I. Kaji, and M. Kito, J. Nucl. Energy, C4:271 (1962). Numerical calculations are made of the boundary of stability of an anisotropic plasma against electrostatic perturbations (§ 10.3).
8. E. G. Harris, J. Nucl. Energy, C2:138 (1962). Rewiew of the instabilities of a non-Maxwellian plasma.
9. L. S. Hall and W. Heckrotte, Phys. Rev., A134:1474 (1964). It is shown that a plasma with $T_\parallel < T_\perp$ is stable against electrostatic perturbations.

10. Y. Shima and L. S. Hall, Phys. Rev., A139:1115 (1965). It is shown that a plasma with $T_{\perp} > 2T_{\parallel}$ is stable against electrostatic perturbations (§ 10.3).
11. L. S. Hall, W. Heckrotte, and T. Kammash, Phys. Rev., A139:1117 (1965). Discussion of electrostatic instabilities of a plasma with anisotropic and transversely non-Maxwellian velocity distribution.
12. R. A. Dory, G. E. Guest, and E. G. Harris, Phys. Rev. Lett., 14:131 (1965). The majority of the results obtained in this paper is given in § 10.4.
13. F. W. Crawford and J. A. Tataronis, Internat. J. Electronics, 19:557 (1965). The stability of a δ-function distribution is analyzed numerically.
14. V. V. Alikaev, V. M. Glagolev, and S. A. Morozov, Plasma Phys., 10:753 (1968). A study is made of the instability of a plasma with anisotropic electrons.
15. A. V. Timofeev and V. I. Pistunovich, In: Reviews of Plasma Physics, Vol. 5, Consultants Bureau, New York (1970), p. 401. A summary is given of results hitherto obtained and some new results are derived on the stability of a plasma with Maxwellian electrons.
16. J. E. McCune, Phys. Rev. Lett., 15:398 (1965). A general instability condition (§ 10.3) is obtained.
17. S. Gruber, M. W. Klein, and P. L. Auer, Phys. Fluids, 8:1504 (1965). The instabilities of a plasma with an anisotropic electron distribution are investigated.

Chapter 11

Plasmas with a Group of Non-Maxwellian Electrons

§ 11.1. Excitation of Oscillations of a Cold Plasma by a Group of Fast Particles with an Anisotropic Velocity Distribution

In contrast to Chapter 10, we shall assume in this chapter that the electron component of the plasma consists of two, and not one, groups of particles: cold electrons and a group of fast electrons with a non-Maxwellian velocity distribution. We shall investigate the excitation of oscillations of the cold plasma by the fast particles.

A group of fast particles with an anisotropic velocity distribution ($T_\perp \neq T_\parallel$) can excite plasma oscillations even if v = 0. The mechanism of anisotropic excitation of oscillations was discussed in § 10.1 and the instability of a plasma that consists entirely of fast particles with $T_\perp > T_\parallel$ was studied in § § 10.2 and 10.3. We shall now assume that there are many more slow than fast particles, $\alpha \equiv n_1/n_0 \ll 1$. We shall now show that, as in the case of a single-component plasma, a mixture of two plasmas of this kind can sustain an instability of hydrodynamic type (for $T_\parallel \ll T_\perp$) or kinetic type (if $T_\parallel$ and $T_\perp$ are comparable).

1. Hydrodynamic Anisotropic Instability. Suppose that the velocity distribution of the fast particles is strongly anisotropic, $T_\parallel/T_\perp \to 0$, and that their mean velocity vanishes, V = 0. Let us consider the interaction of these particles with a

cold dense plasma. We shall proceed from the dispersion equation

$$1-\left(\frac{\omega_p \cos\theta}{\omega}\right)^2-\frac{(\omega_p \sin\theta)^2}{\omega^2-\omega_B^2}-\alpha\omega_p^2\sum_{n=-\infty}^{\infty}\left[\frac{I_n e^{-z_\perp}\cos^2\theta}{(\omega-n\omega_B)^2}+\right.$$

$$\left.+\frac{nI_n e^{-z_\perp}\sin^2\theta}{z_\perp\omega_B(\omega-n\omega_B)}\right]=0. \qquad (11.1)$$

This equation can be obtained by means of Eqs. (7.3) and (7.21). We recall that ω_p is the electron plasma frequency of the cold plasma; ω_B is the electron cyclotron frequency; α is the ratio of the densities of the fast and slow particles; $z_\perp = k_\perp^2 T_\perp / m\omega_B^2$, where m is the electron mass.

From the analysis of § 9.1, we conclude that, provided the instability is possible, the maximal growth rate of the perturbations must be attained at the value of $\cos\theta$ corresponding to the intersection of the branches ω_1 and ω_2 of the plasma oscillations and the beam branches (cf. Fig. 9.1). In the given case, the frequencies of the beam branches are equal to the electron-cyclotron harmonics, $\omega = n\omega_B$, n = 1, 2, ... For n ≠ 1 the beam branches can therefore intersect only the branch ω_1; for n = 1, only the branch ω_2. The conditions for intersection are

$$\left.\begin{aligned}&\left(\frac{\omega_p}{n\omega_B}\right)^2\approx\left(1+\frac{\sin^2\theta}{n^2-1}\right)^{-1}, \quad n\neq 1,\\ &\omega_p^2\approx\omega_B^2, \quad \sin^2\theta\lessapprox 1, \qquad n=1.\end{aligned}\right\} \qquad (11.2)$$

Under these conditions, there is a resonant hydrodynamic instability with the growth rate

$$\left.\begin{aligned}&\gamma\approx\frac{\sqrt{3}}{2}\left[\frac{\alpha I_n e^{-z_\perp}\cos^2\theta}{2}\right]^{1/3} n\omega_B, \quad n\neq 1,\\ &\gamma\approx\frac{\sqrt{3}}{2}\left(\frac{\alpha}{2} I_1 e^{-z_\perp}\right)^{1/3}\omega_B, \quad n=1.\end{aligned}\right\} \qquad (11.3)$$

Perturbations with n = 1 and $z_\perp \approx 1$ are excited only if $\sin\theta < \alpha^{1/3}$. If the density is greater than that allowed by condition (11.2), a nonresonant instability arises with growth rate

$$\gamma \simeq \alpha^{1/2}\omega_B. \qquad (11.4)$$

2. Kinetic Anisotropic Instability. If the ratio $T_\parallel / T_\perp$ is finite, the contribution of the fast particles to (11.1) is given by the right-hand side of (7.20), so that in this case the dis-

persion equation has the form

$$\varepsilon_0 \equiv 1 - \left(\frac{\omega_p \cos\theta}{\omega}\right)^2 - \frac{(\omega_p \sin\theta)^2}{\omega^2 - \omega_B^2} + \frac{1}{(kd_{\parallel})^2} \sum_{n=-\infty}^{\infty} e^{-z_\perp} I_n \times$$

$$\times \left\{1 + i\sqrt{\pi}\, \frac{\omega}{|k_z| v_{T\parallel}} \left[1 - \frac{n\omega_B}{\omega}\left(1 - \frac{T_\parallel}{T_\perp}\right)\right] W\left((\omega - n\omega_B)/k_z v_{T\parallel}\right)\right\}. \quad (11.5)$$

From this we conclude that the instability with growth rate (11.3) occurs if

$$\frac{T_\parallel}{T_\perp} < \begin{cases} \alpha^{2/3}, & n > 1, \\ \alpha^{4/3}, & n = 1. \end{cases} \quad (11.6)$$

For the nonresonant instability it is necessary that

$$\frac{T_\parallel}{T_\perp} < \begin{cases} \alpha, & n < 1, \\ \alpha^2, & n = 1. \end{cases} \quad (11.7)$$

It can be seen that the conditions on the smallness of $T_\parallel/T_\perp$ are more stringent for n = 1 than for n > 1.

If the condition (11.6) is violated, a kinetic instability develops. Its growth rate is

$$\gamma = -\frac{2\sqrt{\pi}\,\alpha\omega_p^2}{v_{T\parallel}^3 k^2 |k_z|} \left\{ \frac{1}{\partial \mathrm{Re}\,\varepsilon_0/\partial\omega} \sum_{n=-\infty}^{\infty} I_n e^{-z_\perp} \times \right.$$

$$\left. \times \exp\left\{-\left(\frac{\omega - n\omega_B}{k_z v_{T\parallel}}\right)^2\right\} \left[\omega - n\omega_B\left(1 - \frac{T_\parallel}{T_\perp}\right)\right]\right\}_{\omega=\omega_{1,2}}. \quad (11.8)$$

Under conditions similar to (11.2) (with n > 1) and

$$\alpha^{2/3} < T_\parallel/T_\perp < \frac{1}{2} \quad (11.9)$$

Eq. (11.8) yields the approximate relationship

$$\gamma \simeq \alpha\omega_B T_\perp/T_\parallel \quad (n > 1). \quad (11.10)$$

If $T_\parallel/T_\perp \approx \alpha^{2/3}$, the expressions (11.3) and (11.10) are of the same order of magnitude. The restriction $T_\parallel/T_\perp < 1/2$ follows from the results of § 10.3.

Let us now consider perturbations with $\omega \approx \omega_B$. If $\left|\frac{\omega}{\omega_B} - 1\right| < \cot\theta$ and $T_\parallel/T_\perp < 1/2$, Eq. (11.5) yields

$$\gamma = -2\alpha\sqrt{\pi}\,\frac{T_\perp}{T_\parallel}\,\frac{I_1 e^{-z_\perp}}{z_\perp}\cdot\frac{(\omega-\omega_B)^2}{\omega_B|k_z|v_{T\parallel}}\left[\omega-\omega_B \times \right.$$

$$\left.\times\left(1-\frac{T_\parallel}{T_\perp}\right)\right]\exp\left\{-\left(\frac{\omega-\omega_B}{k_z v_{T\parallel}}\right)^2\right\}. \qquad (11.11)$$

It can be seen that $\gamma > 0$ if

$$\frac{\omega}{\omega_B} < 1 - \frac{T_\parallel}{T_\perp}. \qquad (11.12)$$

Perturbations with $|\omega| < |\omega_B|$ satisfying the conditions $|\omega-\omega_B| \ll \omega_B$ exist only in a plasma with $\omega_p \geq \omega_B$ and only if $\sin\theta \ll 1$ (this is the branch ω_2; see § 8.1). An instability is possible not only if $\omega_p \approx \omega_B$ but also if ω_p and ω_B differ appreciably, $\omega_p > \omega_B$.

Arguments similar to those adduced in § 10.3 show that the interval of frequencies of the excited oscillations is bounded below by the inequality $\omega_2 > \omega_B/2$ and that excitation is possible only if $T_\parallel/T_\perp < 1/2$. The growth rate of the oscillations is given approximately by (11.10).

§ 11.2. Two-Stream – Anisotropic Instabilities

If a plasma interacts with a group of fast particles for which $T_\perp > T_\parallel$ and $V \neq 0$, instabilities can arise that can be described as a hybrid of the two-stream and anisotropic instabilities. Their growth rates are larger than those in the cases $T_\perp = T_\parallel$, $V \neq 0$, and $T_\perp > T_\parallel$, $V = 0$. Let us consider some examples of instabilities of this kind.

1. Excitation of Oscillations of a Cold Plasma by an Anisotropic Beam of Low Density. Suppose that a low-density beam with $T_\parallel = 0$. passes through a cold plasma. The oscillations of such a system are described by a dispersion equation that is similar to (11.1), except that the substitution $\omega \rightarrow \omega - k_z V$ must be made in the beam term:

$$\varepsilon_0 \equiv 1 - \frac{\omega_p^2\cos^2\theta}{\omega^2} - \frac{\omega_p^2\sin^2\theta}{\omega-\omega_B^2} - \alpha\omega_p^2\sum_{n=-\infty}^{\infty}\times$$

$$\times\left[\frac{I_n e^{-z_\perp}\cos^2\theta}{(\omega-n\omega_B-k_zV)^2} + \frac{nI_n e^{-z_\perp}\sin^2\theta}{z_\perp\omega_B(\omega-n\omega_B-k_zV)}\right] = 0. \qquad (11.13)$$

In contrast to § 11.1, the beam branches $\omega = n\omega_B + k_z V$ can intersect both the branches ω_1 and ω_2 of plasma oscillations at any value of the ratio ω_p/ω_B. The general form of the branches is similar to that shown in Fig. 9.1. There are however two differences: first, beam branches with $n > 1$ are present and also all the beam branches are doubled — the dispersion equation in the limit $\alpha \to 0$ has two solutions $\omega = k_z V + n\omega_B$. For finite α, each of the intersections of the branches therefore corresponds to instability. The maximal growth rate of the oscillations is determined by an expression similar to (9.10):

$$\gamma_{1,2}^{(n)} = \frac{\sqrt{3}}{2}\alpha^{1/3}\left(\frac{\omega_p^2 \cos^2\theta I_n e^{-z_\perp}}{\partial\varepsilon_0^{(0)}/\partial\omega}\right)_{\substack{\omega=\omega_{1,2} \\ k_z=k_{z1,2}^{(n)}}}, \qquad (11.14)$$

where $k_{z1,2}^{(n)}$ satisfy the equation

$$k_{z1,2}^{(n)} = (\omega_{1,2} - n\omega_B)/V. \qquad (11.15)$$

Equations (11.14) and (11.15) can be analyzed as in §§ 9.1 and 9.2. As an example, let us consider the excitation by the first cyclotron resonance ($n = 1$) of the low-frequency branch $\omega = \omega_2$ in a plasma with $\omega_p \ll \omega_B$. In this case $k_{z2} \approx \omega_B/V$, and

$$\gamma_2 = \frac{\sqrt{3}}{2}\left(\frac{\alpha I_1 e^{-z_\perp}}{2}\right)^{1/3}\omega_p \cos\theta. \qquad (11.16)$$

In accordance with (11.16) and (9.30), the growth rate of the hybrid instability for $v_{T_\perp} > V$ exceeds that of the two-stream instability by a factor of $(\omega_B/\omega_p\alpha^{1/3})^{1/2}$.

The applicability of Eq. (11.14) is restricted by the condition $\gamma > k_z v_{T_\parallel}$. If this condition is not satisfied, the hybrid instability becomes a kinetic instability, and its growth rate is determined by the interaction between the resonant particles and the wave. In this case the growth rate is given by Eq. (11.8) if the substitution $\omega \to \omega - k_z V$ is made.

2. Excitation of Electron-Cyclotron Oscillations in a Plasma Consisting of Two Interstreaming Anisotropic Beams. Suppose that the electron component of a plasma consists of two equal-density beams, $n_1 = n_2 \equiv n_0$, moving toward each other, $V_2 = -V_1 \equiv V$, and such that $T_\perp \gg T_\parallel$.

We shall assume that the density of the beams is fairly low, $\omega_p \ll \omega_B$, and consider their cyclotron–cyclotron interaction (cf. § 9.2).

We shall proceed from the dispersion equation

$$1-\omega_p^2 I_n e^{-z_\perp}\cos^2\theta\left[\frac{1}{(\omega-n\omega_B-k_zV)^2}+\frac{1}{(\omega-n\omega_B+k_zV)^2}\right]=0. \quad (11.17)$$

It is similar to Eq. (1.39). Using the results of § 1.5.1, we find that Eq. (11.17) has solutions with $\operatorname{Im}\omega > 0$, these satisfying

$$\left.\begin{aligned} &\operatorname{Re}\omega \approx n\omega_B,\\ &\gamma_{max}=\frac{\omega_p}{2}\cos\theta\,(I_n e^{-z_\perp})^{1/2},\\ &k_{opt}=\frac{\sqrt{3}}{2}\cdot\frac{\omega_p}{V}(I_n e^{-z_\perp})^{1/2}\cos\theta. \end{aligned}\right\} \quad (11.18)$$

The longitudinal thermal spread of the particles in the beams can be neglected if

$$v_{T_\parallel} < V. \quad (11.19)$$

Depending on $k_\perp$, the growth rate is maximal for $k_\perp\rho_\perp \simeq 1$. For perturbations with $k_\perp \simeq k_z$, this condition means

$$v_{T_\perp}/V \simeq \omega_B/\omega_p \gg 1. \quad (11.20)$$

It follows from (11.19) and (11.20) that a plasma in which this instability can arise must be strongly anisotropic:

$$T_\parallel/T_\perp \lesssim (\omega_p/\omega_B)^2. \quad (11.21)$$

A comparison of the growth rates (11.18) and (9.27) shows that the hybrid excitation of electron-cyclotron oscillations must occur more rapidly than the two-stream excitation for $T_\parallel = T_\perp$.

§ 11.3. Excitation of Plasma Oscillations by a Group of Fast Particles with a Nonequilibrium Transverse Velocity Distribution

In § 11.1 we have shown that fast particles with $V = 0$ excite oscillations of a cold plasma if their transverse energy is high compared with the longitudinal energy, $T_\perp > T_\parallel$. We shall now show that the excitation of oscillations by particles with V = 0 is also possible when $\bar{v}_\perp^2/2 \approx \bar{v}_\parallel^2$ if the particles have a nonequilibrium distribution of their transverse velocities, $\partial F/\partial v_\perp > 0$. This effect is similar to the instability of a single-component plasma with $\partial F/\partial v_\perp > 0$ considered in § 10.4.

We shall restrict ourselves to an analysis of perturbations with $k_z = 0$. Using Eqs. (7.3) and (7.16), we find that perturbations of this kind are described by the dispersion equation

$$1 - \frac{\omega_p^{(2)}}{\omega^2 - \omega_B^2} - \alpha\omega_p^2 \sum_{n=-\infty}^{\infty} \frac{n\left\langle 2\frac{J_n J'_n}{\xi}\right\rangle}{\omega_B(\omega - n\omega_B)} = 0. \tag{11.22}$$

1. Excitation of Electron-Cyclotron Harmonics. The plasma oscillation frequency $\omega_1 = (\omega_p^2 + \omega_B^2)^{1/2}$ is near $n\omega_B$ ($|n| > 1$) if the density and the magnetic field are related by the equation [cf. (11.2)]:

$$(\omega_p/\omega_B)^2 = n^2 - 1, \quad |n| > 1. \tag{11.23}$$

A beam leads to the excitation of oscillations of a plasma of this density if

$$\langle J_n J'_n/\xi\rangle < 0. \tag{11.24}$$

If the condition (11.23) is satisfied, the growth rate of the n-th harmonic is

$$\gamma = \alpha^{1/2}\frac{\omega_p^2}{\omega_B}\sqrt{-\left\langle\frac{J_n J'_n}{\xi}\right\rangle}. \tag{11.25}$$

The instability condition (11.24) is satisfied for a larger class of nonequilibrium distributions than in the case of a single-component plasma [cf. (11.24) and (10.66)]. In particular, oscillations can be excited by fast particles with a distribution of the type (10.69):

$$F = \frac{m(T+T_1)}{2\pi T^2}\exp\left(-\frac{mv_\perp^2}{2T}\right)\left[1 - \exp\left(-\frac{mv_\perp^2}{2T_1}\right)\right],$$

whereas a single-component plasma with a distribution function of this kind is stable.

It should be noted that the lowest harmonic of the excited cyclotron harmonics is two, i.e., oscillations with the frequency $\omega \approx \omega_B$ are not excited.

2. Excitation of High-Frequency Oscillations with $\gamma > \omega_B$ in the case $\alpha > \omega_B/\omega_p$. If $\omega_p \gg \omega_B$, the unstable harmonic has a high number, $n \gg 1$. In this case, the growth rate of the instability is approximately

$$\gamma \simeq \alpha^{1/2} \omega_p. \tag{11.26}$$

If $\alpha \gtrsim (\omega_B/\omega_p)^2$, the relation (11.26) would imply $\gamma \gtrsim \omega_B$. In this case, however, several harmonics make an appreciable contribution to the growth rate, so that it is incorrect to treat only a single harmonic. One can take into account many harmonics by going over to the approximation of a continuous spectrum (see § 7.4). It then follows from (11.22) that

$$\varepsilon_0 \equiv 1 - \frac{\omega_p^2}{\omega_B^2} - \frac{\alpha\omega_p^2}{k^2} \left\langle \frac{\partial F}{\partial \varepsilon_\perp} \left(1 - \frac{\omega}{[\omega^2 - (kv_\perp)^2]^{1/2}}\right)\right\rangle = 0. \tag{11.27}$$

In the case of a δ-function distribution, $F \propto \delta(v_\perp - v_0)$, this means

$$1 - \frac{\omega_p^2}{\omega^2} - \frac{i\alpha\omega_p^2\omega}{[(k_\perp v_0)^2 - \omega^2]^{3/2}} = 0. \tag{11.28}$$

The maximal growth rate is attained for $kv_0 \approx \omega_p$, the expression for it differing formally only little from (11.26):

$$\gamma \simeq \alpha^{2/5} \omega_p; \tag{11.29}$$

however, since we have assumed $\alpha > \omega_B/\omega_p$, it does exceed ω_B. High-frequency perturbations with $\gamma > \omega_B$ can be excited not only in the case of a δ-function distribution but also for every other $F(\varepsilon_\perp)$ if the corresponding one-dimensional function

$$f_0(v_x) \equiv \int F(\varepsilon_\perp)\, dv_y$$

has a positive derivative, $\partial f_0/\partial v_x > 0$, at small v_x. Because of the presence of the cold component of the plasma, the total one-dimensional distribution function has a fairly deep minimum and therefore belongs to the class of unstable distributions (see §§ 2.7 and 2.8).

§ 11.4. Excitation of Plasma Oscillations by an Azimuthal Electron Stream

We shall now consider the excitation of plasma oscillations by a cold beam that rotates about a symmetry center of the system (and, possibly, also moves along the magnetic field). It follows from the hydrodynamic equation of motion [see (1.3)] that the transverse velocity of the equilibrium motion of the stream is

$$\mathbf{V}_{\perp}^{(0)} = -r\omega_B \mathbf{e}_{\varphi}, \tag{11.30}$$

where $\mathbf{e}_{\varphi}$ is a unit azimuthal vector.

If the space-time dependence of perturbations is taken in the form $F(r)\exp(il\varphi + ik_z z - i\omega t)$ and $\partial V_z^{(0)}/\partial r = 0$, Eq. (1.3) yields the perturbed density of the stream:

$$n'_1 = \frac{e}{m}\left\{\frac{n_1 k_z^2}{\Omega^2}\psi - \nabla_{\perp}\left(\frac{n_1\nabla_{\perp}\psi}{\Omega^2-\omega_B^2}\right) - \frac{l}{r}\frac{\omega_B}{\Omega}\psi\frac{\frac{\partial n_1}{\partial r}}{\Omega^2-\omega_B^2}\right\}. \tag{11.31}$$

Here, $\Omega = \omega - k_z V_z^{(0)} + l\omega_B$. The effects of the azimuthal motion of the stream can only be manifested if $l \neq 0$, since otherwise (11.31) would be the same as in the case $V_{\perp}^{(0)} = 0$. In what follows we shall assume that this condition is satisfied. Physically, the presence of the term with l in Ω corresponds to a Doppler shift of the effective frequency of the oscillations due to the azimuthal motion of the particles. The expression for Ω can also be represented in the more suggestive form

$$\Omega = \omega - k_z V_z^{(0)} - k_{\varphi} V_{\varphi}^{(0)}, \tag{11.32}$$

where $V_{\varphi}^{(0)}$, which is defined by Eq. (11.30), is the azimuthal component of the stream velocity and $k_{\varphi} \equiv -l/r$ can be interpreted as an azimuthal wave number. Substituting (11.31) into Poisson's equation; using the method of successive approximations adopted in § 9.6.1; and assuming that the short-wavelength condition $k_{\perp}a \gg 1$ is satisfied, we arrive at an equation for the correction to the frequency. This has the form of (9.76) with $\varepsilon_0^{(0)}$ in the previous form and $\varepsilon_{\text{eff}}^{(1)}$ equal to

$$\varepsilon_{\text{eff}}^{(1)} = -\frac{\alpha_{\text{eff}}\,(\omega_p \cos\theta)^2}{(\omega - k_z V_z - k_\varphi V_\varphi)^2} - \frac{\alpha_{\text{eff}}\,(\omega_p \sin\theta)^2}{(\omega - k_z V_z - k_\varphi V_\varphi)^2 - \omega_B^2},$$

$$k_\varphi V_\varphi = -l\omega_B. \tag{11.33}$$

(We have omitted the index 0 from V_z and V_φ).

The expression (11.33) differs from $\varepsilon_0^{(1)}$ of the form (9.3) by the substitution

$$\left.\begin{array}{c} \alpha \to \alpha_{\text{eff}} \\ k_z V_z \to k_z V_z + k_\varphi V_\varphi . \end{array}\right\} \tag{11.34}$$

From this it follows that under conditions when the discrete nature of k_φ is not important the interaction between the plasma and an azimuthal stream, $\mathbf{V} \perp \mathbf{B}_0$, is described by the same equations as in the case $\mathbf{V} \parallel \mathbf{B}_0$.

If $V_z = 0$, Eq. (11.33) is similar to the expression for $\varepsilon_0^{(1)}$ for a group of particles with an anisotropic velocity distribution; for the n-th term of the series (11.1), which is quadratic in $(\omega - n\omega_B)^{-1}$, differs from the first term of the right-hand side of (11.33) only by the substitution

$$\alpha I_n e^{-z} \to \alpha_{\text{eff}} \tag{11.35}$$

By the same token we have discovered an analogy between the properties of an azimuthal stream and an anisotropic plasma; this also follows from § 10.1.2.

Finally, comparing (11.33) and (11.22), we see that there is a similarity between the properties of an azimuthal stream and a group of particles with a nonequilibrium transverse distribution function: if $\cos\theta = 0$ and $|\omega + l\omega_B \pm \omega_B| \ll \omega_B$, the difference between (11.33) and the n-th term of the series (11.22) consists of the substitution

$$\alpha \left\langle -\frac{2nJ_n J'_n}{\xi} f_1 \right\rangle \to \frac{\alpha_{\text{eff}}}{2}. \tag{11.36}$$

These analogies enable us to use a number of the results of §§ 11.1-11.3. In what follows we shall assume $V_z = 0$.

1. Excitation of Oscillations with $\gamma > \omega_B$ of a Plasma in a Weak Magnetic Field. If $\omega_p \gg \omega_B$ and α_{eff} satisfies the opposite condition to (9.13), an azimuthal stream can excite oscillations with $\gamma > \omega_B$. In this approximation

$$\varepsilon_{eff}^{(1)} = -\frac{\alpha_{eff}\,\omega_p^2}{(\omega - k_\varphi V_\varphi)^2}, \tag{11.37}$$

$$\gamma = \frac{\sqrt{3}}{2^{4/3}}\alpha_{eff}^{1/3}\,\omega_p, \tag{11.38}$$

$$\mathrm{Re}\,\omega \approx \omega_p \approx k_\varphi V_\varphi. \tag{11.39}$$

It follows from the resonance condition $k_\varphi V_\varphi = \omega_p$ that the azimuthal number of the excited oscillations is high:

$$|l| \approx \frac{\omega_p}{\omega_B} > \alpha_{eff}^{-1/3}. \tag{11.40}$$

If the radial wavelength of the perturbations does not exceed the width of the beam, the above analysis must be augmented by a consideration of the growth of wave packets (see §§ 4.4 and 5.4). We then obtain equations of the type (11.37)-(11.40), in which α_{eff} is replaced by the ratio of the local densities α, this being greater than α_{eff} by a factor of R/a.

The condition of applicability $\gamma_{loc} \gg \frac{1}{a}\cdot\frac{\partial\omega}{\partial K_r}$ of the local treatment (see § 5.3 and the remarks at the beginning of § 9.6) in the case $K_r \approx l/r$ leads to

$$a/R \gg \alpha^{2/3} \tag{11.41}$$

if allowance is made for a relation of the type (4.11) and the fact that $\frac{\partial\omega}{\partial K_r} \simeq \frac{\omega}{K_r}\left(\frac{\omega_B}{\omega_p}\right)^2$ for $\omega_p \gg \omega_B$.

If (11.41) holds, the most important factor is the growth of wave packets; if not, the growth of the characteristic oscillations of the plasma cylinder.

2. Excitation of Cyclotron Harmonics in a Plasma with $\omega_B < \omega_p < \alpha_{eff}^{-1/3}\omega_B$.

If the condition (9.13) holds, the perturbations excited by the beam have growth rates that are small compared with the cyclotron frequency, $\gamma \ll \omega_B$. At a resonance of the Cherenkov type

($\omega \approx -l\omega_B$) the growth rate of the perturbations is determined by Eq. (11.3) in conjunction with (11.35); at a cyclotron resonance ($\omega \approx -l\omega_B \pm \omega_B$), by Eq. (11.25) with the substitution (11.36).

The condition for the excitation of the n-th harmonic has the form (11.23). Instability is possible only in a plasma with $\omega_p > \omega_B$. At Cherenkov resonance, the serial number of the excited harmonic is equal to the azimuthal number, $n = l$; at cyclotron resonance, it differs by unity, and $|l| = |n| + 1$.

Bibliography

1. A. B. Kitsenko and K. N. Stepanov, In: Plasma Physics and Problems of Controlled Thermonuclear Fusion, Vol. 3 [in Russian], Naukova Dumka, Kiev (1963), p. 143. An investigation is made of the instabilities of a beam with an almost δ -function distribution (§§ 11.1-11.3).
2. L. S. Hall, W. Heckrotte, and T. Kammash, Phys. Rev., A139:1117 (1965).
3. K. Jungwirth, Czech. J. Phys., B17:498 (1967).
4. V. I. Shevchenko, In: Plasma Physics and Problems of Controlled Thermonuclear Fusion, Vol. 2 [in Russian], Naukova Dumka, Kiev (1963), p. 156. In [2-4] the reader can find discussions of the excitation of plasma oscillations by a group of electrons with a non-Maxwellian transverse velocity distribution (§ 11.3)
5. K. Jungwirth and J. Preinhaelter, Czech. J. Phys., B16:228 (1966). A study is made of the cyclotron instability of interstreaming electron beams with an anisotropic velocity distribution (§ 11.2).
6. A. B. Mikhailovskii and K. Jungwirth, Zh. Tekh. Fiz., 36:777 (1966) [Soviet Phys. – Tech. Phys., 11:581 (1966)]. A study is made of the excitation of plasma oscillations by an azimuthal electron stream (§ 11.4) and also the excitation of plasma oscillations by a spatially inhomogeneous stream with a non-Maxwellian velocity distribution (§§ 11.1-11.3).
7. M. Seidl and P. Šunka, Nucl. Fusion, 7:237 (1967). Discussion of the excitation of plasma oscillations by a beam with a non-Maxwellian velocity distribution (§§ 11.1-11.3).
8. K. Jungwirth, Czech. J. Phys., 18:629 (1968). The instability of an azimuthal electron beam (§ 11.4) is investigated.

Chapter 12

Plasmas with a Longitudinal Current

§ 12.1. Plasmas with a High Current Velocity in a Weak Magnetic Field

A relative motion of the electron and ion components – a current in the plasma – can give rise to various instabilities. We have already discussed some of these in the $B_0 = 0$ approximation – the Buneman instability in § 1.6 and the ion-acoustic instability in §§ 3.4, 6.3, and 6.4.

Suppose the current velocity exceeds the electron thermal velocity $(V > v_{T_e})$, and the magnetic field is not too strong, $\omega_{pe} > \omega_{B_e}$. In this case, we know from § 1.6 that the plasma must have a hydrodynamic instability of the Buneman type with the growth rate

$$\gamma \simeq (m_e/m_i)^{1/3}\,\omega_{pe} \tag{12.1}$$

and longitudinal wave number

$$|k_z| = \omega_{pe}/V. \tag{12.2}$$

The real part of the frequency of the growing perturbations is of the same order as the growth rate:

$$\mathrm{Re}\,\omega \simeq (m_e/m_i)^{1/3}\,\omega_{pe}. \tag{12.3}$$

The generalization of the dispersion equation (1.60) to the case of nonvanishing $\omega_{B_e}/\omega_{p_e}$ is

$$1-\frac{(\omega_{p_e}\cos\theta)^2}{(\omega-k_zV)^2}-\frac{(\omega_{p_e}\sin\theta)^2}{(\omega-k_zV)^2-\omega_{B_e}^2}-\left(\frac{\omega_{p_i}}{\omega}\right)^2=0. \tag{12.4}$$

In the zeroth approximation in the parameter m_e/m_i, the nonvanishing roots of (12.4) are simply the electron branches ω_1 and ω_2 shifted by k_zV [see (8.2)]. The interaction of the electrons with the ions leads to excitation of these branches, the maximum of the growth rate being attained when

$$k_z=-\omega_{1,2}/V. \tag{12.5}$$

In the limit of large $\omega_{p_e}/\omega_{B_e}$, the growth rate of the branch ω_1 is given by Eq. (12.1) and that of the branch ω_2 is

$$\gamma_2=\frac{\sqrt{3}}{2^{4/3}}\left(\frac{m_e}{m_i}\cdot\frac{\sin^2\theta}{\cos\theta}\right)^{1/3}\omega_{B_e}. \tag{12.6}$$

If $\omega_{p_e}\simeq\omega_{B_e}$, both growth rates are of the same order.

§ 12.2. Plasma with a High Current Velocity in a Strong Magnetic Field

Suppose, as in § 12.2, that $V>v_{T_e}$, but $\omega_{p_e}\ll\omega_{B_e}$, i.e., we have a strong magnetic field. The frequency of the branch ω_1 is then near the cyclotron frequency, $\omega_1\approx\omega_{B_e}$, and the resonant k_z is given by

$$k_z^{(1)}=\omega_{B_e}/V. \tag{12.7}$$

Equation (12.4) shows that this branch has the growth rate

$$\gamma_1=\frac{\sqrt{3}}{2^{4/3}}\left(\frac{m_e}{m_i}\cdot\frac{\omega_{p_e}\sin^2\theta}{\omega_{B_e}}\right)^{1/3}\omega_{p_e}. \tag{12.8}$$

It must however be borne in mind that the growth of the electron-cyclotron oscillations is very sensitive to the thermal spread of the electron velocities and the inequality

$$\frac{v_{T_e}}{V}\lesssim\left(\frac{m_e}{m_i}\right)^{1/3}\left(\frac{\omega_{p_e}}{\omega_{B_e}}\right)^{4/3} \tag{12.9}$$

must be satisfied.

The growth of the branch ω_2 for large ω_{B_e} is described by the dispersion equation

$$1-\left(\frac{\omega_{p_e}\cos\theta}{\omega-k_zV}\right)^2-\left(\frac{\omega_{p_i}}{\omega}\right)^2=0. \tag{12.10}$$

If $\cos\theta > (m_e/m_i)^{1/2}$, this equation can be investigated by the method of successive approximations, the ion contribution being assumed small. We then find that the instability occurs if

$$k \leqslant \omega_{p_e}/V. \tag{12.11}$$

This shows that an instability cannot arise in a spatially bounded plasma if its density is too low (cf. § 9.7). In the case of a long plasma, $k_z \ll k_\perp$, the condition (12.11) means

$$\omega_{p_e}^2 \gg (k_\perp V)^2 \gtrsim (V\pi/a_\perp)^2. \tag{12.12}$$

This is the desired lower bound for the density. It can be expressed as a restriction on the current at which the instability can develop:

$$j_0 \equiv en_0V \gtrsim \frac{V^3}{4\pi e}\left(\frac{\pi}{a_\perp}\right)^2. \tag{12.13}$$

If the instability condition (12.11) is satisfied, the corresponding growth rate must be

$$\gamma=\frac{\sqrt{3}}{2^{4/3}}\left(\frac{m_e}{m_i}\right)^{1/3}\cos\theta\,\omega_{p_e}. \tag{12.14}$$

The value of $\operatorname{Re}\omega$ is of the same order.

If $\cos\theta \lessgtr (m_e/m_i)^{1/2}$, the instability condition is [cf. (1.58)]

$$k_\perp^2 < \frac{\omega_{p_e}^2}{V^2}\left[1+\left(\frac{m_e}{m_i\cos^2\theta}\right)^{1/3}\right]^3. \tag{12.15}$$

Thus, if the plasma is sufficiently long, an instability can occur even if the condition (12.13) is not satisfied.

§ 12.3. Excitation of High-Frequency Ion-Acoustic Oscillations by a Current

If the velocity of the current is low, $v_{T_i} \ll V < v_{T_e}$, and $T_e \gg T_i$, the high-frequency ion-acoustic branch (§ 8.3) can be excited.

If $k_\perp = 0$, a magnetic field does not affect this instability and it is described by the equations of § 3.4. We shall now assume that $k_\perp \approx k_z$.

If the ions are cold and $\omega \gg \omega_{B_i}$, the dispersion equation is

$$1+\frac{1}{(kd_e)^2}\left\{1+i\sqrt{\pi}\,\frac{\omega-k_zV}{|k_z|\,v_{T_e}}\sum_{n=-\infty}^{\infty} I_n(z_e)\times\right.$$
$$\left.\times\exp\left[-z_e-\left(\frac{\omega-n\omega_{B_e}-k_zV}{k_zv_{T_e}}\right)^2\right]\right\}-\left(\frac{\omega_{p_i}}{\omega}\right)^2=0. \qquad (12.16)$$

Since the characteristic k are of order ω_{p_e}/v_{T_e}, we conclude that in a weak magnetic field, $\omega_{p_e} \gg \omega_{B_e}$, it is possible to assume $z_e \gg 1$, $k_zv_{T_e} \gg \omega_{B_e}$. The series can then be summed and it is then seen that the magnetic field drops out of (12.16) and the latter reduces to the equation considered in § 3.4.

If $\omega_{B_e} \gg \omega_{p_e}$, only the term with n = 0 is important in the sum (12.16). In this case, the growth rate is

$$\gamma=\left(\frac{\pi}{8}\cdot\frac{m_e}{m_i}\right)^{1/2}\frac{k\left[V(1+k^2d_e^2)^{1/2}-(\cos\theta)^{-1}\left(\frac{T_e}{m_i}\right)^{1/2}\right]}{(1+k^2d_e^2)^2}. \qquad (12.17)$$

It is greater than the growth rate in the case $\omega_{p_e} \gg \omega_{B_e}$, by the factor $1/\cos\theta$. The growth rate (12.17) is positive if

$$V>V_{\mathrm{cr}}=(T_e/m_i)^{1/2}(\cos\theta)^{-1}(1+k^2d_e^2)^{-1/2}. \qquad (12.18)$$

If $V \gg V_{\mathrm{cr}}$, then γ does not depend on $\cos\theta$. The maximum of γ is of the order $(m_e/m_i)^{1/2}\,\omega_{p_i}$, as in the case $B_0 = 0$.

§ 12.4. Excitation of High-Frequency Ion-Acoustic Oscillations by Runaway Electrons

Suppose there is a group of electrons in a plasma with velocities $v_1 > v_{T_e}$ and distribution function in the form of a plateau [Eq. (9.58)]. If $\omega_{p_e} \ll \omega_{B_e}$, these electrons can excite oscillations of the branch $\omega_2 = \omega_{p_e}\cos\theta$ if their velocity is such that $v_1/v_{T_e} > \omega_{B_e}/\omega_{p_e}$ [see condition (9.65)]. Suppose that condition (9.65) is not satisfied, i.e.,

$$1 < v_1/v_{T_e} \lesssim \omega_{B_e}/\omega_{p_e}. \qquad (12.19)$$

If $T_e \gg T_i$, fast electrons can excite high-frequency ion-acoustic oscillations in this case. The dispersion equation describing this instability has the form [cf. (9.59)]

$$1 + \frac{1}{(kd_e)^2} - \left(\frac{\omega_{p_i}}{\omega}\right)^2 + i\pi\alpha \frac{(\omega_{p_e} \sin\theta)^2}{2\omega_{B_e} |k_z| v_1} = 0. \qquad (12.20)$$

The imaginary term in this equation is due to the cyclotron resonance between the fast electrons and the wave. The condition of this resonance when $\omega \ll \omega_{B_e}$ is $v_z = \omega_{B_e}/k_z$. Since wave numbers satisfying $k_z \simeq \omega_{p_e}/v_{T_e}$ correspond to ion-acoustic oscillations, the velocity of the resonant particles is of the order $v_z \simeq (\omega_{B_e}/\omega_{p_e})v_{T_e}$, which does not contradict the condition (12.19). The growth rate of the perturbations is

$$\gamma \simeq \alpha\omega_{p_i} (\omega_{p_e}/\omega_{B_e})^2. \qquad (12.21)$$

§ 12.5. Ion-Cyclotron Instability

If $V \neq 0$, the ion-cyclotron oscillations considered in the case $V = 0$ in § 8.4.5 can be excited in the plasma. If T_e/T_i is not too small and $\cos\theta > v_{T_i}/v_{T_e}$, these oscillations have a phase velocity that is small compared with the electron thermal velocity, $\omega/k_z < v_{T_e}$, and therefore – as in the case of ion-acoustic oscillations – an instability is possible even if $V < v_{T_e}$. In contrast to the ion-acoustic instability, it is not essential for T_i/T_e to be small for excitation of the cyclotron oscillations.

If $\omega/k_z \ll v_{T_e}$, the real part of the frequency of the cyclotron oscillations is determined by Eq. (8.52). We can find the growth rate by means of Eq. (2.17), substituting into it $\mathrm{Re}\,\varepsilon_0$ in the form (8.54) and $\mathrm{Im}\,\varepsilon_0 = \mathrm{Im}\,\varepsilon_0^{(e)} + \mathrm{Im}\,\varepsilon_0^{(i)}$, where $\mathrm{Im}\,\varepsilon_0^{(e)} = \sqrt{\pi}(kd_e)^{-2}(\omega - k_z V)/|k_z| v_{T_e}$, and $\mathrm{Im}\,\varepsilon_0^{(i)}$ is given by the expression (8.55):

$$\gamma = -\frac{\sqrt{\pi}\Delta^2}{|k_z| v_{T_e}} \frac{T_i}{T_e} \left[1 - \frac{k_z V}{n\omega_{B_i}} + \left(\frac{T_e}{T_i}\right)^{3/2} \left(\frac{m_i}{m_e}\right)^{1/2} \times \right.$$
$$\left. \times I_n(z_i) e^{-z_i} \exp\left\{-\left(\frac{\Delta_0}{k_z v_{T_i}}\right)^2 - 1\right\}\right]. \qquad (12.22)$$

TABLE 1

T_i/T_e	V_{cr}/v_{T_i}	$z_{i\,opt}$	$(\cos\theta)_{opt}$
1.00	23	1.1	0.05
0.33	15	0.8	0.12
0.10	12	0.8	0.15

Here, as in (8.56), $\Delta = \mathrm{Re}\,\omega - n\omega_{B_i}$; $\Delta_0 = \lim\limits_{T_i/T_e \to 0} \Delta$. The growth rate is positive if

$$V > V_{cr} = \frac{n\omega_{B_i}}{k_z}\left[1 + \left(\frac{T_e}{T_i}\right)^{3/2}\left(\frac{m_i}{m_e}\right)^{1/2} \times \right.$$
$$\left. \times I_n(z_i)\, e^{-z_i} \exp\left\{-1 - \left(\frac{\Delta_0}{k_z v_{T_i}}\right)^2\right\}\right]. \qquad (12.23)$$

The ion-cyclotron instability was first considered by Drummond and Rosenbluth. Lominadze and Stepanov have made a numerical calculation of the critical velocity, minimized with respect to k_z and z_i. The results of the numerical calculation, which are given in Table I, show that oscillations grow in a plasma with $T_e \approx T_i$ if the directed velocity V is not much smaller than the electron thermal velocity.

However, under these conditions the ion-acoustic instability is completely absent.

The value of m_i/m_e is taken equal to 3680 (deuterium).

§ 12.6. Excitation of Low-Frequency Ion-Acoustic Oscillations

If a plasma carries a current, the low-frequency ion-acoustic branch $\omega_5 = k_z (T_e/m_i)^{1/2}$ (§ 8.3) can also be excited. If binary collisions between the particles are unimportant, the growth rate of these oscillations is proportional to the derivative $\partial f_{0e}/\partial v_z$. It is then small compared with the increment of the high-frequency ion-acoustic oscillations:

$$\frac{\gamma_{lf}}{\gamma_{hf}} \simeq \frac{\omega_{B_i}}{\omega_{p_i}}. \qquad (12.24)$$

Under these conditions, low-frequency ion-acoustic oscillations may not be excited in practice since the rapid growth of the high-frequency ion-acoustic oscillations leads to a relaxation of the electron distribution function and the derivative $\partial f_{0e}/\partial v_z$ is then not positive for any v_z.

The low-frequency ion-acoustic instability plays a more important role in a collisional plasma when the electron mean free path does not exceed the length of the system:

$$\lambda_{\mathrm{coll}} \equiv v_{T_e}/\nu_e < L. \tag{12.25}$$

At the same time, the collisional instability mechanism discussed in §§ 6.3 and 6.4 may be manifested. In contrast to the case of a collisionless plasma, an instability can arise even if $T_e = T_i$.

Bibliography

1. I. B. Bernstein et al., Phys. Fluids, 3:136 (1960). The low-frequency ion-acoustic instability (§ 12.6) is discussed.
2. W. E. Drummond and M. N. Rosenbluth, Phys. Fluids, 5:1507 (1962). It is shown that a longitudinal current may excite ion-cyclotron oscillations (§ 12.5).
3. D. G. Lominadze and K. N. Stepanov, Zh. Tekh. Fiz., 34:1823 (1964) [Soviet Phys. — Technical Physics, 9:1408 (1965)]. The instabilities in a plasma with a longitudinal current — the cyclotron and low-frequency instabilities (§§ 12.5 and 12.6) — are investigated.
4. K. N. Stepanov, in: Plasma Physics and Problems of Controlled Thermonuclear Fusion [in Russian], Naukova Dumka (1963), p. 164. A study is made of the instability of a plasma whose electrons move relative to the ions along $\mathbf{B}_0$ with velocity $V > v_{T_e}$ (§§ 12.1 and 12.2).
5. L. V. Korablev and L. I. Rudakov, Zh. Eksp. Teor. Fiz., 50:220 (1966) [Soviet Phys. — JETP, 23:145 (1966)]. The high-frequency ion-acoustic instability (§ 12.3) is discussed.
6. B. B. Kadomtsev and O. P. Pogutse, Zh. Eksp. Teor. Fiz., 53:2025 (1967) [Soviet Phys. — JETP, 26:1146 (1968)]. The excitation of high-frequency ion-acoustic oscillations by runaway electrons (§ 12.4) is discussed.
7. D. G. Lominadze and K. N. Stepanov, Zh. Tekh. Fiz., 35:148 (1965) [Soviet Phys. — Tech. Phys., 10:113 (1965)]. The excitation of magnetoacoustic waves is discussed in the case when two equal-density plasma streams moving along the magnetic field collide.
8. D. G. Lominadze and K. N. Stepanov, Zh. Tekh. Fiz., 35:205 (1965) [Soviet Phys. — Tech. Phys., 10:169 (1965)]. The excitation of electrostatic (ion-acoustic) oscillations is discussed in the case when equal-density plasmas moving toward each other along the magnetic field collide.

Chapter 13

Plasmas with a Transverse Current

§ 13.1. Instability of a Cold Plasma

Suppose a plasma is cylindrically symmetric and that all the ions or some of them move azimuthally with respect to the electrons. It is then possible for electron-ion instabilities similar to the electron instabilities considered in § 11.4 to arise. If the plasma density is not too low, the growth rate of the electron-ion oscillations may be greater than the cyclotron frequency of the ions. It is these high-frequency instabilities that we shall now consider. Instabilities with $\gamma \lesssim \omega_{B_i}$ are discussed in Chapter 17.

In this section we shall assume that all the ions move azimuthally with respect to the electrons with a velocity V. If thermal effects are ignored, the perturbations of such a plasma are described by the dispersion equation

$$1+\left(\frac{\omega_{p_e}}{\omega_{B_e}}\sin\theta\right)^2-\left(\frac{\omega_{p_e}}{\omega}\cos\theta\right)^2-\left(\frac{\omega_{p_i}}{\omega-k_\varphi V}\right)^2=0. \tag{13.1}$$

This equation shows that if $\cos\theta \approx (m_e/m_i)^{1/2}$, when the "effective" densities of the electrons and ions are of the same order, the perturbations have the growth rate

$$\gamma \simeq \operatorname{Re}\omega \simeq \omega_{p_i}\left(1+\frac{\omega_{p_e}^2}{\omega_{B_e}^2}\right)^{-1/2}. \tag{13.2}$$

The wave number k_φ is then of order $(\omega_{p_i}/V)(1+\omega_{p_e}^2/\omega_{B_e}^2)^{-1/2}$. The temperature of the particles can be neglected in these perturbations if

$$v_{T_i} < V, \tag{13.3}$$

$$(T_e/m_i)^{1/2} < V. \tag{13.4}$$

Perturbations with $\cos\theta \gg (m_e/m_i)^{1/2}$ are characterized by the relations

$$\left.\begin{aligned} \mathrm{Re}\,\omega &\simeq k_{\varphi\,\mathrm{opt}} V \simeq \omega_{p_e} \cos\theta\,(1+\omega_{p_e}^2/\omega_{Be}^2)^{-1/2}, \\ \gamma_{\max} &\simeq (m_e/m_i \cos^2\theta)^{1/3}\,\mathrm{Re}\,\omega. \end{aligned}\right\} \tag{13.5}$$

The growth rate (13.5) is greater than (13.2). However, perturbations with large $\cos\theta$ can be excited only if there is a sufficiently small thermal spread of the ions and the electrons [cf. (13.3) and (13.4)]:

$$v_{T_i}/V < (m_e/m_i \cos^2\theta)^{1/3}, \tag{13.6}$$

$$T_e/m_iV^2 < m_e/m_i \cos^2\theta. \tag{13.7}$$

The dependence of the growth rate on $\cos\theta$ is shown schematically in Fig. 13.1.

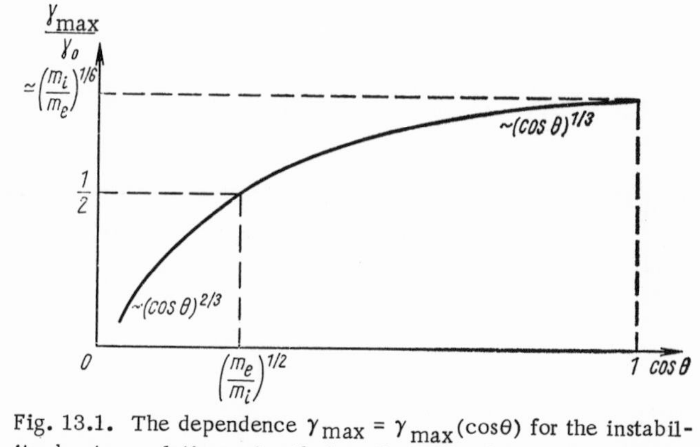

Fig. 13.1. The dependence $\gamma_{\max} = \gamma_{\max}(\cos\theta)$ for the instability due to a relative azimuthal motion of cold ions and cold electrons. Here, $\gamma_0 \equiv \omega_{p_i}(1+\omega_{pe}^2/\omega_{B_e}^2)^{-1/2}$.

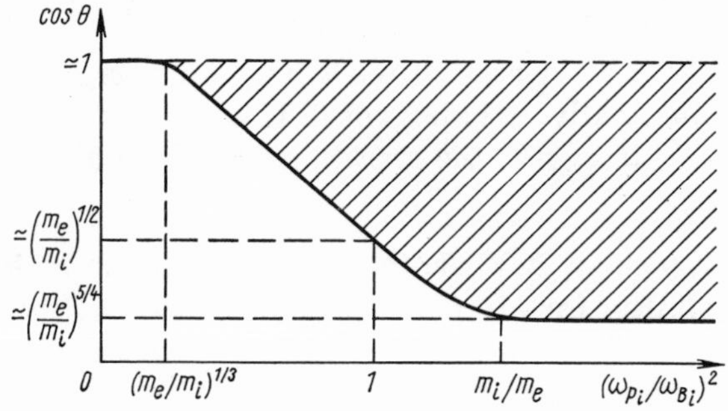

Fig. 13.2. Region of cos θ values for which $\gamma > \omega_{B_i}$ (shaded region).

As follows from (13.2) and (13.5), perturbations with $\gamma > \omega_{B_i}$, are possible only if the plasma density is not too low. The region of cos θ values corresponding to perturbations with $\gamma > \omega_{B_i}$, is shown as a function of the parameter $(\omega_{p_i}/\omega_{B_i})^2$ in Fig. 13.2.

§ 13.2. Excitation of Electron-Acoustic Oscillations in a Plasma with Hot Ions

It follows from the inequality (13.6) that the thermal spread of the ions has its greatest effect on the excitation of oscillations with maximal cos θ.

In particular, the hydrodynamic instability is impossible for $\cos\theta \approx 1$ if $v_{T_i}/V \gtrsim (m_e/m_i)^{1/3}$. For $(m_e/m_i)^{1/3} < v_{T_i}/V < 1$ the kinetic growth rate is approximately

$$\gamma_{\text{kin}} \simeq (m_e/m_i \cos^2\theta)^{1/2} (V/v_{T_i})^2 \omega_{p_i} (1 + \omega_{p_e}^2/\omega_{B_e}^2)^{-1/2}. \qquad (13.8)$$

This expression is valid for cos θ values that satisfy the opposite condition to (13.6). For given V/v_{T_i}, the largest growth rate corresponds to hydrodynamically unstable perturbations with maximal cos θ, i.e., with

$$\cos\theta \simeq (m_e/m_i)^{1/2} (V/v_{T_i})^{3/2}. \qquad (13.9)$$

Approximately

$$\gamma_{\text{max}} \simeq \omega_{p_i} (1 + \omega_{p_e}^2/\omega_{B_e}^2)^{-1/2} (V/v_{T_i})^{1/2}. \qquad (13.10)$$

The transverse wave number k_φ is determined by Eq. (13.5).

If $v_{T_i}/V \simeq 1$, all perturbations with $\cos\theta > (m_e/m_i)^{1/2}$ are unstable only kinetically. Their growth rate is less than in the case $\cos\theta \simeq (m_e/m_i)^{1/2}$; if $v_{T_i} \simeq V$ the growth rate is still approximately equal to (13.2).

Now suppose $v_{T_i} > V$. In this case the kinetic instability is the only possibility for all $\cos\theta$. It is described by the dispersion equation

$$1+\left(\frac{\omega_{p_e}}{\omega_{B_e}}\right)^2-\left(\frac{\omega_{p_e}}{\omega}\cos\theta\right)^2+\frac{1}{(kd_i)^2}\left(1+i\sqrt{\pi}\,\frac{\omega-k_\varphi V}{|k|\,v_{T_i}}\right)=0. \quad (13.11)$$

There is excitation of the branch of electron-acoustic oscillations with frequency $\omega = \omega_6$, where ω_6 is determined by Eq. (8.47). The maximal growth rate of the perturbations is approximately

$$\gamma_{\max} \simeq \omega_{p_i}\left(1+\frac{\omega_{p_e}^2}{\omega_{B_e}^2}\right)^{-1/2}\left(\frac{V}{v_{T_i}}\right)^2. \quad (13.12)$$

These perturbations correspond to

$$\cos\theta \simeq \left(\frac{m_e}{m_i}\right)^{1/2}\frac{V}{v_{T_i}}. \quad (13.13)$$

The transverse wave number is approximately $k_\perp \simeq \frac{1}{d_i}(1+\omega_{p_e}^2/\omega_{B_e}^2)^{-1/2}$.

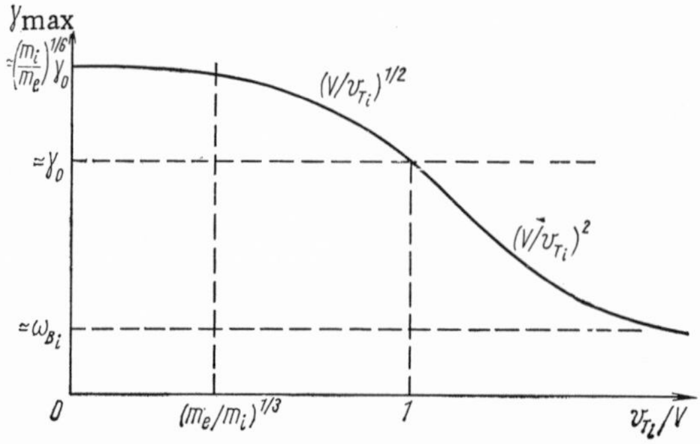

Fig. 13.3. Maximal growth rate of the instability of a dense plasma with V ≠ 0 as a function of the thermal spread of the ions.

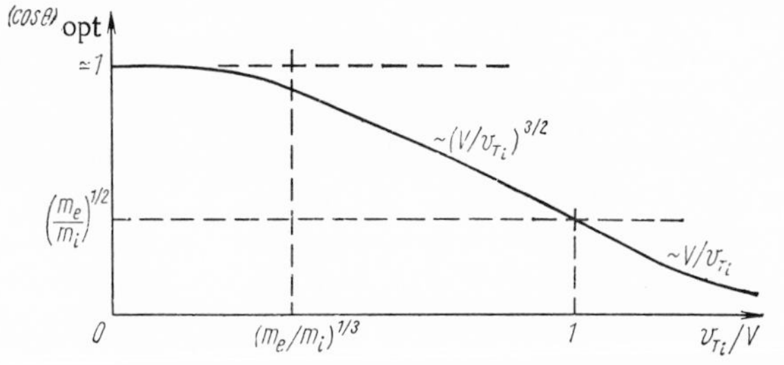

Fig. 13.4. Dependence of the parameter cos θ corresponding to the maximal growth rate on the thermal spread of the ions.

The dependences $\gamma_{max} = \gamma_{max}(v_{T_i}/V)$ and $(\cos\theta)_{opt} = (\cos\theta)_{opt}(v_{T_i}/V)$ are shown schematically in Figs. 13.3 and 13.4.

§ 13.3. Excitation of Ion-Acoustic Oscillations in a Plasma with Hot Electrons

If $T_e > m_e V^2$, the electrons behave as if they were "hot" in perturbations with $\cos\theta \approx 1$; this follows from (13.7). In this case the hydrodynamic instability is impossible for $\cos\theta \approx 1$. The maximal growth rate of the instability is then determined by Eq. (13.5) with $\cos\theta$ at the limit permitted by the inequality (13.7), i.e.

$$\gamma_{max} \simeq \frac{\omega_{p_i}}{(1+\omega_{p_e}^2/\omega_{B_e}^2)^{1/2}} (m_i V^2/T_e)^{1/6}, \tag{13.14}$$

$$(\cos\theta)_{opt} \simeq (m_e V^2/T_e)^{1/2}. \tag{13.15}$$

The transverse wave number of the corresponding perturbations is given approximately by

$$k_\varphi \rho_e \simeq \frac{1}{(1+\omega_{B_e}^2/\omega_{p_e}^2)^{1/2}}. \tag{13.16}$$

It can be seen that $z_e \approx 1$ if $\omega_{p_e} > \omega_{B_e}$ and $z_e < 1$ if $\omega_{p_e} < \omega_{B_e}$.

The results (13.14)-(13.16) are valid for T_e right up to $T_e \approx m_i V^2$. If T_e is greater, the hydrodynamic instability is not possible for any $\cos\theta$ and it is then necessary to take into account the interaction between the wave and the resonant electrons.

Assuming $|\omega - k_\varphi V| \gg kv_{T_i}$ and $kd_e \gg 1$ and regarding $\frac{\omega}{k_z v_{T_e}} W\left(\frac{\omega}{k_z v_{T_e}}\right) I_0(z_e) e^{-z_e}$ as a small quantity (because $1/z_e$ or $\omega/k_z v_{T_e}$, is small), we find that under these conditions ion plasma oscillations (the branch ω_4) are excited with a Doppler-shifted frequency:

$$|\omega - k_\varphi V| \approx \omega_{p_i}. \tag{13.17}$$

The characteristic $k_\perp$ are of order ω_{p_i}/V, and the characteristic k_z are of order ω_{p_i}/v_{T_e}. To estimate the growth rate we use the formula

$$\gamma = -\frac{\omega_{p_i}\sqrt{\pi}(\omega - |k_\varphi V|)}{2(kd_e)^2 |k_z| v_{T_e}} I_0(z_e) e^{-z_e}. \tag{13.18}$$

Since $z_e \approx (\omega_{p_e}/\omega_{B_e})^2 (T_e/m_i V^2)$, we must assume $z_e \gg 1$ in (13.18) if $\omega_{p_e} > \omega_{B_e}$ and we then have approximately

$$\gamma \simeq \left(\frac{m_i V^2}{T_e}\right)^{3/2} \left(\frac{m_i}{m_e}\right)^{1/2} \omega_{B_i}. \tag{13.19}$$

This result is still valid if $m_i V^2/T_e < (\omega_{p_e}/\omega_{B_e})^2 < 1$. If the density is even lower, $(\omega_{p_e}/\omega_{B_e})^2 < m_i V^2/T_e$ we must assume $z_e < 1$ in (13.18), and then

$$\gamma \simeq \frac{m_i V^2}{T_e} \omega_{p_i}. \tag{13.20}$$

The dependence $\gamma_{max} = \gamma_{max}(m_i V^2/T_e)$ is shown schematically in Fig. 13.5.

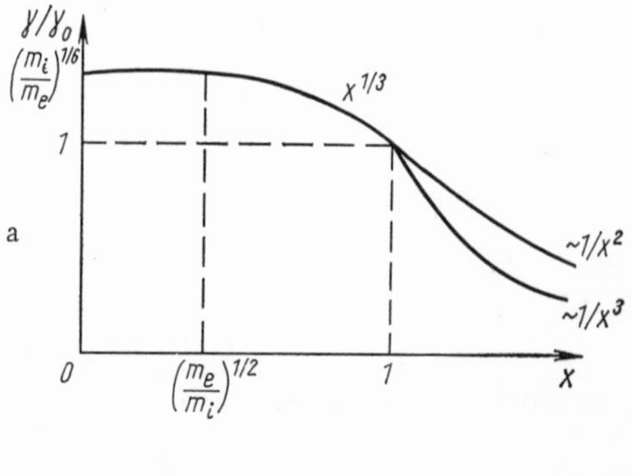

Fig. 13.5. Growth rate γ_{max} as a function of $x \equiv (T_e/m_i V^2$. If $x > 1$ in a dense plasma, than $(\omega_{p_e} > \omega_{B_e})$, then $\gamma_{max} \propto 1/x^3$; in a less dense plasma $(\omega_{p_e} < \omega_{B_e})$, we have $\gamma_{max} \propto 1/x^2$.

§ 13.4. Excitation of Plasma Oscillations by a Transverse Ion Stream

1. Cold Stream in a Cold Plasma. We shall assume that the plasma density is such that $\omega_{B_i} \ll \omega_{p_i} \ll \omega_{B_e}$. In such a plasma, we shall consider the excitation of oscillations by a beam of ions with density $n_1 = \alpha n_0 \ll n_0$ which move at right angles to the magnetic field, $\mathbf{V} \perp \mathbf{B}_0$. The thermal spread of the beam will be assumed to be negligibly small; the plasma, cold.

Under these assumptions, the dispersion equation reduces to the type (9.2) with

$$\varepsilon_0^{(0)} = 1 + \left(\frac{\omega_{p_e} \sin\theta}{\omega_{B_e}}\right)^2 - \left(\frac{\omega_{p_e} \cos\theta}{\omega}\right)^2 - \left(\frac{\omega_{p_i}}{\omega}\right)^2; \quad (13.21)$$

$$\varepsilon_0^{(1)} = -\alpha \left(\frac{\omega_{p_i}}{\omega - k_\varphi V}\right)^2. \quad (13.22)$$

The equation $\varepsilon_0^{(0)} = 0$ is the same as (8.7). The frequency $\omega^{(0)}$ of the plasma oscillations excited by the stream is determined by (8.4) or (8.9). The resonant wave number is $k_\varphi = \omega_2/V$. The maximal growth rate corresponds to this wave number:

$$\gamma \simeq \alpha^{1/3} [1 + (m_i/m_e)\cos^2\theta]^{-1/3}\, \omega_2. \quad (13.23)$$

The cold-plasma approximation, $kv_{T_1} < \gamma$, is valid if

$$v_{T_1}/V < \alpha^{1/3} [1 + (m_i/m_e)\cos^2\theta]^{-1/3}. \quad (13.24)$$

The ions of the plasma are effectively cold $\omega > kv_{T_i}$, if $v_{T_i} < V$; the electrons $z_e < 1$, $\omega > k_z v_{T_e}$, if

$$v_{T_e}/V < \min[\omega_{B_e}/\omega_2, (\cos\theta)^{-1}]. \quad (13.25)$$

The condition $\gamma > \omega_{B_i}$ means

$$\alpha^{1/3} > (\omega_{B_i}/\omega_2)[1 + (m_e/m_i)\cos^2\theta]^{1/3}. \quad (13.26)$$

2. Hot Beam in a Cold Plasma. If the particles of the beam have a Maxwellian distribution, $f^{(1)} \propto \exp[-m_i(\mathbf{v} - \mathbf{V})^2/$

$2T_1$], Eq. (13.22) is replaced by

$$\varepsilon_0^{(1)} = \frac{2\alpha\omega_{p_i}^2}{(kv_{T_1})^2}\left[1 + i\sqrt{\pi}\,\frac{\omega - k_\varphi V}{|k|v_{T_1}}\,W\left(\frac{\omega - k_\varphi V}{|k|v_{T_1}}\right)\right]. \tag{13.27}$$

If the condition opposite to (13.24) holds, Eqs. (13.27) and (13.21) show that oscillations of the frequency (8.4) or (8.9) are excited with the growth rate

$$\gamma = \frac{2\alpha\sqrt{\pi}\,\omega_{p_i}^2}{(kv_{T_1})^3}\left\{\frac{k_\varphi V - \omega}{\partial\varepsilon_0^{(0)}/\partial\omega}\exp\left[-\frac{(\omega - k_\varphi V)^2}{(kv_{T_1})^2}\right]\right\}_{\omega=\omega_2}. \tag{13.28}$$

If the thermal spread of the beam is not very great, $v_{T1} < V$, the maximal growth rate as a function of k_φ is attained at $k_\varphi \approx \omega_2/V$ [cf. § 3.2]:

$$\gamma_{\max} \simeq \alpha\left(\frac{\omega_{p_i}}{\omega_2}\cdot\frac{V}{v_{T_1}}\right)^2\left(\frac{\partial\varepsilon_0^{(0)}}{\partial\omega}\right)^{-1}. \tag{13.29}$$

The growth rate γ depends on the thermal spread of the stream in the same way as in the case $B_0 = 0$ [cf. Fig. 3.2].

3. <u>Ion Stream in a Plasma with Hot Electrons.</u> If neither of the conditions (13.25) is satisfied, the electrons must be regarded as hot. In this case, $\varepsilon_0^{(1)}$ retains the previous form (13.22) and

$$\varepsilon_0^{(0)} = 1 - \left(\frac{\omega_{p_i}}{\omega}\right)^2 + \frac{1}{(kd_e)^2}. \tag{13.30}$$

In accordance with § 8.3, the equation $\varepsilon_0^{(0)} = 0$ with $\varepsilon_0^{(0)}$ of the form (13.30) describes oscillations which correspond to high-frequency ion sound if $\omega < k_z v_{T_e}$ and the short-wavelength part of the branch ω_2 if $z_e > 1$.

If a beam is present, the dispersion equation reduces to

$$1 + \frac{1}{(kd_e)^2} - \left(\frac{\omega_{p_i}}{\omega}\right)^2 - \alpha\left(\frac{\omega_{p_i}}{\omega - k_\varphi V}\right)^2 = 0. \tag{13.31}$$

In this limiting case the magnetic field disappears from the expression and the problem of the excitation of oscillations by a transverse beam is found to be similar to that of the excitation of oscillations by a longitudinal beam (see § 3.5). The results that follow from (13.31) are similar to those given in § 3.5 if the substitutions $k_z \to k_\varphi$ and $k_\perp^2 \to k_r^2 + k_z^2$ are made in the latter.

Bibliography

1. B. B. Kadomtsev, In; Plasma Physics and the Problems of Controlled Thermonuclear Reactions, Vol. 4, Pergamon Press, Oxford (1960), p. 431.
2. A. A. Vedenov, E. P. Velikhov, and R. Z. Sagdeev, Usp. Fiz. Nauk, 73:701 (1961) [Soviet Phys. — Uspekhi, 4:332 (1961)]. In [1, 2] a study is made of the stability of a plasma in which there are streams of ions moving at right angles to a magnetic field (§ 13.4).
3. R. Z. Sagdeev, Zh. Tekh. Fiz., 31:1185 (1961) [Soviet Phys. — Tech. Phys., 6:867 (1962)]. It is noted that a plasma with an unstable multivelocity distribution of the ions can be formed as a result of the breaking of a shock wave (§ 13.4).
4. M. F. Forbatenko, In: Plasma Physics and Problems of Controlled Thermonuclear Fusion, Vol. 1 [in Russian], Naukova Dumka, Kiev (1962), p. 39. A study is made of the high-frequency instability of a cold ion stream moving in a plasma at right angles to a magnetic field.
5. O. Buneman, J. Nucl. Energy, C4:111 (1962). A study is made of the instability due to a transverse relative motion of the electrons and ions (§ 13.1).
6. M. V. Babykin et al., Zh. Eksp. Teor. Fiz., 46:511 (1964) [Soviet Phys. — JETP, 19:349 (1964)]. A study of the high-frequency instability of relative azimuthal motion of electrons and ions (§§ 13.1 and 13.3).
7. A. B. Mikhailovskii and V. S. Tsypin, ZhETF Pis. Red., 3:247 (1966) [JETP Letters, 3:158 (1966)]. A study is made of the high-frequency instability of a plasma in a radial electric field (§ 13.1).
8. V. L. Sizonenko and K. N. Stepanov, Nucl. Fusion, 7:131 (1967). The excitation of oscillations by a transverse current is investigated with allowance for the thermal motion of the ions and electrons (§§ 13.1-13.3).

Chapter 14

High-Frequency Instability of Plasmas with a Non-Maxwellian Transverse Velocity Distribution of the Ions

§ 14.1. Instability of a Plasma with δ-Function Transverse Velocity Distribution of the Ions

The transverse velocity distribution of the ions may be strongly non-Maxwellian because of, for example, injection into a trap of fast particles. Another mechanism leading to the same effect is the depletion of cold ions in a plasma due to their escape through the loss cone.

If the ion distribution is such that $\partial f_0/\partial v_\perp > 0$ for small $v_\perp$, the ions possess properties similar to those of a beam moving through the ion component of the plasma. It is then possible for a kind of two-stream instability to arise. Such is the high-frequency instability, $(\gamma, \omega) \gg \omega_{B_i}$, which we discuss below. It can develop in a plasma of not too low density, $\omega_{p_i} > \omega_{B_i}$.

Suppose that the distribution of the ions over the transverse velocities has the form $f_\perp \propto \delta(v_\perp^2 - v_0^2)$, i.e., all the ions possess the same transverse energy. Suppose the electron velocity distribution is Maxwellian with a temperature appreciably less than the ion energy, $T_e \ll m_i v_0^2/2$. We shall show that if the density of the plasma is not too low, $\omega_{p_i} \gg \omega_{B_i}$, a plasma of this kind must be unstable against perturbations with $\omega \gg \omega_{B_i}$ and $k_z \ll k$.

Under the assumptions we have made, Eqs. (7.3) and (7.28) yield the dispersion equation

$$1+\frac{\omega_{p_e}^2}{\omega_{B_e}^2}-\left(\frac{\omega_{p_e}\cos\theta}{\omega}\right)^2-\frac{\omega\omega_{p_i}^2}{(\omega^2-k^2v_0^2)^{3/2}}=0. \tag{14.1}$$

It is simplest to demonstrate the possibility of an instability by considering perturbations with $\cos\theta > (m_e/m_i)^{1/2}$. In this case the ions make a small contribution to (14.1). If we neglected their contribution, we should have purely electron oscillations with the real frequency (8.4). In the subsequent approximation, arguments similar to those used in § 1.5 show that the interaction of these oscillations with the ions leads to an instability whose maximal growth rate is approximately

$$\gamma\simeq\left(\frac{m_e}{m_i\cos^2\theta}\right)^{2/5}\frac{\omega_{p_e}\cos\theta}{[1+\omega_{p_e}^2/\omega_{B_e}^2]^{1/2}}. \tag{14.2}$$

In this case the quantity $\alpha\equiv m_e/m_i\cos^2\theta$ plays the role of a small parameter. It follows that (14.2) is qualitatively correct until $\cos\theta\approx(m_e/m_i)^{1/2}$. At these values of $\cos\theta$, the relation (14.2) becomes

$$\gamma\simeq\frac{\omega_{p_i}}{[1+\omega_{p_e}^2/\omega_{B_e}^2]^{1/2}}. \tag{14.3}$$

The transverse and longitudinal wave numbers are then of the same order:

$$\left.\begin{aligned} k_\perp\simeq k_{\perp\,\mathrm{opt}}&\simeq\left(\frac{\omega_{p_i}}{v_0}\right)\left(1+\frac{\omega_{p_e}^2}{\omega_{B_e}^2}\right)^{-1/2},\\ k_z\simeq k_{z\,\mathrm{opt}}&\simeq\left(\frac{m_e}{m_i}\right)^{1/2}k_{\perp\mathrm{opt}}\quad . \end{aligned}\right\} \tag{14.4}$$

We see that if the plasma has a high density, $\omega_{p_e}\gtrsim\omega_{B_e}$, the growth rate attains values of the order of the hybrid frequency, $\gamma\approx(m_i/m_e)^{1/2}\,\omega_{B_i}$. If $\omega_{p_i}\lesssim\omega_{B_i}$, the approximation $\gamma>\omega_{B_i}$ ceases to be valid and this means that the high-frequency instability does not arise in a plasma whose density is so low.

If $\omega \approx kv_0$, the high-frequency instability is hydrodynamic in nature – all the ions participate in this buildup of the oscillations. If the perturbations have a small phase velocity, $\omega/k \ll v_0$, they interact with only a small fraction of the ions – those whose phase velocities are almost perpendicular to the wave vector, $\mathbf{kv} = \omega \ll kv$. The instability is then essentially kinetic and its growth rate is determined by the derivative $\partial f_0(v_x)/\partial v_x$ of the one-dimensional distribution function

$$f_0(v_x) \equiv \int f_\perp(v_\perp)\, dv_y$$

(for $\mathbf{k} \| \mathbf{x}$). Using (14.1), we can show that the growth rate of the kinetic instability for $\cos\theta \approx (m_e/m_i)^{1/2}$ is somewhat less than (14.3). As ω/k decreases, the growth rate falls off as $1/k^3$.

§ 14.2. Instability of a Plasma with a Broadened Transverse Velocity Distribution of the Ions

If the distribution $f_{\perp i}(\varepsilon_\perp)$ has an arbitrary form, the dispersion relation (14.1) is replaced by

$$1 + \frac{\omega_{p_e}^2}{\omega_{B_e}^2} - \frac{\omega_{p_e}^2}{\omega^2}\cos^2\theta - \frac{\omega_{p_i}^2}{k^2}\left\langle \frac{\partial f_\perp}{\partial \varepsilon_\perp}\left[1 - \frac{\omega}{(\omega^2 - k^2 v_\perp^2)^{1/2}}\right]\right\rangle = 0. \quad (14.5)$$

This is derived using Eq. (7.27).

It follows from (14.5) that the instability considered in § 14.1 is reduced if the velocity spread of the particles about the mean velocity v_0 is increased. This velocity spread has its most pronounced effect on perturbations with $\omega \approx kv_0$. In particular, if

$$f_{\perp i} = \frac{1}{v_0}(\pi v_T^2)^{-1/2} \exp\left[-(v_\perp - v_0)^2/v_T^2\right], \quad (14.6)$$

it follows from (14.5) that the hydrodynamic resonant instability with the growth rate (14.2) disappears if

$$\frac{v_T}{v_0} < \left(\frac{m_e}{m_i \cos^2\theta}\right)^{2/5}. \quad (14.7)$$

If the velocity spread satisfies this condition, only the kinetic instability with a growth rate that is several times smaller than (14.3) remains, and the phase velocity of the oscillations is $\omega/k < v_0$. At the same time, the kinetic instability occurs even if $v_T \approx v_0$ provided the one-dimensional ion distribution has a minimum. If $f_{\perp i}$ has an arbitrary form, the frequency Re ω of the oscillations is determined by the expression (8.4) and the growth rate by

$$\gamma = \zeta \int\limits_{\mathrm{Re}\,\frac{\omega}{k}}^{\infty} \frac{\partial f_\perp}{\partial v_\perp} \cdot \frac{dv_\perp}{[(v_\perp k/\mathrm{Re}\,\omega)^2 - 1]^{1/2}} = \zeta \left(\frac{1}{v_x} \cdot \frac{\partial f_0(v_x)}{\partial v_x} \right)_{v_x = \frac{\mathrm{Re}\,\omega}{k_x}},$$

$$\zeta \equiv \frac{\mathrm{Re}\,\omega}{2} \cdot \frac{\omega_{p_i}^2}{k^2} \left(1 + \frac{\omega_{p_e}^2}{\omega_{B_e}^2} \right)^{-1}. \tag{14.8}$$

In § 10.4, we considered the nonequilibrium distributions (10.69) and (10.71) as examples. These two-dimensional distributions correspond to the one-dimensional distributions

$$f_0(v_x) = \left(\frac{m}{2\pi T} \right)^{1/2} \left(1 + \frac{T_1}{T} \right) \exp\left(- \frac{mv_x}{2T} \right) \times$$
$$\times \left[1 - \left(\frac{T_1}{T + T_1} \right)^{1/2} \exp\left(- \frac{mv_x^2}{2T_1} \right) \right], \tag{14.9}$$

$$f_0(v_x) = \frac{2}{\sqrt{\pi}\, j!} (-1)^j \alpha^{j+1} \left(\frac{\partial}{\partial \alpha} \right)^j [\alpha^{-1/2} e^{-\alpha v_x^2}],$$
$$j = 1, 2, \ldots \tag{14.10}$$

In both cases $\partial f_0/\partial v_x > 0$ if v_x is sufficiently small. In accordance with (14.8), such distributions are therefore unstable.

§ 14.3. Instability of a Stationary Ion Distribution in an Adiabatic Trap

A plasma confined in an adiabatic trap is depleted of ions with a low transverse energy because of the escape of such ions through the loss cone. It is then possible for instabilities of the type considered in §§ 14.1 and 14.2 to arise. To justify this assertion, we proceed as follows.

We first calculate the stationary distribution function when the problem is simplified to the maximum extent. We shall assume that for $|z| < L/2$ the magnetic field B_0 is homogeneous and directed along the z axis and increases in the region $|z| - L/2| \ll L/2$ to some maximum value $B_{0\,\max} > B_0$.

The motion of a charged particle in a field of this form is governed by the laws of conservation of its energy, $\varepsilon \equiv (v_\perp^2 + v_\parallel^2)/2 = \text{const}$, and magnetic moment, $\mu \equiv v_\perp^2/2\omega_B = \text{const}$.

Suppose that for $|z| < L/2$ the particle has the velocity components $v_\perp^{(0)}$ and $v_\parallel^{(0)}$. As it approaches the region with the stronger field, its transverse velocity increases but the longitudinal velocity decreases, so that

$$\left.\begin{aligned} v_\perp(z) &= v_\perp^{(0)} \left(\frac{B_0(z)}{B_0(0)}\right)^{1/2}, \\ v_\parallel(z) &= \sqrt{(v^{(0)})^2 - (v_\perp^{(0)})^2 \frac{B_0(z)}{B_0(0)}}. \end{aligned}\right\} \tag{14.11}$$

The second equation shows that the particle penetrates into the region where the field satisfies $B_0 = B_{0\max}$ if

$$\sin\theta \equiv \frac{v_\perp^{(0)}}{v^{(0)}} \leqslant \frac{B_0(0)}{B_{0\max}} \equiv \sin\theta_0, \tag{14.12}$$

and that otherwise it is reflected into the region with $B_0 = B_0(0)$. It follows that a field with this configuration – usually called a field with magnetic stoppers or magnetic mirrors – can contain only particles whose velocities are inclined at a sufficiently large angle to the magnetic field, $\theta > \theta_0$ (in other words if θ does not lie in the loss cone). The stationary distribution function in such a field must therefore satisfy the condition

$$f_0(\mathbf{v}) = 0 \text{ for } \theta \leqslant \theta_0. \tag{14.13}$$

We shall assume that the escape of particles through the loss cone is compensated by the injection of other particles from outside the system. Let q be the number of particles injected in unit time in unit interval of the velocities and the space. Then the distribution function $f_0(\mathbf{v})$ for $\theta > \theta_0$ is determined by the Boltzmann equation (I) with q added to the collision term (6.2) on the right-hand side. We shall solve this equation under the following assumptions.

We shall assume that q depends only on $v_\perp$ and $v_\parallel$ (and does not depend on the coordinates or the azimuthal angle in the space of transverse velocities). We shall take into account only ion–ion collisions – it is the ion distribution with which we are concerned. In the collision term C_{ii}, we shall neglect the departure of the function $f(\mathbf{v}')$ from a Maxwellian distribution (this assumption is not, of course, entirely rigorous). We shall assume that $q(v, \theta)$ can be chosen such that the distribution function has the form

$$f_0 = \Psi(\theta) \exp\left(-\frac{m_i v^2}{2T_i}\right). \tag{14.14}$$

The Boltzmann equation then reduces to

$$\frac{1}{\sin\theta} \cdot \frac{\partial}{\partial\theta}\left(\sin\theta \frac{\partial\Psi}{\partial\theta}\right) = -\frac{qv^2}{a(v)} \exp\left(\frac{m_i v^2}{2T_i}\right), \tag{14.15}$$

where $a(v)$ is some function of v whose explicit form is not required.

The left-hand side of (14.15) does not depend on v. It follows that q has the form

$$q = C \frac{a(v)}{v^2} \exp\left(-\frac{m_i v^2}{2T_i}\right). \tag{14.16}$$

The solution of (14.15) with the boundary conditions $\Psi(\theta_0) = 0$ and $\Psi(\theta) = \Psi(-\theta)$ has, with allowance for (14.16), the form

$$\Psi(\theta) = C \ln \frac{\sin\theta}{\sin\theta_0}. \tag{14.17}$$

Finally, from (14.14) and (14.17)

$$f_0 = C \ln \frac{\sin\theta}{\sin\theta_0} \exp\left(-\frac{m_i v^2}{2T_i}\right). \tag{14.18}$$

The constant C can be expressed in terms of the plasma density, $n_0 = \int f_0 d\mathbf{v}$.

We shall use (14.18) to find the transverse velocity distribution of the particles, $f_\perp = \int f_0 dv_\parallel$. For simplicity we shall assume that the "stopper ratio" is small, i.e., $(B_{0\,max} - B_0) \ll B_0$; in this case

$$\frac{\pi}{2} - \theta_0 \ll 1, \quad v^2 - v_\perp^2 \ll v_\perp^2. \tag{14.19}$$

Then

$$\ln \frac{\sin \theta}{\sin \theta_0} \simeq \frac{1}{2} \left[\left(\frac{\pi}{2} - \theta_0 \right)^2 - \left(\frac{v_{\parallel}}{v_{\perp}} \right)^2 \right], \tag{14.20}$$

where $|v_{\parallel}| < \left(\frac{\pi}{2} - \theta_0\right) v_{\perp}$. Substituting (14.20) into (14.18), integrating the result over the transverse velocity, and expressing C in terms of n_0, we obtain

$$f_{\perp} = \frac{n_0}{2} \left(\frac{m_i}{2\pi T_i} \right)^{3/2} v_{\perp} \exp\left(- \frac{m_i v_{\perp}^2}{2T_i} \right). \tag{14.21}$$

It can be seen that the presence of the loss cone leads to the setting up of a stationary state with a nonequilibrium transverse velocity distribution, $\partial f_{\perp}/\partial v_{\perp} > 0$.

We shall now show that the distribution (14.21) is unstable. To within a positive constant, Eq. (14.21) yields

$$\frac{\partial f_0(v_x)}{\partial v_x} \propto v_x \int_{-\infty}^{\infty} \frac{\exp\left[- \frac{m_i (v_x^2 + v_y^2)}{2T_i} \right]}{\sqrt{v_x^2 + v_y^2}} \left(1 - \frac{m_i (v_x^2 + v_y^2)}{T_i} \right) dv_y. \tag{14.22}$$

For small v_x/v_T we find

$$\frac{\partial f_0(v_x)}{\partial v_x} \propto v_x \ln \frac{v_T}{|v_x|} \tag{14.23}$$

to within logarithmic accuracy. It can be seen that $v_x \frac{\partial f_0(v_x)}{\partial v_x} > 0$. Thus, in accordance with (14.8), we have instability.

More detailed calculations of the high-frequency instability of a plasma in an adiabatic trap can be found in the paper of Post and Rosenbluth.

§ 14.4. Stabilizing Role of the Electron Temperature

In studying the influence of the electron temperature on the high-frequency instability, we shall restrict ourselves to a δ-function ion distribution. Using (7.23) and (7.28), we obtain a dispersion equation that takes into account the finite electron temperature:

$$\varepsilon_0(\omega) \equiv 1 + \frac{1}{(kd_e)^2}\left[1 + \frac{i\sqrt{\pi}\,\omega}{|k_z|\,v_{T_e}} W\left(\frac{\omega}{|k_z|\,v_{T_e}}\right) I_0(z_e)\,e^{-z_e}\right] - \frac{\omega\omega_{p_i}^2}{(\omega^2 - k_\perp^2 v_0^2)^{3/2}} = 0. \tag{14.24}$$

The necessary condition for instability – the Nyquist condition (§ 2.7) – is that Im $\varepsilon_0(\omega)$ should vanish for some $\omega = \omega^*$. It is evident from (14.24) that this can happen only if

$$\omega^* < k_\perp v_0. \tag{14.25}$$

In this case

$$\operatorname{Re}\varepsilon_0(\omega^*) = 1 + \frac{1}{(kd_e)^2}\left[1 - \frac{\sqrt{\pi}\,\omega^*}{|k_z|\,v_{T_e}} \operatorname{Im} W\left(\frac{\omega^*}{|k_z|\,v_{T_e}}\right) I_0(z_e)\,e^{-z_e}\right]. \tag{14.26}$$

For instability it is necessary that Re $\varepsilon_0(\omega^*) < 0$, i.e.,

$$\frac{\sqrt{\pi}\,\omega^*}{|k_z|\,v_{T_e}} \operatorname{Im} W\left(\frac{\omega^*}{|k_z|\,v_{T_e}}\right) I_0(z_e)\,e^{-z_e} > 1 + (kd_e)^2. \tag{14.27}$$

Using the tables of the functions W and I_0, we can show that this condition can be satisfied only if

$$\left.\begin{aligned} &\omega^*/|k_z|\,v_{T_e} > 0.925, \\ &z_e < 1, \quad kd_e < 1. \end{aligned}\right\} \tag{14.28}$$

It follows from the inequalities (14.25) and $\omega^* > k_z v_{T_e}$ that in a plasma with a finite electron temperature the values of $\cos\theta$ for growing perturbations are bounded above by the condition

$$\cos\theta < \frac{v_0}{v_{T_e}}. \tag{14.29}$$

In accordance with § 14.1 the characteristic values of $\cos\theta$ are $\sim (m_e/m_i)^{1/2}$. The restriction (14.29) is therefore important if $v_0/v_{T_e} < (m_e/m_i)^{1/2}$, i.e.,

$$T_e > m_i v_0^2. \tag{14.30}$$

If the condition (14.30) is satisfied, the instability with the growth rate (14.3) cannot develop. Using (14.24), we find that in this case the maximum possible growth rate is approximately

$$\gamma \simeq \left(\frac{m_i v_0^2}{T_e}\right)^{1/2} \frac{\omega_{p_i}}{\left(1+\frac{\omega_{p_e}^2}{\omega_{B_e}^2}\right)^{1/2}}. \tag{14.31}$$

The condition $\gamma > \omega_{B_i}$ can be satisfied provided the plasma density is such that

$$\left(\frac{\omega_{p_i}}{\omega_{B_i}}\right)^2 > \frac{T_e}{m_i v_0^2}. \tag{14.32}$$

The transverse and longitudinal wave numbers of growing perturbations are bounded above by the condition [cf. (14.4)]:

$$\left.\begin{aligned} k_\perp &\lesssim k_{\perp\mathrm{opt}}\left(\frac{m_i v_0^2}{T_e}\right)^{1/2}, \\ k_z\rho_i &\lesssim \frac{m_i v_0^2}{T_e}\min\left(1, \frac{\omega_{p_e}}{\omega_{B_e}}\right). \end{aligned}\right\} \tag{14.33}$$

The inequality (14.32) shows that an increase in the electron temperature shifts the boundary of the instability to higher densities, $n_0 \sim T_e$.

It follows from (14.33) that for $T_e > m_i v_0^2$ the growing perturbations must have long wavelengths, both transversely and longitudinally, $\lambda_\parallel \propto 1/k_z \propto T_e$. If the plasma is of finite length, $\lambda_\parallel$ may be greater than this length L and the instability is then impossible. All this demonstrates clearly that a finite electron temperature is a stabilizing factor.

§ 14.5. Plasmas with Two Groups of Ions

Suppose that a plasma consists of two groups of ions – one cold, $T \to 0$, and the other hot with $\partial f_\perp/\partial v_\perp > 0$. A plasma of this kind may be unstable because of the collective interaction between these groups of ions. Let us consider some examples of this kind of instability.

1. Plasma with a Low Fraction of Monoenergetic Ions and Cold Electrons. In this case the dispersion equation is

$$1+\left(\frac{\omega_{p_e}}{\omega_{B_e}}\right)^2-\left(\frac{\omega_{p_i}}{\omega}\right)^2\left(1+\frac{m_i}{m_e}\cos^2\theta\right)-\frac{\alpha\omega\omega_{p_i}^2}{(\omega^2-k^2v_0^2)^{3/2}}=0. \tag{14.34}$$

It follows from this equation that "resonant" oscillations, $\omega \approx kv_0$, are excited with the growth rate

$$\gamma\simeq\alpha^{2/5}\frac{\omega_{p_i}}{\left(1+\frac{\omega_{p_e}^2}{\omega_{B_e}^2}\right)^{1/2}}. \tag{14.35}$$

This is subject to the condition

$$\cos\theta\leqslant\left(\frac{m_e}{m_i}\right)^{1/2},\quad k\simeq\frac{\omega_{p_i}}{v_0\left(1+\frac{\omega_{p_e}^2}{\omega_{B_e}^2}\right)^{1/2}}.$$

The approximation $\gamma > \omega_{B_i}$ is valid if

$$\left(\frac{\omega_{p_i}}{\omega_{B_i}}\right)^2>\alpha^{-4/5}. \tag{14.36}$$

At lower plasma densities, an instability with frequencies near the cyclotron harmonics (see below, § 16.6) is possible.

2. Plasma with a Small Fraction of Hot Ions with Finite Thermal Spread about a Mean Transverse Energy. Suppose the distribution function of the hot ions has the form (14.6). The results of the foregoing subsection are valid provided

$$v_{T_1}<\alpha^{2/5}v_0. \tag{14.37}$$

If this condition is not satisfied, the resonant oscillations are not excited. This can be seen as follows. Suppose $|\omega - kv_0| < kv_{T_1}$. Then

$$\varepsilon_0^{(1)}=\frac{\alpha\omega_{p_i}^2(1+i)\,\Gamma\left(\frac{3}{4}\right)}{2\sqrt{\pi}\,k^2(2v_0)^{1/2}v_{T_1}^{3/2}} \tag{14.38}$$

The real part of this expression does not depend on the difference $\omega - kv_0$, so that an instability of the hydrodynamic type is impossible. Since $\mathrm{Im}\,\varepsilon_0^{(1)} > 0$, a kinetic instability is also impossible, as follows from Eq. (2.17).

If the opposite condition to (14.37) is satisfied, nonresonant oscillations with $\omega < kv_0$ can be excited. This is the kinetic instability of a beam with a large thermal spread of the kind considered in § 3.3. The corresponding growth rate is approximately

$$\gamma_{\mathrm{kin}} \simeq \alpha \,\mathrm{Re}\,\omega, \tag{14.39}$$

where $\mathrm{Re}\,\omega$ is given by Eq. (8.9). The approximation $\gamma > \omega_{B_i}$ is valid if

$$\alpha > \max\left(\frac{\omega_{B_i}}{\omega_{p_i}}, \left(\frac{m_e}{m_i}\right)^{1/2}\right). \tag{14.40}$$

3. Group of Cold Ions of Low Density in a Plasma with Hot Nonequilibrium Ions. In §§ 14.5.1 and 14.5.2 we have assumed that there are many more cold than hot ions. We shall now make the opposite assumption. We shall assume that the plasma density is fairly high, $\omega_{p_i}\alpha^{1/2} > \omega_{B_i}$, where α is the ratio of the densities of the cold and hot ions. The high-frequency approximation can then be employed, as in §§ 14.5.1 and 14.5.2. The main effect due to the cold ions is the excitation of oscillations with $k_z = 0$. The dispersion equation for this instability has the form ($\omega \ll k_\perp v_0$):

$$1 + \frac{\omega_{p_e}^2}{\omega_{B_e}^2} - \alpha\left(\frac{\omega_{p_i}}{\omega}\right)^2 - \frac{i\omega\omega_{p_i}^2}{(k_\perp v_0)^3} = 0. \tag{14.41}$$

The frequency and the growth rate are approximately [cf. (14.35)]:

$$\gamma \simeq \mathrm{Re}\,\omega \simeq \frac{\alpha^{1/2}\omega_{p_i}}{\left(1 + \dfrac{\omega_{p_e}^2}{\omega_{B_e}^2}\right)^{1/2}}. \tag{14.42}$$

The transverse wave number is

$$k_\perp \simeq \frac{\alpha^{1/6}\omega_{p_i}}{v_0\left(1 + \dfrac{\omega_{p_e}^2}{\omega_{B_e}^2}\right)^{1/2}}. \tag{14.43}$$

Recalling that this is an absolute and not a convective (see § 4.2) instability, we see that a small addition of cold ions worsens the stability of hot ions in an adiabatic trap.

§ 14.6. Plasmas of Finite Length

We shall now consider how the results we have obtained for the stability of an idealized, spatially homogeneous plasma can be extended to the case of a plasma of finite length with a density that is not uniform along its length. We shall have in mind the high-frequency instability of a plasma with cold electrons (§ 14.1).

Assuming that the plasma density depends on z and using the method of § 5.4, we obtain the dispersion equation

$$\frac{d}{dz}\left(\varepsilon_{\parallel}\frac{d\psi}{dz}\right)-k_{\perp}^{2}\varepsilon_{\perp}\psi=0, \tag{14.44}$$

where

$$\varepsilon_{\parallel}=1-\frac{\omega_{p_e}^2}{\omega^2}, \quad \varepsilon_{\perp}=1+\frac{\omega_{p_e}^2}{\omega_{B_e}^2}-\frac{i\omega_{p_i}^2\omega}{(k_{\perp}^2v_0^2-\omega^2)^{3/2}}. \tag{14.45}$$

In the limiting cases of abrupt and smooth decreases of the density, this equation yields the results of the next two subsections.

1. Plasma with an Abrupt Decrease of the Density. Suppose the plasma is uniform for $|z| < L/2$ and that its density drops abruptly to zero at $|z| = L/2$. The solution of (14.44) for $|z| < L/2$ has the form

$$\psi=C\cos(k_z z+\alpha), \tag{14.46}$$

where

$$k_z^2=-\frac{\varepsilon_{\perp}}{\varepsilon_{\parallel}}k_{\perp}^2, \tag{14.46}$$

and C and α are arbitrary constants.

If $\omega_{p_e}\gg\omega$ the boundary condition for the potential ψ has the form

$$\psi'(L/2)=0. \tag{14.48}$$

Equations (14.47) and (14.48) yield the dispersion equation

$$\left(\frac{\pi n}{L}\right)^2 \varepsilon_{\parallel} + k_{\perp}^2 \varepsilon_{\perp} = 0, \tag{14.49}$$

which is the same as (14.1) subject to the substitution

$$k_z \longleftrightarrow \pi n/L. \tag{14.50}$$

Exploiting the analogy between (14.49) and (14.1), we obtain an estimate of the length of an unstable plasma:

$$L \gtrsim \pi \begin{cases} \rho_i \text{ for } \omega_{p_e} > \omega_{B_e}, \\ \left(\frac{m_i}{m_e}\right)^2 d_0 \text{ for } \omega_{p_e} < \omega_{B_e}. \end{cases} \tag{14.51}$$

2. Plasma with a Smooth Decrease of the Density. We shall now assume that the plasma is confined in a trap whose length is large compared with the length defined by the right-hand side of (14.51). A large number of wavelengths can then be fitted into the length of the plasma. We shall assume that the plasma density varies smoothly over a distance of one wavelength and that, in contrast to § 14.6.1, there is no abrupt decrease of the density at the ends of the trap.

The solution of (14.44) can then be sought in the quasiclassical form $\psi \propto e^{i\int K_z(z)dz}$. An expression similar to (14.47) is obtained for $k_z(z)$:

$$K_z^2(z) = -\frac{\varepsilon_{\perp}(z)}{\varepsilon_{\parallel}(z)} k_{\perp}^2. \tag{14.52}$$

If the density is decreased, $\varepsilon_{\parallel}(z)$ decreases and $K_z(z)$ increases. This means that Eq. (14.44) does not have localized solutions in the adopted approximation. The plasma may nevertheless be unstable but the stability problem must now be solved in a different manner by investigating the behavior of a wave packet (see §§ 4.4, 5.3, and 5.4).

Suppose that a wave packet that has arisen somewhere in the plasma has a real frequency and propagates toward the edge of the trap. Equation (14.52) gives the law of variation of the wave number K_z in space. The amplitude of the packet then changes in accordance with the law

$$|\psi| \propto \exp\{-\int \operatorname{Im} K_z(z)\, dz\}. \tag{14.53}$$

Using (14.52), we can show that if $\varepsilon_\perp$ has the form (14.45) the wave is amplified, $\operatorname{Im} K\ (z) < 0$ (see § 4.5), so that the amplitude increases in space. An instability arises if no other factors prevent the packet amplitude from attaining large values.

If the packet propagates into the region $z \approx L/2$, where the plasma density is low, the function $K_z(z)$ increases appreciably and may attain values of the order

$$K_z^{\max} \simeq \omega / v_{T_e}. \tag{14.54}$$

Equation (14.44) then ceases to hold and a more exact equation taking into account the thermal motion of the electrons – an equation of the type (14.24) – shows that the spatial growth of the amplitude is now replaced by damping. This is, in fact, the main factor that prevents the growth of the packet to large amplitudes.

Estimating the argument of the exponential function in (14.53) by $L|\operatorname{Im} K_z|$, we write the instability condition in the approximate form

$$L|\operatorname{Im} K_z| \simeq N, \tag{14.55}$$

where N is some number that exceeds unity (for example, $N = 10$). Equation (14.52) shows that $|\operatorname{Im} K_z| \simeq (m_e/m_i)^{1/2} k_\perp^{(0)}$. From (14.55) we then obtain an estimate of the type (14.51), but with the coefficient N on the right-hand side.

Thus, if there is a smooth decrease of the density, the length of the unstable plasma is greater than in the case of an abrupt change.

Numerical calculations of the critical length can be found in the paper of Post and Rosenbluth. Aamodt and Book have shown that the reflection of waves from inhomogeneities must be taken into account in the calculation of the critical length (this in the second approximation of the quasiclassical treatment, for which the reader is referred to Ginzburg's monograph).

Bibliography

1. Yu. N. Dnestrovskii, Nucl. Fusion, 3:259 (1963). It is noted that plasmas in adiabatic traps are depleted of ions of low transverse energy, with the consequence that the transverse ion distribution is nonmonotonic and has a beam nature.
2. V. B. Krasovitskii and K. N. Stepanov, Zh. Tekh. Fiz., 34:1013 (1964) [Soviet Phys. — Tech. Phys., 9:786 (1964)]. It is shown that the high-frequency instability (§ 14.1) can develop.
3. A. B. Mikhailovskii, Nucl. Fusion, 5:125 (1965). A study is made of the high-frequency instability of a plasma with nonequilibrium ions and cold electrons (§ 14.1).
4. M. N. Rosenbluth and R. F. Post, Phys. Fluids, 8:547 (1965).
5. R. F. Post and M. N. Rosenbluth, Phys. Fluids, 9:730 (1966). In [4, 5] a study is made of the high-frequency instability in a plasma with cold electrons (§§ 14.1 and 14.2). A study is made of the excitation of wave packets in a plasma of finite length (§ 14.6).
6. R. E. Aamodt and L. Book, Phys. Fluids, 9:143 (1966). The behavior of wave packets is considered in an unstable plasma with a δ-function distribution of the ions and cold electrons. The critical length of a trap at which the development of the high-frequency instability is not dangerous (§ 14.6) is considered.
7. L. S. Hall and W. Heckrotte, Proceedings of the Seventh International Conference on Phenomena in Ionized Gases. Vol. 2, Gradevinska Knjiga, Belgrad, (1966), p. 624.
8. G. I. Budker, in: Plasma Physics and the Problem of Controlled Thermonuclear Reactions, Vol. 3, Pergamon Press, Oxford (1959), p. 1. The stationary distribution of ions in an adiabatic trap is found (§ 14.3).
9. D. G. Lominadze and K. N. Stepanov, Zh. Tekh. Fiz., 35:441 (1965) [Soviet Phys. — Tech. Phys., 10:347 (1965)].
10. A. B. Mikhailovskii and É. A. Pashitskii, Zh. Tekh. Fiz., 35:1961 (1965) [Soviet Phys. — Tech. Phys., 10:1507 (1966)].
11. L. D. Pearlstein, M. N. Rosenbluth, and D. B. Chang, Phys. Fluids, 9:953 (1966). In [9-11] a study is made of the excitation of plasma oscillations by a group of ions with a non-Maxwellian transverse velocity distribution (§ 14.5).
12. V. L. Ginzburg, The Propagation of Electromagnetic Waves in Plasmas, Pergamon Press, Oxford (1964).

Chapter 15

Plasmas with Anisotropic Ions

§ 15.1. Low-Density Strongly Anisotropic Plasma with Cold Electrons

An anisotropic distribution of the ion velocities is characteristic of a plasma confined in adiabatic traps. The ion anisotropy leads to instabilities whose mechanism is similar to the electron anisotropic instability mechanisms considered in Chapter 10. The situation is, however, complicated by the fact that both ions and electrons participate in the oscillations.

In the case of a low-pressure plasma, $\beta \ll 1$, to which we shall restrict ourselves, all the ion anisotropic instabilities have frequencies near the ion cyclotron frequency or its harmonics and are electrostatic.

The growth rates and the conditions under which the instabilities arise depend strongly on the relative density $(\omega_{p_i}/\omega_{B_i})^2$ of the plasma, on the ratio of the electron temperature to the ion energy, $T_e/T_{\perp i}$, and, of course, on the degree of anisotropy of the ions.

In the present section we shall consider the excitation of ion-cyclotron oscillations under the assumptions $T_{\parallel i}/T_{\perp i} \to 0$, $T_e/T_{\perp i} \to 0$. The finite value of $T_{\parallel i}/T_{\perp i}$ will be taken into account in § 15.2; that of $T_e/T_{\perp i}$, in §§ 15.3 and 15.4. We shall assume that the plasma has a low density, $\omega_{p_i}/\omega_{B_i} \ll 1$. The oscillations of a plasma with $\omega_{p_i} \gtrsim \omega_{B_i}$ will be considered in § 15.4.

The dispersion equation for the ion oscillations can be obtained by means of (10.29), by replacing the electron indices by

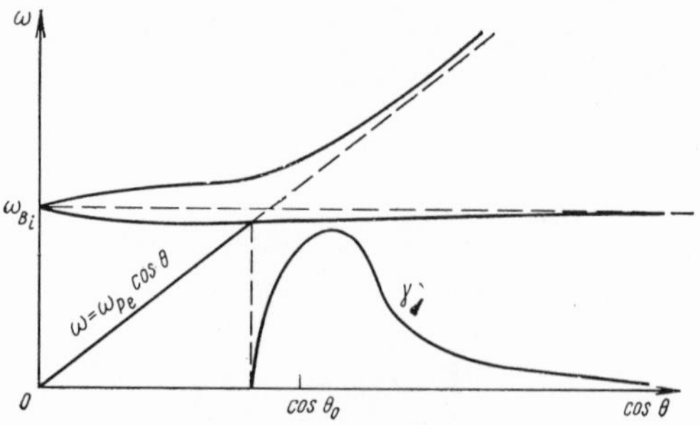

Fig. 15.1. Oscillation branches of a plasma with anisotropic ions when $\omega \simeq \omega_{B_i}$.

ion indices and then adding to the left-hand side of the resulting equation the electron term

$$\varepsilon_0^{(e)} = -(\omega_{p_e}\cos\theta/)\omega^2.$$

Thus,

$$1-\left(\frac{\omega_{p_e}\cos\theta}{\omega}\right)^2-\frac{\omega_{p_i}^2 I_n(z_i)\,e^{-z_i}\cos^2\theta}{(\omega-n\omega_{B_i})^2}-\frac{\omega_{p_i}^2\sin^2\theta}{\omega_{B_i}(\omega-n\omega_{B_i})}\cdot\frac{nI_n e^{-z_i}}{z_i}=0 \tag{15.1}$$

Here, terms of order $(\omega_{p_i}/\omega_{B_i})^2$ have been neglected.

The oscillation branches $\omega = \omega(\cos\theta)$ are shown in Fig. 15.1. This figure is similar to the corresponding figure Fig. 10.1 for the electron branches in a plasma with $T_{\perp e} \gg T_{\parallel e}$.

In a plasma with $T_{\perp i} = T_{\parallel i}$, Fig. 15.1 is replaced by Fig. 8.2. A comparison of these figures shows that in a strongly anisotropic plasma there is one and not two branches with $\omega \approx n\omega_{B_i}$. These branches are similar to the beam branches of § 1.5.2.

At $\theta = \theta_0$ defined by Eq. (8.45), the cyclotron branches "intersect" the branch of magnetized plasma oscillations (see Fig. 15.1). Near this value of θ a resonant instability with growth rate

$$\gamma = \frac{\sqrt{3}}{2}\,|n\omega_{B_i}|\left[\frac{m_e}{2m_i}I_n(z_i)\,e^{-z_i}\right]^{1/3} \tag{15.2}$$

can arise. The growth rate of the resonant oscillations has a fairly smooth maximum at $z_i \approx 1$, its order of magnitude being

$$\gamma \simeq \left(\frac{m_e}{m_i}\right)^{1/3}\omega_{B_i}. \tag{15.3}$$

If $\cos\theta > \cos\theta_0$, a nonresonant instability arises, the characteristic growth rate being of the order

$$\gamma \simeq \left(\frac{m_e}{m_i}\right)^{1/2} \omega_{B_i}. \tag{15.4}$$

The results (15.2)-(15.4) and the schematic diagram of Fig. 15.1 were obtained by neglecting the last term of the left-hand side of Eq. (15.1). This is justified if

$$\frac{\gamma}{\omega_{B_i}} \ll z_i \cot^2\theta. \tag{15.5}$$

If $z_i \approx 1$, this condition is satisfied for all $\cos\theta \gtrless \cos\theta_0$ provided the plasma density is not too high:

$$\left(\frac{\omega_{p_i}}{\omega_{B_i}}\right)^2 < \left(\frac{m_e}{m_i}\right)^{2/3}. \tag{15.6}$$

In a plasma with a density satisfying this condition, only the resonant instability with the increment (15.3) is possible. If this condition is not satisfied, only the nonresonant instability with the growth rate (15.4) can arise.

The oscillation branch $\omega = \omega_{p_e} \cos\theta$ intersects the cyclotron branches only if the plasma density is not too low:

$$\left(\frac{\omega_{p_i}}{\omega_{B_i}}\right)^2 > \frac{m_e}{m_i}. \tag{15.7}$$

This inequality is a necessary condition for instability; otherwise all the perturbations described by Eq. (15.1) have a real frequency.

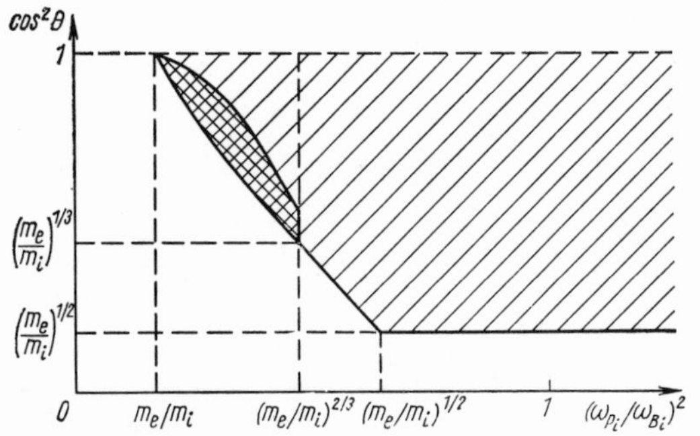

Fig. 15.2. Stability diagram of a plasma with anisotropic in the limit $T_{\parallel i} \to 0$, $T_e \to 0$. The instability region is shaded. The region of resonant instability is shaded more darkly.

The stability diagram on the plane $(\omega_{p_i}/\omega_{B_i})^2$, $\cos^2\theta$ is shown in Fig. 15.2. If $(\omega_{p_i}/\omega_{B_i})^2 \simeq m_e/m_i$, only perturbations with $\cos^2\theta \approx 1$ are excited. As the density increases right up to the point at which the condition (15.5) is satisfied, the stability boundary is determined by the approximate equation

$$(\cos^2\theta)_{\text{bound}} \approx \cos^2\theta_0 = \left(\frac{m_e}{m_i}\right)\left(\frac{\omega_{B_i}}{\omega_{p_i}}\right)^2. \tag{15.8}$$

The condition (15.5) for $\cos\theta = \cos\theta_0$ is violated at the density

$$\left(\frac{\omega_{p_i}}{\omega_{B_i}}\right)^2 \gtrsim \left(\frac{m_e}{m_i}\right)^{1/2}. \tag{15.9}$$

In this case only the nonresonant instability is possible [cf. (15.6)] and then only if (see also Fig. 15.2)

$$\cos^2\theta > (\cos^2\theta)_{\min} \simeq \left(\frac{m_e}{m_i}\right)^{1/2}. \tag{15.10}$$

The dependence of the largest possible growth rate on the parameter $b_i \equiv (\omega_{p_i}/\omega_{B_i})^2$ is shown in Fig. 15.3. For $m_e/m_i < b_i < (m_e/m_i)^{2/3}/\gamma_{ma}$ this growth rate refers to the resonant instability [see (15.2) and (15.3)]; for $b_i > (m_e/m_i)^{2/3}$, to the nonresonant instability.

The approximation adopted in this section $T_e \to 0$, $T_{\parallel i} \to 0$ is valid if $\omega \gg k_z v_{T_e}$, and $|\omega - n\omega_{B_i}| \gg k_z v_{T_{\parallel i}}$. In the case of the resonant

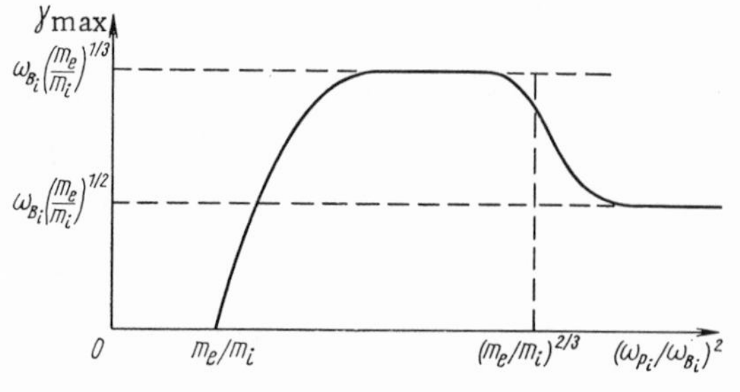

Fig. 15.3. Maximum possible growth rate of the instability of an anisotropic plasma as a function of $(\omega_{p_i}/\omega_{B_i})^2$ as $T_e \to 0$ and $T_{\parallel i} \to 0$.

instability and when $z_i \approx 1$, this means

$$\eta_e < b_i \lesssim \mu^{2/3}, \qquad (15.11)$$

$$\eta_i < b_i/\mu^{1/3} \lesssim \mu^{1/3} \qquad (15.12)$$

Here, $\eta_e \equiv T_e/T_{\perp i}$, $\eta_i \equiv T_{\parallel i}/T_{\perp i}$, $\mu \equiv m_e/m_i$. The right-hand inequalities take into account the upper bound on the density [the condition (15.6)]. For the nonresonant instability, (15.11) and (15.12) are replaced by

$$(\eta_e, \eta_i) < \frac{\mu}{\cos^2\theta}, \qquad (15.13)$$

where $\cos^2\theta$ lies in the shaded region of Fig. 15.2. The maximum of the right-hand side of (15.13) is attained for $\cos^2\theta$ adjoining the instability boundary. In the case of a plasma with $b_i < \mu^{1/2}$, this boundary is given by Eq. (15.8) and it then follows from (15.13) that

$$(\eta_e, \eta_i) < b_i. \qquad (15.14)$$

At larger values of the parameter b_i, we find that $(\cos\theta)_{min}$ is independent of the density [see (15.10)]; in this case, (15.14) is replaced by the condition

$$(\eta_e, \eta_i) < \mu^{1/2}. \qquad (15.15)$$

It is evident from the inequalities (15.11)-(15.15) that the hydrodynamic excitation of ion-cyclotron oscillations considered above is possible only if the distribution of the ion energies is sufficiently anisotropic and the electrons are much colder than the ions.

§ 15.2. Low-Density Plasma with Cold Electrons and Finite $T_{\parallel i}/T_{\perp i}$

As in § 15.1, we shall assume that the electrons are cold, $T_e \to 0$, but we shall take into account the longitudinal ion temperature. Using an expression for $\varepsilon_0^{(i)}$ of the form (7.20), we obtain the dispersion equation

$$\varepsilon_0 = 1 - \left(\frac{\omega_{p_e}\cos\theta}{\omega}\right)^2 + \frac{I_n e^{-z_i}}{k^2 d_{\parallel i}^2}\left\{1 + \right.$$

$$\left. + \frac{i\sqrt{\pi}\,\omega}{|k_z| v_{T_{\parallel i}}}\left[1 - \frac{n\omega_{B_i}}{\omega}\left(1 - \frac{T_{\parallel i}}{T_{\perp i}}\right)\right] W\left(\frac{\omega - n\omega_{B_i}}{|k_z| v_{T_{\parallel i}}}\right)\right\} = 0. \qquad (15.16)$$

Suppose that the longitudinal temperature does not satisfy the conditions (15.12)-(15.15). Then the instabilities of hydrodynamic type considered in § 15.1 are absent. Under these conditions a kinetic instability can arise. Its growth rate can be found from (15.16):

$$\gamma = -\frac{\sqrt{\pi}}{2}\cdot\frac{(n\omega_{B_i})^2}{|k_z|\, v_{T_{\parallel i}}}\cdot\frac{I_n e^{-z_i}}{k^2 d^2_{\parallel i}}\left[1-\frac{n\omega_{B_i}}{\omega_k}\left(1-\frac{T_{\parallel i}}{T_{\perp i}}\right)\right] e^{-\left(\frac{\omega_k - n\omega_{B_i}}{k_z v_{T_{\parallel i}}}\right)^2}, \tag{15.17}$$

where $\omega_k = \omega_{p_e}\cos\theta$. The maximum of the growth rate is attained for

$$\frac{\omega - n\omega_{B_i}}{n\omega_{B_i}} = -\frac{1}{2}\left\{\frac{T_{\parallel i}}{T_{\perp i}} + \sqrt{\left(\frac{T_{\parallel i}}{T_{\perp i}}\right)^2 + 2\left(\frac{k_z v_{T_{\parallel i}}}{n\omega_{B_i}}\right)^2}\right\}. \tag{15.18}$$

This maximum is not exponentially small if

$$\eta_i \lesssim k_z \rho_{\parallel i}/n,\quad \rho_{\parallel i} \equiv v_{T_{\parallel i}}/\omega_{B_i}. \tag{15.19}$$

Using (15.19) and the opposite inequality to (15.12), we find the conditions under which a kinetic instability with a growth rate that is not exponentially small can arise in the plasma:

$$b_i/\mu^{1/3} < \eta_i < \mu/b_i. \tag{15.20}$$

The range of the parameter $T_{\parallel i}/T_{\perp i}$ determined by the inequalities (15.20) is shown in Fig. 15.4. This figure also indicates

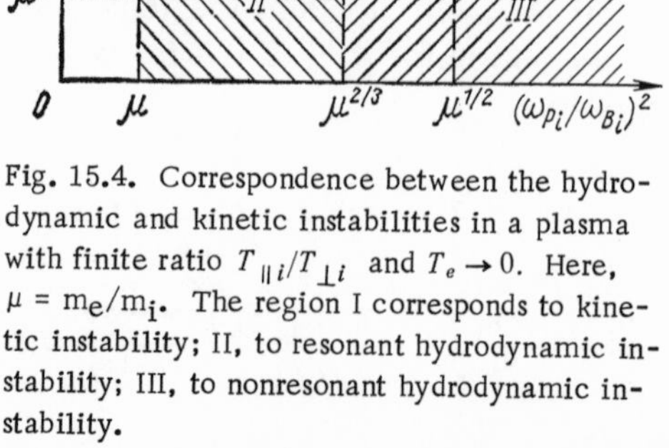

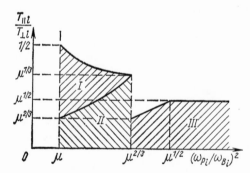

Fig. 15.4. Correspondence between the hydrodynamic and kinetic instabilities in a plasma with finite ratio $T_{\parallel i}/T_{\perp i}$ and $T_e \to 0$. Here, $\mu = m_e/m_i$. The region I corresponds to kinetic instability; II, to resonant hydrodynamic instability; III, to nonresonant hydrodynamic instability.

the correspondence between the kinetic instability considered here and the hydrodynamic instabilities considered in § 15.1. It can be seen that the kinetic instability can arise only at the densities at which the resonant hydrodynamic instability is also possible. This accords with the comments made in § 3.2 concerning the relationship between the hydrodynamic and kinetic instabilities.

The kinetic growth rate is given approximately by

$$\gamma \simeq (b_i/\eta_i)\,\omega_{B_i}. \tag{15.21}$$

The parameters η_i and b_i in this relation lie in the limits (15.20) or, equivalently, in the corresponding shaded region of Fig. 15.4.

The approximation of cold electrons is justified if the condition (15.11) is satisfied. If the condition opposite to (15.11) holds, a plasma with $b_i \ll 1$ and not too small η_i – the condition opposite to (15.12) – is stable.

§ 15.3. Low-Density, Strongly Anisotropic Plasma with Finite Electron Temperature

If $k_z v_{T_e}/\omega$ is finite and $k_z v_{T_{\parallel i}}/(\omega - n\omega_{B_i}) \to 0$, the dispersion equation has the form

$$\varepsilon_0 = 1 + \frac{1}{k^2 d_e^2}\left[1 + i\sqrt{\pi}\,\frac{\omega}{|k_z|\,v_{T_e}}\,W\left(\frac{\omega}{|k_z|\,v_{T_e}}\right)\right] - \\ - \frac{\omega_{p_i}^2 I_n e^{-z_i}\cos^2\theta}{(\omega - n\omega_{B_i})^2} - \frac{\omega_{p_i}^2 \sin^2\theta}{\omega_{B_i}(\omega - n\omega_{B_i})}\,\frac{n I_n e^{-z_i}}{z_i} = 0. \tag{15.22}$$

As in (15.16), this equation does not take into account the low-frequency longitudinal inertia of the ions (the term with $I_0 e^{-z_i}/\omega^2$), which is important in the case $T_e \gtrsim T_{\perp i}$ and $\omega_{p_i} \gtrsim \omega_{B_i}$. This effect will be taken into account below in § 15.4. In addition, we have omitted an ion term $\simeq (kd_{\perp i})^{-2}$ in (15.22). If $T_e \lesssim T_{\perp i}$ and $\omega \lesssim k_z v_{T_e}$, it is small compared with the electron term. If $z_i \approx 1$ and $k_z \lessdot k_\perp$, the electron contribution to (15.22) is of the order of b_i/η_e. Suppose

$$b_i \lesssim \eta_e. \tag{15.23}$$

In this case it is sufficient to take into account only the imaginary part in $\varepsilon_0^{(e)}$. It is small compared with the real terms of the equation and the frequency and growth rate of the oscillations can therefore be determined by means of (2.16) and (2.17).

If $\cos\theta > \omega_{p_i}/\omega_{B_i}$, when the last term of the left-hand side of (12.22) is not important, we have

$$\omega_k^{(\pm)} = n\omega_{B_i} \pm |\omega_{p_i} \cos\theta| (I_n e^{-z_i})^{1/2}, \tag{15.24}$$

$$\gamma_k^{(\pm)} = \pm \frac{\sqrt{\pi}}{2} \cdot \frac{n\omega_{B_i} |\omega_{p_i} \cos\theta|}{|k_z| v_{T_e} k^2 d_e^2} \exp\left[-\left(\frac{n\omega_{B_i}}{k_z v_{T_e}}\right)^2\right]. \tag{15.25}$$

For perturbations with $\omega = \omega^{(+)}$ we have damping ($\gamma < 0$); with $\omega = \omega^{(-)}$, growth ($\gamma > 0$) [to be specific, we assume $n\omega_{B_i} > 0$)]. The damping and growth is due to the interaction of the oscillations with the resonant electrons. In both cases, the electrons take energy from the wave. This follows from the energy balance equation (2.18) of the oscillations and the expression (15.22) for ε_0. It is the negative energy W_k of the $\omega = \omega^{(-)}$ branch – proved by (15.22), (15.24), and (2.19) – that makes it possible for the electrons to take energy from these oscillations and yet cause them to grow. (A detailed analysis of the energy balance of oscillations with $W_k < 0$ was given in § 2.3).

The growth rate of the anisotropic instability as a function of $\eta_e \equiv T_e/T_{\perp i}$ is shown schematically in Fig. 15.5. If η_e is very small, γ is determined by the hydrodynamic relations given in § 15.1. Figure 15.5b corresponds to densities at which both the resonant and nonresonant hydrodynamic instabilities are possible. This explains its similarity to the corresponding figure (Fig. 3.4) for the two-stream instability. If $b_i < \mu$, the only type of instability is kinetic (Fig. 15.5a). In a plasma with $b_i < \mu^{2/3}$ – the condition opposite to (15.6) – and $\eta_e \to 0$, only the nonresonant hydrodynamic instability is possible. This is why Fig. 15.5c, which corresponds to this range of densities, differs from Fig. 15.5b.

If the condition opposite to (15.23) holds, the solution of (15.22) is

$$\omega = n\omega_{B_i} \pm k_z \left(\frac{T_e}{m_i}\right)^{1/2} \frac{(I_n e^{-z_i})^{1/2}}{\left[1 + i\sqrt{\pi}\, \frac{n\omega_{B_i}}{|k_z| v_{T_e}} W\left(\frac{n\omega_{B_i}}{|k_z| v_{T_e}}\right)\right]^{1/2}}. \tag{15.26}$$

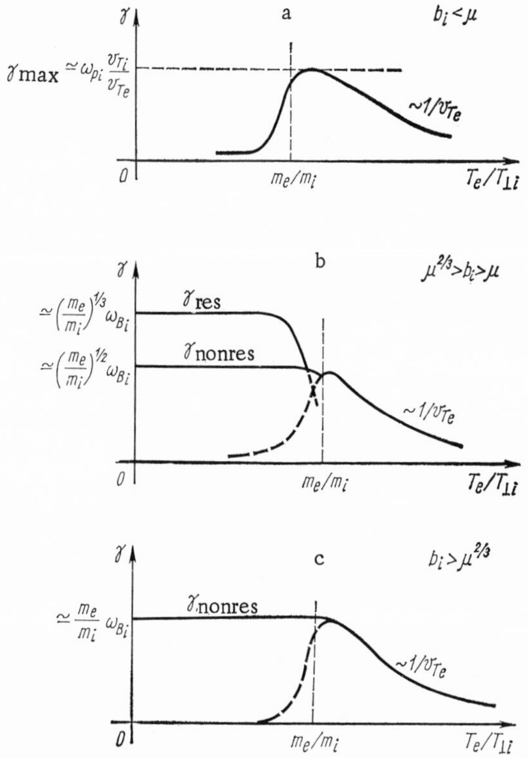

Fig. 15.5. Growth rates of the instabilities of a low-density plasma with a finite electron temperature for different values of the parameter $\left(\frac{\omega_{p_i}}{\omega_{B_i}}\right)^2 \equiv b_i$. $\mu = m_e/m_i$.

It will be seen that ω is independent of the density in this limit. The frequency and growth rate of perturbations with $k_z > n\omega_{B_i}/v_{T_e}$ are

$$\left.\begin{aligned} \operatorname{Re}\omega &= n\omega_{B_i} - k_z\left(\frac{T_e}{m_i}\right)^{1/2}(I_n e^{-z_i})^{1/2},\\ \gamma &= \left(\frac{\pi}{8}\cdot\frac{m_e}{m_i}\right)^{1/2} n\omega_{B_i}(I_n e^{-z_i})^{1/2}. \end{aligned}\right\} \quad (15.27)$$

The approximate dependence of the growth rate on the density predicted by (15.25) and (15.27) is shown in Fig. 15.6.

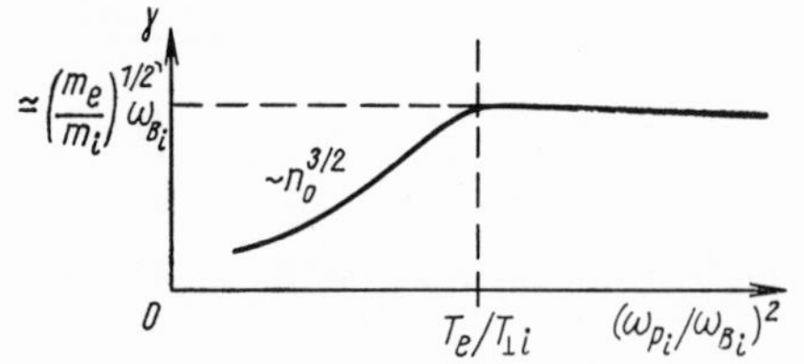

Fig. 15.6. Kinetic growth rate as a function of the density in a plasma with finite $T_e/T_{\perp i}$.

The instability considered above is possible only in a plasma with a sufficiently small η_i. If the condition (15.23) is satisfied, when the frequency of the oscillations satisfies (15.24), the condition for η_i to be small is identical with (15.14). In the opposite case, i.e., when $\eta_e < b_i$, Eq. (15.27) shows that the inequality $|\omega - n\omega_{B_i}| \gg k_z v_{T\parallel i}$ entails

$$\eta_i \ll b_i. \tag{15.28}$$

In deriving (15.22) we have assumed $b_i \ll 1$ and it therefore follows from (15.28) that $\eta_i \ll 1$.

§ 15.4. Dense Plasma with Finite Electron Temperature

1. Instability of a Strongly Anisotropic Ion Distribution, $T_{\parallel i}/T_{\perp i} \to 0$. If $T_{\parallel i}/T_{\perp i} \to 0$ and $\omega \ll k_z v_{T_e}$, the dispersion equation is

$$1 + \frac{1}{k^2 d_e^2} - \omega_{p_i}^2 \sum_{n=-\infty}^{\infty} \left[\frac{I_n e^{-z_i} \cos^2\theta}{(\omega - n\omega_{B_i}^2)} + \frac{n I_n e^{-z} \sin^2\theta}{z\omega_{B_i}(\omega - n\omega_{B_i})} \right] = 0. \tag{15.29}$$

Except for the notation, this equation is identical with (10.29). If $\cos\theta$ is not too small and $\omega \approx n\omega_{B_i}$, Eq. (15.29) simplifies to [cf. (10.40)]

$$1 + (kd_e)^{-2} + \frac{1 - I_0 e^{-z_i}}{k^2 d_{\perp i}^2} - \left(\frac{\omega_{p_i}\cos\theta}{\omega}\right)^2 I_0 e^{-z_i} - \frac{\omega_{p_i}^2 I_n e^{-z_i}\cos^2\theta}{(\omega - n\omega_{B_i})^2} = 0. \tag{15.30}$$

In the approximation $I_n/I_0 \ll 1$, which is valid for $z_i \lessapprox 1$, Eq. (15.30) describes four oscillation branches—two cyclotron and two ion-acoustic. If $\omega - n\omega_{B_i}$ is not too small, the frequency of the latter satisfies [cf. (8.28)]

$$\omega^2 = \frac{k_z^2 T_e}{m_i} \cdot \frac{I_0 e^{-z_i}}{1 + k^2 d_e^2 + \eta_e (1 - I_0 e^{-z_i})}. \tag{15.31}$$

Because of the condition $\eta_i \ll 1$, the perturbations of the type (15.31) are weakly damped, $\omega/k_z \gg v_{T\parallel i}$, even in a plasma with a not very high electron temperature, $T_e < T_{\perp i}$ (but it is still necessary that $T_e \gg T_{\parallel i}$).

If the branches (15.31) intersect the cyclotron branches, the frequency of the perturbations is complex and one of the intersecting branches has $\gamma > 0$. The maximal growth rate of perturbations with $k_\perp \rho_{\perp i} \simeq 1$ is of the order of the cyclotron frequency [cf. (10.33)]:

$$\gamma_{\max} \lesssim \frac{\omega_{B_i}}{2}. \tag{15.32}$$

The frequency (15.31) can attain the ion-cyclotron frequency only if [cf. (10.35)]

$$b_i > 1. \tag{15.33}$$

This is the necessary condition for instability. The dependence of the critical density on the parameter $k_\perp \rho_{\perp i}$ is similar to the dependence shown in Fig. 10.2.

Thus, the hydrodynamic instability of a plasma with anisotropic ions and hot electrons is possible only if the density is not too low, this condition being expressed by (15.33).

The lowest density at which an instability is still possible corresponds to perturbations with $(k_\perp \rho_{\perp i})^2 \gg T_{\perp i}/T_e$. In this case, Eq. (15.29) or (15.31) shows that the electron terms disappear completely from the dispersion equation, which becomes identical – subject to the substitution $m_i \to m_e$ and $e_i \to e_e$ – with that in the case of electron-cyclotron oscillations. The necessary condition (15.33) for the ion instability is therefore identical with (10.35) except for the notation. Numerically, these conditions differ by the mass ratio.

2. Instability of a Plasma with Finite $T_{\parallel i}/T_{\perp i}$. If $T_{\parallel i}/T_{\perp i}$ is finite, it is necessary to take into account the inter-

action of the oscillations with the resonant particles. Using (7.20) and assuming $|\omega - n\omega_{B_i}| \gg k_z v_{T\parallel i}$, $k_z v_{T\parallel i} \ll \omega \ll k_z v_{T_e}$, we find a dispersion equation which is identical with (15.30) in the zeroth approximation in the given expansion parameters and differs from (15.30) when allowance is made for the finite value of the latter by the presence on the left-hand side of the term

$$\mathrm{i}\,\mathrm{Im}\,\varepsilon_0 = \frac{\mathrm{i}\sqrt{\pi}\,\omega}{|k_z|\,v_{T_e} k^2 d_e^2} + \frac{\mathrm{i}\sqrt{\pi}\,\omega I_0 e^{-z_i} \exp[-(\omega/k_z v_{T\parallel i})^2]}{k^2 d_{\parallel i}^2 |k_z|\,v_{T\parallel i}} +$$

$$+ \frac{\mathrm{i}\sqrt{\pi}\left[\omega - n\omega_{B_i}\left(1 - \frac{T_{\parallel i}}{T_{\perp i}}\right)\right]}{|k_z|\,v_{T\parallel i} k^2 d_i^2} I_n e^{-z_i} \exp\left[-\left(\frac{\omega - n\omega_{B_i}}{k_z v_{T\parallel i}}\right)^2\right]. \quad (15.34)$$

The first term corresponds to resonant electrons; the second, to a Cherenkov resonant interaction with the ions; the last, to ion-cyclotron resonance. If the solutions of (15.30) are real, it is the sign of $\mathrm{Im}\,\varepsilon_0$ that determines whether the oscillations grow or are damped (cf. § 10.3). If electron resonance is ignored, Eq. (15.34) yields the instability condition [cf. (10.49)]

$$T_{\perp i} > 2T_{\parallel i}. \quad (15.35)$$

A comparison of the first and last terms of the right-hand side of (15.34) shows that instability is possible (i.e., electron resonance is unimportant) if

$$\eta_e > \mu^{1/3}. \quad (15.36)$$

The growth rate of the kinetic ion instability is given by an expression similar to (10.50). If $T_{\perp i} \gg 2T_{\parallel i}$, it is of the order (15.32) but if $T_{\perp i} \simeq 2T_{\parallel i}$ it is exponentially small.

Bibliography

1. E. G. Harris, Phys. Rev. Lett., 2:34 (1959). It is shown that an instability can develop in an anisotropic plasma of zero pressure.
2. A. V. Timofeev, Zh. Eksp. Teor. Fiz., 39:397 (1960) [Soviet Phys. — JETP, 12:281 (1961)]. A study of the ion-acoustic instability of an anisotropic plasma (§ 15.4).
3. A. B. Kitsenko and K. N. Stepanov, Zh. Tekh. Fiz., 31:176 (1961) [Soviet Phys. — Tech. Phys., 6:127 (1961)]. A study of the instability of a plasma with a strongly anisotropic ion distribution and cold electrons (§§ 15.1-15.3).

4. V. I. Pistunovich, At. Energ., 14:72 (1963). The maximal growth rate of the ion-cyclotron instability of a low-density plasma [Eq. (15.3)] is calculated.
5. Yu. N. Dnestrovskii, D. P. Kostomarov, and V. I. Pistunovich, Nucl. Fusion, 3:30 (1963). An analytic and numerical investigation of the ion-cyclotron instability of a plasma with cold electrons.
6. V. B. Krasovitskii and K. N. Stepanov, Zh. Tekh. Fiz., 34:1013 (1964) [Soviet Phys. — Tech. Phys., 9:786 (1964)].
7. V. I. Pistunovich and A. V. Timofeev, Dokl. Akad. Nauk SSSR, 159:779 (1964) [Soviet Phys. — Doklady, 9:1083 (1965)]. In [6, 7] a study is made of the instability of a plasma with anisotropic ions due to the absorption of energy by resonant electrons (§ 15.3).
8. B. B. Kadomtsev, A. B. Mikhailovskii, and A. V. Timofeev, Zh. Eksp. Teor. Fiz., 47:2266 (1964) [Soviet Phys. — JETP, 20:1517 (1965)]. It is shown that the oscillations of a plasma with strongly anisotropic ions may have negative energy (§ 15.3).
9. L. S. Hall, W. Heckrotte, and T. Kammash, Phys. Rev., A139:1117 (1965). A detailed investigation of the principal types of instability of an anisotropic plasma.
10. G. K. Soper and E. G. Harris, Phys. Fluids, 8:984 (1965). A numerical analysis of the limits of the anisotropic instability.
11. G. E. Guest and R. A. Dory, Phys. Fluids, 8:1853 (1965). The instabilities of an anisotropic plasma are investigated.
12. A. V. Timofeev and V. I. Pistunovich, In: Reviews of Plasma Physics, Vol. 5, Consultants Bureau, New York (1970), p. 401. A detailed review of the main types of instability of an anisotropic plasma.

Chapter 16

Growth of Ion-Cyclotron Oscillations in a Plasma with Transversely Non-Maxwellian Ions

§ 16.1. Instabilities of Plasmas with Cold Electrons

One of the causes of excitation of ion-cyclotron oscillations is a nonequilibrium distribution of the transverse velocities of the ions. In this chapter we shall consider some examples of such excitation. We shall first consider a plasma with cold electrons; subsequently, with a finite electron temperature.

1. Hydrodynamic Excitation of Oscillations with $\omega \simeq n\omega_{B_i}$. In § 14.1, we have considered the instability of a high-density plasma, $\omega_{p_i}^2/\omega_{B_i}^2 \equiv b_i \gg 1$. We shall now assume $b_i \lessgtr 1$. We shall neglect the perturbed motion of the ions along the magnetic field (it will be taken into account subsequently). In this approximation, $\varepsilon_0^{(i)}$ has the form (7.16) with $k_z = 0$. We shall assume that the electrons are cold, $\omega/k_z \gg v_{T_e}$. If the plasma density is low $(\omega_{p_e}^2/\omega_{B_e}^2 \ll 1)$, then $\varepsilon_0^{(e)} = -(\omega_{p_e}\cos\theta/\omega)^2$. The basic dispersion equation is therefore

$$1 - \frac{\omega_{p_e}^2}{\omega^2}\cos^2\theta - \frac{\omega_{p_i}^2}{\omega_{B_i}} \sum_{n=-\infty}^{\infty} \frac{n\left\langle \frac{2J_n J'_n}{\xi} \right\rangle}{\omega - n\omega_{B_i}} = 0. \tag{16.1}$$

An equation of a similar kind [see (11.22)] was investigated in § 11.3.1. Exploiting this analogy, we conclude that the plasma is unstable if for some value of the wave number $k_\perp$ the following condition holds [cf. (11.23)]:

$$\cos\theta \approx n\frac{\omega_{B_i}}{\omega_{p_e}} \equiv n\left(\frac{m_e}{m_i}\right)^{1/2}\frac{\omega_{B_i}}{\omega_{p_i}}. \tag{16.2}$$

The growth rate of the oscillations is

$$\gamma = \omega_{p_i}\sqrt{-\left\langle\frac{J_n J_n'}{\xi}\right\rangle}. \tag{16.3}$$

In contrast to § 11.3.1, the first harmonic ($\omega \approx \omega_{B_i}$) and not only the higher harmonics ($\omega \approx n\omega_{B_i}$, $n = 2, 3, \ldots$), can be excited.

The function $\langle -J_n J'_n/\xi\rangle$ can be positive only if $\xi \equiv \bar{v}_\perp k_\perp/\omega_{B_i} > 1$, where $\bar{v}_\perp$ is some effective value of the transverse velocity. It follows that only short-wavelength perturbations, $k_\perp > 1/\rho_i$, can be excited. This condition is not essential in the case of the anisotropic excitation considered in Chapter 15. A δ-function distribution is the most strongly unstable distribution. Ion distributions of the type (10.69) and (10.71) are also unstable.

In accordance with (16.3), the growth rate increases with the density, being $\gamma \simeq \omega_{B_i}$ at

$$b_i \simeq \frac{1}{\left|\left\langle\frac{(J_n^2)'}{\xi}\right\rangle\right|} \gtrsim 1. \tag{16.4}$$

At higher densities, Eq. (16.3) predicts $\gamma > \omega_{B_i}$. However, Eq. (16.3) ceases to be valid in this case, since several terms of the sum in the original equation (16.1) are then important (cf. §11.3.2). A treatment which takes into account the large number of terms needed when $b_i \gg 1$ corresponds to the high-frequency approximation of § 14.1.

In the above, we have neglected the longitudinal motion of the ions. In the case of a plasma with an approximately isotropic distribution of both the longitudinal and transverse velocities, $\bar{v}_\perp^2/2 \approx \bar{v}_\parallel^2$, this is valid if $|\omega - n\omega_{B_i}| > k_z v_{T_i}$. Using (16.2) and (16.3), we find that our results for a plasma of this kind are valid provided

$$b_i > (m_e/m_i)^{1/2}. \tag{16.5}$$

In § 16.1.2 we shall consider a less dense plasma with $\bar{v}_\perp^2/2 \approx \bar{v}_\parallel^2$.

In the case of a strongly anisotropic plasma $(\bar{v}^2_{\|i} \to 0)$, the neglect of the longitudinal motion of the ions entails $\gamma/\omega_{B_i} \gg \cos^2 \theta$. This is the opposite condition to (15.5), for which the longitudinal motion of the ions is important. It can be seen that in the case of a strongly anisotropic plasma the region of applicability of (16.2) and (16.3) is larger than in the case of an isotropic plasma [cf. (16.5)]:

$$b_i > (m_e/m_i)^{2/3}. \tag{16.6}$$

An anisotropic plasma with $\partial f_0/\partial v_\perp > 0$ for which the opposite condition to (16.6) is satisfied will be considered in § 16.1.3.

Equation (16.1) can also have complex solutions for values of $\cos \theta$ less than those allowed by condition (16.2). This question is discussed in § 16.2.

The cold-electron approximation adopted above is valid if $\omega > k_z v_{T_e}$. Taking into account (16.2) and assuming n $\approx$ 1 and $k_\perp \approx 1/\rho_i$, we find that we must then have

$$T_e/m_i v_0^2 \ll b_i. \tag{16.7}$$

At the limit of applicability of the approximation of a discrete spectrum of the cyclotron harmonics (for $b_i \approx 1$) the condition (16.7) is identical with the high-frequency condition (14.32). We have also encountered a condition of the type (16.7) in our study of the stability of an anisotropic plasma [see the first inequality in (15.14)].

The stability of a plasma for which the condition opposite to (16.7) holds is discussed in § 16.2.

2. Kinetic Instability of a Plasma with $\bar{v}^2_\perp/2 \simeq \bar{v}^2_\|$. Suppose the condition opposite to (16.5) holds:

$$b_i < (m_e/m_i)^{1/2}. \tag{16.8}$$

It is then necessary to take into account the longitudinal thermal spread of the ions and to assume $|\omega - n\omega_{B_i}| < k_z v_{T_{\|i}}$. We find the permittivity of the ions by means of (7.16). The oscillation growth rate, which is maximal if (16.2) is satisfied, is given by the formula

$$\gamma^{\text{kin}}_{\text{max}} = \sqrt{\pi}\, \frac{\omega^2_{p_i}}{|k_z| v_{T_{\|i}}}\, n^2 \sin^2 \theta \left\langle -\frac{J_n J'_n}{\xi} \right\rangle. \tag{16.9}$$

Approximately,

$$\gamma_{\max}^{\text{kin}} = \left(\frac{m_i}{m_e}\right)^{1/2} b_i^{3/2}\omega_{B_i}. \tag{16.10}$$

At the limit of applicability [for $k_z v_{T_{\parallel i}} \simeq \omega_{p_i}$, $b_i \simeq (m_e/m_i)^{1/2}$], the expressions (16.3) and (16.10) are approximately the same; at the same time

$$\gamma \simeq \left(\frac{m_e}{m_i}\right)^{1/4} \omega_{B_i}. \tag{16.11}$$

The density dependence of the growth rate that follows from (16.3) and (16.10) is shown in Fig. 16.1. This figure also takes into account the absence of an instability for $b_i < m_e/m_i$ [in this case the condition (16.2) is not satisfied]. If $b_i > 1$, the growth rate is determined by Eq. (14.3) of the high-frequency approximation (§ 14.1).

The results of this subsection are valid provided the electron temperature is sufficiently low, i.e., the condition (16.7) must hold.

3. Instability of an Anisotropic Plasma with $\partial f_{\perp i}/\partial v_\perp > 0$. If allowance is made for the perturbed longitudinal motion of the ions, the dispersion equation for a strongly anisotropic plasma ($v_{T_{\parallel i}} \to 0$) has the form [cf. (15.1)]

$$1 - \frac{\omega_{p_e}^2}{\omega^2}\cos^2\theta - \omega_{p_i}^2 \sum_{n=-\infty}^{+\infty}\left[\frac{n\left\langle \frac{2J_n J'_n}{\xi}\right\rangle}{\omega_{B_i}(\omega - n\omega_{B_i})} + \frac{\langle J_n^2\rangle \cos^2\theta}{(\omega - n\omega_{B_i})^2}\right] = 0. \tag{16.12}$$

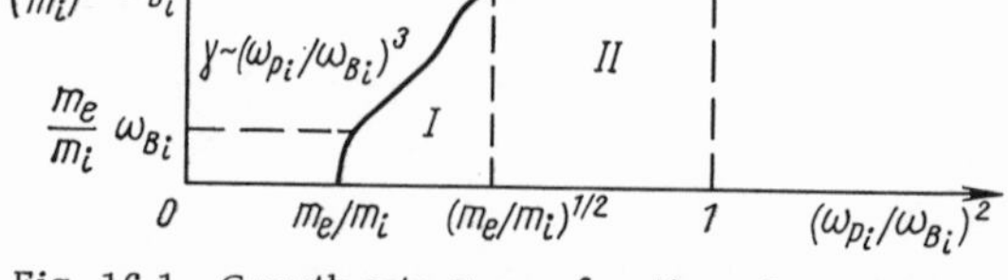

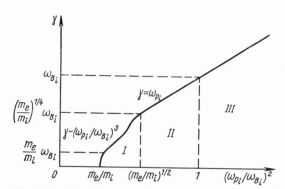

Fig. 16.1. Growth rate γ as a function of $b_i = (\omega_{p_i}/\omega_{B_i})^2$ for a plasma with $(\partial f_{\perp i}/\partial v_\perp) > 0$, $\frac{\bar{v}_\perp^2}{2} \simeq \bar{v}_\parallel^2$, $T_e \to 0$: I) kinetic instability, § 16.2; II) hydrodynamic excitation of cyclotron oscillations, § 16.1.1; III) high-frequency instability, § 14.1.

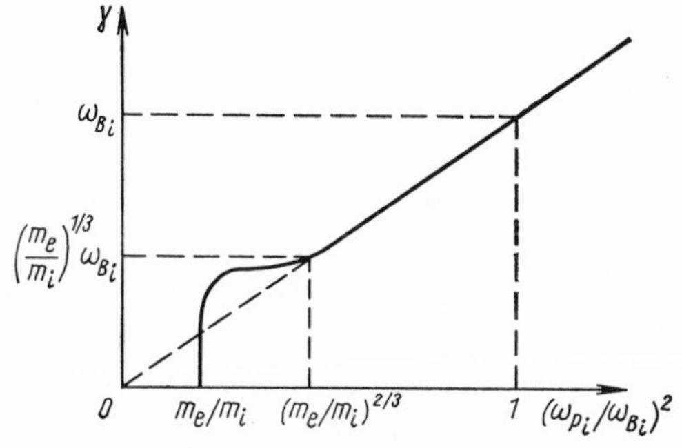

Fig. 16.2. The dependence $\gamma = \gamma(b_i)$ for a strongly anisotropic plasma, $\frac{\partial f_{\perp i}}{\partial v_\perp} > 0$, $T_e \to 0$.

For a plasma with the condition (16.6) this equation was investigated in § 16.1.1; with the opposite condition, in § 15.1 [see (15.6)]. For a plasma with an equilibrium transverse velocity distribution, $\partial f_{\perp i}/\delta v_\perp < 0$ (considered in § 15.1), the condition (16.5) means that hydrodynamic excitation of oscillations with $\omega \approx \omega_{p_e} \cos\theta \approx n\omega_{B_i}$ is impossible. (We recall that the maximal growth rate corresponds to such oscillations.) It can be seen that for $\partial f_{\perp i}/\partial v_\perp > 0$ [or rather, for $f_{\perp i}$ satisfying (11.24)] the condition (15.6) has a different meaning: for a plasma satisfying this condition the excitation of oscillations with $\omega \approx n\omega_{B_i}$ $\omega \approx \omega_{p_e} \cos\theta$ is due to ion anisotropy; for a plasma satisfying the opposite condition, to a nonequilibrium transverse ion distribution.

The dependence of the growth rate on the density for a strongly anisotropic plasma with $\partial f_{\perp i}/\partial v_\perp > 0$ is shown in Fig. 16.2. If $b_i < (m_e/m_i)^{2/3}$, this dependence is identical with that in Fig. 15.3. If the anisotropy of the plasma is not very great, the law $\gamma \approx (m_e/m_i)^{2/3}$ is violated at a density greater than that corresponding to the value $b_i = m_e/m_i$ (see § 15.2) shown in Fig. 15.3.

§ 16.2. Instabilities of a Dense Plasma with Hot Electrons

1. The Approximation $T_e \to \infty$. Even if the electron temperature is very high, a dense plasma may still be unstable against perturbations with $\omega \lesssim \omega_{B_i}$. In the limit of infinitely hot

electrons, $T_e \gg m_i v_0^2$, we have the dispersion equation

$$1-\frac{\omega_{p_i}^2}{\omega_{B_i}}\sum_{n=-\infty}^{\infty}\frac{n\left\langle 2\frac{J_n J'_n}{\xi}\right\rangle}{\omega-n\omega_{B_i}}=0. \tag{16.13}$$

This equation differs from the equation investigated in §10.4 [the equation $\varepsilon_0 = 0$ with ε_0 of the form (10.53)] only by the subscripts labeling the charge species. It describes growing perturbations, $\gamma > 0$, if, in accordance with (10.68),

$$b_i > \frac{1}{\left\langle \frac{(J_0^2)'}{\xi}\right\rangle} > 0. \tag{16.14}$$

The growth rate of the perturbations does not exceed ω_{B_i}, $\gamma \lesssim \omega_{B_i}$.

The class of unstable distributions that satisfy the condition (16.4) was discussed in § 10.4. It is smaller than the class of distributions that are unstable against high-frequency perturbations, which develop in a dense plasma in the limit $T_e \to 0$ (see § 14.1).

In accordance with (10.57), an ion distribution of δ-function type is unstable if

$$b_i > 17. \tag{16.15}$$

The instability limits of smoother distributions are shown in Fig. 10.5.

The low-frequency instability considered here can also develop in a plasma with cold electrons if $k_z = 0$. However, in view of the small value of the growth rate the low-frequency instability can play a role in a plasma of this kind only if some additional factors lead to stabilization of the high-frequency perturbations. This could be the case, in particular, if the plasma does not have sufficient length, unstable perturbations with $k_z = 0$ then being the only possibility.

2. Excitation of Cyclotron Oscillations by Resonant Electrons. Under conditions when the dispersion equation (16.13) does not have complex roots, it is necessary to take into account the interaction of the oscillations with resonant electrons in order to obtain a true picture of the stability.

Let us assume $\omega \ll k_z v_{T_e}$, $\omega \approx n\omega_{B_i}$. The dispersion equation can be written in the form

$$\varepsilon_0 = 1 + \frac{1}{(kd_e)^2}\left(1 + \frac{i\sqrt{\pi}\,\omega}{|k_z|\,v_{T_e}}\right) + \\ + \left(\frac{\omega_{p_i}}{\omega_{B_i}}\right)^2 \left[\left\langle \frac{2J_0J_1}{\xi}\right\rangle - \frac{n\omega_{B_i}\left\langle \dfrac{2J_nJ'_n}{\xi}\right\rangle}{\omega - n\omega_{B_i}}\right] = 0. \qquad (16.16)$$

We shall neglect the terms $1+\frac{1}{(kd_e)^2}$ in ε_0, assuming $b \gg 1$, $T_e \gg m_i(\bar{v}_\perp^2)_i$. (The estimates are made under the assumption $\xi \approx 1$; it should however be borne in mind that ξ may differ from unity by a factor of, say, five.)

Using (16.6) we find

$$\left.\begin{aligned} \operatorname{Re}\omega &= n\omega_{B_i}\left(1 + \left\langle \frac{J_nJ'_n}{\xi}\right\rangle \Big/ \left\langle \frac{J_0J_1}{\xi}\right\rangle\right), \\ \gamma &= \frac{\sqrt{\pi}}{2}\cdot\frac{(n\omega_{B_i})^2}{|k_z|\,v_{T_e}\,(k^2T_e/m_i\omega_{B_i}^2)}\,\frac{\left\langle -\dfrac{J_nJ'_n}{\xi}\right\rangle}{\left\langle \dfrac{J_0J_1}{\xi}\right\rangle^2}. \end{aligned}\right\} \qquad (16.17)$$

Here we must assume $\langle J_0J_1/\xi\rangle > 0$, since otherwise the stronger instability considered in the foregoing subsection can arise. In finding the growth rate we assumed $\gamma \ll |n\omega_{B_i} - \omega|$ and used the relation (2.17).

The growth rate is positive if $\langle J_nJ'_n/\xi\rangle < 0$ [the condition (11.24)]. In this case the oscillations have negative energy, $\omega_k \partial\, \operatorname{Re}\varepsilon_0/\partial\omega_k < 0$ [see § 2.3]. The oscillations grow by having energy taken from them by resonant electrons (cf. § 15.3).

The maximum of the growth rate as a function k_z is attained for $k_z \simeq n\omega_{B_i}/v_{T_e}$, i.e., at the limits of applicability of the approximation $\omega \ll k_z v_{T_e}$. In order of magnitude

$$\gamma_{\max} \simeq \frac{m_i v_{\perp i}^2}{T_e}\,\omega_{B_i}. \qquad (16.18)$$

This result refers to a plasma with $T_e \gg m_i v_{\perp i}^2$, $b_i \gg 1$. At the limits of applicability of these inequalities, i.e., for $T_e \approx m_i v_{\perp i}^2$,

$b_i \approx 1$, the growth rate (16.18) and the growth rate (16.3) obtained in the opposite limiting case are of the same order.

§ 16.3. Excitation of Cyclotron Oscillations by Resonant Electrons in a Low-Density Plasma

If $\omega_{p_i} < \omega_{B_i}$ and the electrons are sufficiently cold [the condition (16.7)], the instabilities considered in § 16.1 can develop in the plasma. Suppose however that the electron temperature satisfies the opposite condition to (16.7):

$$T_e > m_i \frac{\overline{v_{\perp i}^2}}{2} b_i . \tag{16.19}$$

Under these conditions the instabilities of § 16.1 are impossible. It is however possible for oscillations of the type considered in § 16.2.2 to be excited by resonant electrons. Let us proceed to investigate this instability.

We shall proceed from the dispersion equation (16.16) and take into account in ε_0 unity and neglect the term $(kd_e)^{-2}$ and the terms of order b_i. Under these conditions, Eq. (16.17) is replaced by

$$\left.\begin{aligned} \operatorname{Re}\omega &= n\omega_{B_i}\left[1 + b_i\left\langle \frac{2J_n J'_n}{\xi}\right\rangle\right], \\ \gamma &= \frac{\sqrt{\pi}}{|k_z| v_{T_e}}\left(\frac{n\omega_{B_i}}{kd_e}\right)^2\left\langle -\frac{2J_n J'_n}{\xi}\right\rangle. \end{aligned}\right\} \tag{16.20}$$

Instability occurs for the same class of ion distributions as § 16.2, i.e., $\langle J_n J'_n/\xi\rangle < 0$. The instability mechanism is also similar to that discussed in § 16.2.

The growth rate is maximal for $k_z \simeq n\omega_{B_i}/v_{T_e}$, and is of the order

$$\gamma_{\max} \simeq \omega_{B_i} \frac{m_i \overline{v_{\perp i}^2}}{T_e} b_i^2 . \tag{16.21}$$

As b_i decreases, the longitudinal motion of the ions becomes important and for $\overline{v_\perp^2}/2 \approx \overline{v_\parallel^2}$ this leads to stabilization of the plasma. This occurs at a density whose order of magnitude satisfies the condition

$$b_i \leqslant \left(\frac{m_e}{m_i}\right)^{1/2} \left(\frac{m_i v_{\perp i}^2}{T_e}\right)^2. \qquad (16.22)$$

§ 16.4. Excitation of Oscillations by Resonant Electrons in a Plasma of Intermediate Density

If the above inequality is satisfied, the instabilities of hydrodynamic type considered in §§ 14.1 and 16.1 can arise in the plasma. However, if this is to happen, the plasma must be relatively long (small k_z). If the plasma is shorter, one must also take into account instabilities with $\omega \lesssim k_z v_{T_e}$. The dispersion relation (16.6) then reduces to

$$\frac{1}{(k\,d_e)^2}\left(1 + i\frac{\sqrt{\pi}\,\omega}{|k_z|\,v_{T_e}}\right) - \left(\frac{\omega_{p_i}}{\omega_{B_i}}\right)^2 \frac{n\omega_{B_i}}{\omega - n\omega_{B_i}} \left\langle \frac{2J_n J'_n}{\xi} \right\rangle = 0. \qquad (16.23)$$

Hence,

$$\left.\begin{aligned} \operatorname{Re}\omega &= n\omega_{B_i}\left(1 + \frac{k^2 T_e}{m_i \omega_{B_i}^2}\left\langle \frac{2J_n J'_n}{\xi}\right\rangle\right), \\ \gamma &= \frac{\sqrt{\pi}\,(n\omega_{B_i})^2}{|k_z|\,v_{T_e}} \cdot \frac{k^2 T_e}{m_i (\omega_{B_i})^2} \left\langle -\frac{2J_n J'_n}{\xi}\right\rangle. \end{aligned}\right\} \qquad (16.24)$$

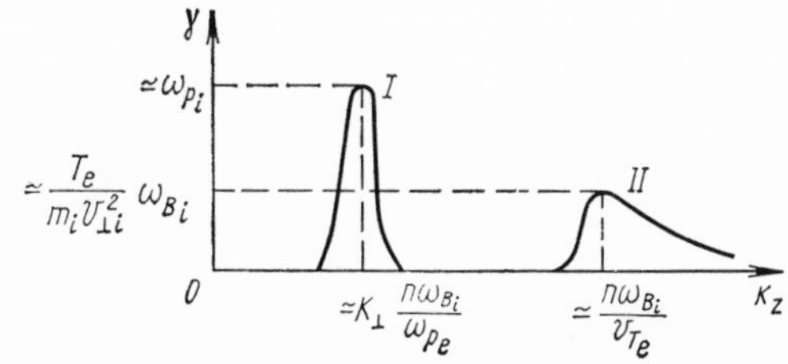

Fig. 16.3. Instabilities of a plasma with $T_e/m_i v_{\perp i}^2 < \left(\frac{\omega_{p_i}}{\omega_{B_i}}\right)^2 < 1$: I) hydrodynamic cyclotron instability, § 16.1.1; II) kinetic cyclotron instability, § 16.4.

We estimate the maximal growth rate by setting $k_z \simeq n\omega_{B_i}/v_{T_e}$, $k_\perp \approx 1/\rho_i$. Then

$$\gamma_{max}^{kin} \simeq \frac{T_e}{m_i v_{\perp i}^2}\,\omega_{B_i}. \tag{16.25}$$

The relationship between the kinetic and the hydrodynamic instabilities is illustrated in Fig. 16.3 (in the case $\omega_{p_i} < \omega_{B_i}$).

§ 16.5. Review of Results Obtained in §§ 16.1-16.4

The correspondence between the principal types of instability of a plasma with $\partial f_{0i}/\partial v_\perp > 0$, and $(\bar{v}_\perp^2/2)_i \simeq (\bar{v}_\parallel^2)_i$ is illustrated in Fig. 16.4. Along the abscissa we have plotted the relative density $(\omega_{p_i}/\omega_{B_i})^2$; along the ordinate, the relative electron temperature $T_e/m_i\bar{v}_{\perp i}^2$.

If the plasma density is high and the electron temperature is low, the high-frequency instability considered in § 14.1 can develop (see the right-hand lower part of the figure). As the density decreases, this instability is first replaced by the hydrodynamic cyclotron instability (§ 16.1.1) and then by the kinetic instability (§ 16.1.2). This all occurs at a fairly low electron temperature.

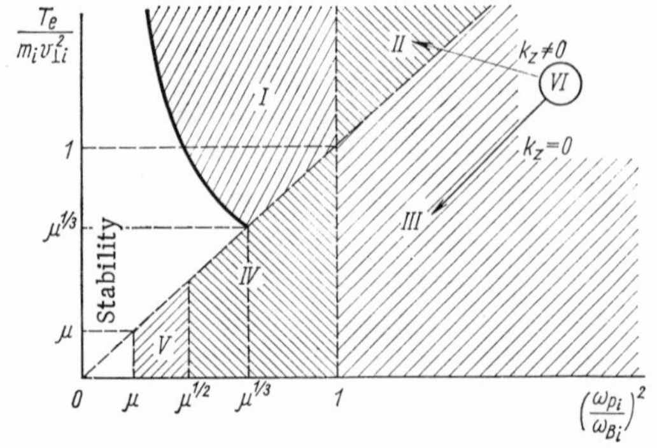

Fig. 16.4. Diagram of the instabilities of a plasma with transversely non-Maxwellian ions: I) cyclotron instability, § 16.3; II) cyclotron instability, § 16.4; III) high-frequency instability, § 14.1; IV) hydrodynamic cyclotron instability, § 16.1.1; V) kinetic cyclotron instability, § 16.2.2; VI) instability considered in § 16.2.1. Here $\mu = m_e/m_i$.

As the electron temperature is increased (going upwards in the figure), the high-frequency instability is replaced by the kinetic cyclotron instability considered in § 16.3. The kinetic instabilities discussed in § 16.1.2 (in the lower left-hand corner of the figure) are completely stabilized when the electron temperature is increased.

These comments refer to a fairly long plasma. If the plasma has a limited length, the kinetic instabilities discussed in § 16.4 may play a role. They are situated in Fig. 16.4 in the region occupied by the hydrodynamic instabilities. If the ion distribution is sufficiently near a δ-function distribution and the plasma density is not low, one can have an "instability of the zeroth harmonic" (see § 16.2.1; this is the right-hand part of Fig. 16.4). For a cold-electron plasma this instability is possible for $k_z = 0$ perturbations; for a hot-electron plasma, for $k_z \neq 0$ perturbations as well.

§ 16.6. Instability of a Plasma Consisting of Two Groups of Ions

Suppose the ion component of the plasma consists of a group of hot particles with $\partial f_\perp / \partial v_\perp > 0$ and a group of particles with temperature T that is low compared with the energy of the hot particles. If the density of a plasma of this kind is high, the excitation of high-frequency perturbations with $\gamma > \omega_{B_i}$ investigated in § 14.5 is possible. We shall now consider perturbations with $\omega \approx n\omega_{B_i}$, $\gamma < \omega_{B_i}$, assuming that the plasma density is not too high.

If the fraction of hot ions is small, they may lead to excitation of oscillations due to the motion of the cold ions. If $k_z = 0$, this effect is described by the dispersion equation

$$1 - \frac{\omega_{p_i}^2}{\omega^2 - \omega_{B_i}^2} - \alpha \frac{\omega_{p_i}^2}{\omega_{B_i}} \sum_{n=-\infty}^{\infty} \frac{n \langle 2J_n J'_n / \xi \rangle}{\omega - n\omega_{B_i}} = 0. \tag{16.26}$$

Here, $\alpha \equiv n_1/n_0$ is the ratio of the densities of the hot and cold ions. From (16.26) we find that the ion-cyclotron harmonics with the numbers n = 2, 3, . . . are excited. Their growth rate is approximately

$$\gamma \simeq \alpha^{1/2} \omega_{B_i}. \tag{16.27}$$

The density at which this occurs is such that

$$\left(\frac{\omega_{p_i}}{\omega_{B_i}}\right)^2 = n^2 - 1, \; n = 2, 3, \ldots . \tag{16.28}$$

As the temperature of the cold ions is increased, this instability is suppressed.

§ 16.7. Role of a Longitudinal Inhomogeneity of the Magnetic Field. Modified Negative Mass Instability

In §§ 16.1-16.6 we have adopted the approximation of a homogeneous magnetic field. If we allow an inhomogeneity of the magnetic field, some further different forms of the ion-cyclotron instabilities of a plasma with transversely non-Maxwellian ions are possible. In the present section we shall consider an instability associated with longitudinal inhomogeneity of the magnetic field and then consider the effect of transverse inhomogeneity in the next section.

Suppose a plasma is confined in an adiabatic trap, so that the particles move between the stoppers with a certain characteristic time τ that depends on the particle velocity, $\tau = \oint dl/v_\parallel$. If this time is short compared with the reciprocal of the growth rate, $\tau \ll 1/\gamma$, the ion-cyclotron frequency ω_{B_i} in the dispersion equation must be replaced by its mean over the period τ:

$$\bar{\omega}_{B_i} = \frac{1}{\tau} \oint \omega_{B_i}(l) \frac{dl}{v_\parallel}; \tag{16.29}$$

$\bar{\omega}_{B_i}$ is some function of the energy and the magnetic moment of the particle, $\bar{\omega}_{B_i} = \omega_{B_i}(\varepsilon, \mu)$, or equivalently,

$$\bar{\omega}_{B_i} = \omega_{B_i}(v_\perp^{(0)}, v_\parallel^{(0)}), \tag{16.30}$$

where $v_\perp^{(0)}$ and $v_\parallel^{(0)}$ are the velocity components of a particle at some fixed point of space, for example, in the center of the trap. In the special case of a field of the form

$$B_0(z) = B_0^{(0)}\left(1 + \frac{z^2}{L^2}\right) \tag{16.31}$$

it follows from (16.29) that

$$\overline{\omega}_{B_i} = \omega_{B_i}^{(0)}\left[1 + \frac{1}{2}\left(\frac{v_{\parallel}^{(0)}}{v_{\perp}^{(0)}}\right)^2\right]. \quad (16.32)$$

We shall take into account the dependence of the cyclotron frequency on the velocity and consider perturbations with $k_{\parallel} = 0$. If the plasma density is sufficiently low, $\omega_{p_e} \ll \omega_{B_e}$, the electrons are not important in such perturbations so that, in accordance with (7.14), the dispersion equation has the form

$$1 - \frac{4\pi e^2}{m_i k^2}\left\langle \frac{\partial F}{\partial \varepsilon_{\perp}^{(0)}}\left[1 - \sum_{n=-\infty}^{\infty} \frac{\omega J_n^2(\xi)}{\omega - n\overline{\omega}_{B_i}(v_{\perp}^{(0)}, v_{\parallel}^{(0)})}\right]\right\rangle = 0. \quad (16.33)$$

Suppose $F \propto \delta(\varepsilon_{\perp}^{(0)} - \varepsilon_{\perp 0})\,\delta(v_{\parallel}^{(0)2} - v_{\parallel 0}^2)$, i.e., we have a plasma with monoenergetic ions. It then follows from (16.33) that for perturbations with $\omega \propto n\overline{\omega}_{B_i}$

$$1 - \frac{\omega_p^2 J_n^2}{(\omega - n\overline{\omega}_{B_i})^2}\cdot\frac{n^2\overline{\omega}_{B_i}}{k^2}\cdot\frac{\partial\overline{\omega}_{B_i}}{\partial\varepsilon_{\perp}} - \frac{\omega_{p_i}^2}{\omega_{B_i}}\cdot\frac{n\,(J_n^2)'}{\xi\,(\omega - n\overline{\omega}_{B_i})} = 0. \quad (16.34)$$

If the density is sufficiently low, the frequency is very near $n\overline{\omega}_{B_i}$ and the last term on the left-hand side of (16.34) is unimportant. Then the expression for $\Delta \equiv \omega - n\overline{\omega}_{B_i}$ has the form

$$\Delta = \pm\left|\omega_{p_i} J_n \frac{n\overline{\omega}_{B_i}}{k}\right|\sqrt{\frac{\partial \ln \overline{\omega}_{B_i}}{\partial \varepsilon_{\perp}}}. \quad (16.35)$$

From (16.32) we find that $\partial \ln \overline{\omega}_{B_i}/\partial\varepsilon_{\perp} < 0$, i.e., the solution (16.35) corresponds to instability. This instability is sometimes called the *modified negative mass instability*.

§ 16.8. Role of a Transverse Inhomogeneity of the Magnetic Field. Instability due to Magnetic Drift

In a zero-pressure plasma, a transverse inhomogeneity of the magnetic field is uniquely related to the curvature of the lines of force. Under the influence of the inhomogeneity and the curvature the particles drift at right angles to the magnetic field with the velocity

$$\mathbf{V}_{\mathrm{m}}=\frac{1}{\omega_B}\left(v_{\parallel}^2+\frac{v_{\perp}^2}{2}\right)\frac{1}{R}\,\mathbf{e}_b, \tag{16.36}$$

where R is the radius of curvature of the lines of force; $\mathbf{e}_b$ is a unit vector in the direction of the binormal to the line of force. As a result, the effective frequency of the perturbed field that acts on the ions is

$$\omega_{\mathrm{eff}}=\omega-n\omega_B-\mathbf{k}\mathbf{V}_{\mathrm{m}}-k_{\parallel}v_{\parallel}. \tag{16.37}$$

To describe this effect, the dispersion equation for perturbations with $k_{\parallel}=0$ must have the form

$$1-\frac{4\pi e^2}{m_i k^2}\left\langle\frac{\partial F}{\partial\varepsilon_{\perp}}-\sum_{n=-\infty}^{\infty}\frac{\omega J_n^2(\xi)\dfrac{\partial F}{\partial\varepsilon_{\perp}}}{\omega-n\omega_{B_i}-\mathbf{k}\mathbf{V}_{\mathrm{m}}}\right\rangle=0. \tag{16.38}$$

As in the case considered in § 16.7, the effective frequency is a function of the velocity, so that if the ions have a non-Maxwellian distribution the magnetic drift plays a destabilizing role. If the ions have a δ-function distribution, (16.8) yields a result similar to (16.35):

$$1+\frac{n\omega_{B_i}}{k^2}\cdot\frac{\partial\omega_{\mathrm{eff}}}{\partial\varepsilon_{\perp}}J_n^2\left(\frac{\omega_{p_i}}{\omega_{\mathrm{eff}}}\right)^2-\frac{\omega_{p_i}^2}{\omega_{\mathrm{eff}}\,\omega_{B_i}}n\frac{(J_n^2)'}{\xi}=0. \tag{16.39}$$

The growth rate is approximately

$$\gamma\simeq\omega_{p_i}(\mathbf{k}\mathbf{V}_{\mathrm{m}}/\omega_{B_i})^{1/2}. \tag{16.40}$$

It is nonvanishing for $\mathbf{k}\mathbf{V}_{\mathrm{m}}/\omega > 0$.

Bibliography

1. Yu. N. Dnestrovskii, Nucl. Fusion, 3:259 (1963). It is shown that a nonmonotonic transverse ion distribution can lead to an ion-cyclotron instability (§ 16.1). A study is made of the stabilizing influence of electron heating and a finite length of the plasma.
2. V. B. Krasovitskii and K. N. Stepanov, Zh. Tekh. Fiz., 34:1013 (1964) [Soviet Phys. – Tech. Phys., 9:786 (1964)].

3. A. B. Mikhailovskii, Nucl. Fusion, 5:125 (1965).
4. L. S. Hall, W. Heckrotte, and T. Kammash, Phys. Rev., A139:1117 (1965).
5. L. S. Hall, W. Heckrotte, and T. Kammash, Phys. Rev. Lett., 13:603 (1965).
6. M. N. Rosenbluth, Plasma Physics, IAEA, Vienna (1965). In [2-6] studies are made of the ion-cyclotron instability in a plasma with one group of ions (§§ 16.1-16.5); in [4] this instability is also considered in the case of a plasma with two groups of ions (§ 16.6).
7. D. G. Lominadze and K. N. Stepanov, Zh. Tekh. Fiz., 35:441 (1965) [Soviet Phys. — Tech. Phys., 10:347 (1965)].
8. A. B. Mikhailovskii and É. A. Pashitskii, Zh. Tekh. Fiz., 35:1961 (1965) [Soviet Phys. — Tech. Phys., 10:1507 (1966)]. In [7, 8] a study is made of the excitation of plasma oscillations by a group if ions with a non-Maxwellian velocity distribution (§ 16.6).
9. J. F. Clarke and G. G. Kelly, Phys. Rev. Lett., 21:1041 (1968).
10. B. B. Kadomtsev and O. P. Pogutse, Report at the Novosibirsk Conference of the International Atomic Energy Agency, N CN-24/G-10, 1968. In [9, 10] a study is made of the instability of a plasma in a longitudinally inhomogeneous magnetic field (§ 16.7).
11. L. V. Mikhailovskaya, ZhETF Pis. Red., 5:339 (1967) [JETP Letters, 5:279 (1967)]. A study of the instability of a plasma in a transversely inhomogeneous magnetic field (§ 16.8).

Chapter 17

Excitation of Ion-Cyclotron Oscillations by Ion Beams

§ 17.1. Cyclotron Instability due to Relative Azimuthal Motion of Ions and Electrons

The interaction of ion beams with a plasma is of interest in several respects. For example, an interaction of this kind can lead to part of the energy of the beam being transferred to the ions of the plasma (this has a bearing on the problem of turbulent heating of the ion component of the plasma). From the point of view of application, one is also interested in the capture of ions of the stream in a trap as a result of scattering by the oscillations excited by the stream.

Suppose a plasma is cylindrically symmetric and all the ions move on the average in a azimuthal direction relative to the electrons. In such a plasma instabilities similar to the electron instabilities studied in § 11.4 can develop.

If the plasma density is not too low, $\omega_{p_i} \gtrsim \omega_{B_i}$, the growth rate of the ion-electron oscillations may exceed the ion-cyclotron frequency, $\gamma > \omega_{B_i}$. In this case the stability problem reduces to the problem considered in Chapter 13. In this section we shall assume that the plasma density is fairly low so that oscillations with $\gamma > \omega_{B_i}$ cannot be excited.

We shall assume that the ions and the electrons are cold. The original dispersion equation obtained by using formula (11.33) then takes the form

$$1-\left(\frac{\omega_{p_e}}{\omega}\cos\theta\right)^2-\frac{(\omega_{p_i}\cos\theta)^2}{(\omega+l\omega_{B_i})^2}-\frac{(\omega_{p_i}\sin\theta)^2}{(\omega+l\omega_{B_i})^2-\omega_{B_i}^2}=0. \tag{17.1}$$

If $\gamma<\omega_{B_i}$, one can consider separately the Cherenkov ($\omega+l\omega_{B_i}\approx 0$) and cyclotron ($\omega+l\omega_{B_i}\approx\pm\omega_{B_i}$) interactions of the ions with the oscillations.

Perturbations with $\omega\simeq -l\omega_{B_i}$. In this case the last term on the left-hand side of (17.1) is not important, so that

$$1-\left(\frac{\omega_{p_e}\cos\theta}{\omega}\right)^2-\frac{(\omega_{p_i}\cos\theta)^2}{(\omega+l\omega_{B_i})^2}=0. \tag{17.2}$$

We shall analyze this equation in the same way as Eq. (1.46) in § 1.5.2. In this case the small parameter α is $\mu\equiv m_e/m_i$.

We find that the maximal growth rate is attained for

$$\cos\theta=\mu^{1/2}l\,\frac{\omega_{B_i}}{\omega_{p_i}}. \tag{17.3}$$

At the same time

$$\left.\begin{aligned}\gamma&=\frac{\sqrt{3}}{2^{4/3}}\mu^{1/3}|l\omega_{B_i}|=\frac{\sqrt{3}}{2^{4/3}}\mu^{1/3}\omega_{p_e}\cos\theta,\\ \operatorname{Re}\omega&=-l\omega_{B_i}(1-\mu^{1/3}/2^{4/3}).\end{aligned}\right\} \tag{17.4}$$

It follows from (17.3) that an instability is possible only if the density is not too low:

$$b_i\equiv\left(\frac{\omega_{p_i}}{\omega_{B_i}}\right)^2\gtrsim\mu. \tag{17.5}$$

For given b_i the number of excited harmonics is, in accordance with (17.3), approximately

$$l_{\max}\simeq\left(\frac{b_i}{\mu}\right)^{1/2}. \tag{17.6}$$

This is when $\cos\theta \approx 1$. In accordance with (17.4) and (17.6), all these harmonics have a growth rate γ that is less than ω_{B_i} if

$$b_i < \mu^{1/3}. \tag{17.7}$$

This condition is the opposite of the condition of the high-frequency approximation for $\cos\theta \approx 1$ (see Fig. 13.2). It can be seen that if $\cos\theta \approx 1$ the high-frequency approximation and the approximation that takes into account only Cherenkov excitation lead to results of the same order at the limits of applicability.

If $\cos\theta \ll 1$, then at the limit of the high-frequency approximation (see Fig. 13.2) perturbations with $\omega \approx -l\omega_{B_i}$ possess a growth rate that is much less than ω_{B_i}. It then follows from (17.4) that

$$\gamma_{\lim} \simeq \left(\frac{m_e}{m_i}\right)^{1/3} \left(\frac{\omega_{B_i}}{\omega_{p_i}}\right)^2 \omega_{B_i}, \tag{17.8}$$

where the ratio $(\omega_{p_i}/\omega_{B_i})^2$ is taken on the curve in Fig. 13.2 that bounds the shaded region for $(m_e/m_i)^{1/2} < \cos\theta < 1$. It can be seen that at small $\cos\theta$ there is no smooth transition from the high-frequency to the $\omega \approx -l\omega_{B_i}$ approximation. This comes about because the principal role in this case is not played by the Cherenkov ($\omega \approx -l\omega_{B_i}$), but the cyclotron resonance ($\omega \approx -l\omega_{B_i} \pm \omega_{B_i}$), to which we now turn.

Perturbations with $\omega + l\omega_{B_i} = \pm\omega_{B_i}$. Assuming that ω is near $-l\omega_{B_i} \pm \omega_{B_i}$, and neglecting the penultimate term of the left-hand side of (17.1) and the terms of order $(\omega + l\omega_{B_i} \mp \omega_{B_i})/\omega_{B_i}$, we have

$$1 - \frac{\omega_{p_e}^2}{\omega^2}\cos^2\theta \mp \frac{\omega_{pi}^2 \sin^2\theta}{2\omega_{B_i}(\omega + l\omega_{B_i} \mp \omega_{B_i})} = 0. \tag{17.9}$$

Assuming that the ion contribution in (17.9) is small, we find that the maximal growth rate is attained for

$$\operatorname{Re}\omega \approx \omega_{p_e}\cos\theta \approx \omega_{B_i}(|l| - 1). \tag{17.10}$$

The growth rate is

$$\gamma = \frac{\omega_{p_i}}{2}\sin\theta\,(|l|-1)^{1/2} \approx \frac{\omega_{p_i}}{2}\sin\theta\left(\frac{\omega_{p_e}}{\omega_{B_i}}\cos\theta\right)^{1/2}. \tag{17.11}$$

If $\sin\theta \simeq \cos\theta \simeq 1$, it follows that

$$\gamma \simeq \omega_{B_i}\left(\frac{m_i}{m_e}\right)^{1/4}\left(\frac{\omega_{p_i}}{\omega_{B_i}}\right)^{3/2}. \tag{17.12}$$

At the limit of applicability of the high-frequency approximation this expression is of the same order as (13.5) (for $\cos\theta \approx 1$) and (17.4). The expressions (17.11) and (13.5) are also approximately equal at the limit of the high-frequency approximation for small $\cos\theta$. Thus, if $\cos\theta \approx 1$ both the cyclotron and Cherenkov interactions are responsible for the excitation of oscillations with $\gamma \simeq \omega_{B_i}$; if $\cos\theta \ll 1$, only the cyclotron interaction is operative.

It follows from a comparison of the Cherenkov and cyclotron growth rates in the case $\omega \simeq \omega_{B_i}$ that $\gamma_{\text{cycl}} > \gamma_{\text{Cher}}$ for

$$\left(\frac{\omega_{p_i}}{\omega_{B_i}}\right)^2 > \left(\frac{m_e}{m_i}\right)^{2/3} \tag{17.13}$$

and $\gamma_{\text{Cher}} > \gamma_{\text{cycl}}$ in the opposite case. A similar condition has already been encountered in the comparison of the anisotropic and δ-function $(\partial f_{\perp 0}/\partial v_\perp > 0)$ mechanisms of excitation in § 16.1 [the condition (16.6)].

<u>Limits of Applicability of the Approximation of Cold Ions and Cold Electrons.</u> The cold-electron approximation assumed in Eq. (17.1) requires $\omega > k_z v_{T_e}$. If $\omega \simeq |l|\omega_{B_i}$ and $k_z \simeq k_\perp \cos\theta \simeq (m_e/m_i)^{1/2}(l\omega_{B_i})^2/(V\omega_{p_i})$, we obtain the condition (13.7), in which $\cos\theta$ is related to the azimuthal number l by Eq. (17.3) or (17.10).

The ions can be assumed to be cold if their longitudinal thermal spread satisfies the condition $k_z v_{T_{\parallel i}} < \gamma$, and the transverse spread the condition $k_\perp v_{T\perp i} < \omega_{B_i}$. The first inequality entails approximately:

for perturbations with $\omega \approx -l\omega_{B_i}$

$$\frac{v_{T\|i}}{V} < \frac{\omega_{p_i}}{l\omega_{B_i}} \left(\frac{m_i}{m_e}\right)^{1/6}; \tag{17.14}$$

for perturbations with $\omega \approx -l\omega_{B_i} \pm \omega_{B_i}$

$$\frac{v_{T\|i}}{V} < \left(\frac{\omega_{p_i}}{\omega_{B_i}}\right)^2 \left(\frac{m_i}{m_e}\right)^{1/2} |l|^{-3/2}. \tag{17.15}$$

The small transverse thermal spread of the ions leads in both cases to the requirement

$$\frac{v_{T\perp i}}{V} < \frac{1}{|l|}. \tag{17.16}$$

Note that if $l = 1$ the condition (17.14) is identical with (15.12), which corresponds to anisotropic excitation. Similarly, (17.15) with $l = 2$ ($\omega \approx \omega_{B_i}$) has the same meaning as (16.5).

If one of the particle species – the electrons or ions – has a large thermal spread in a plasma with an azimuthal motion of the components, kinetic instabilities of the type considered in Chapters 15 and 16 can develop.

§ 17.2. Excitation of Plasma Oscillations by an Azimuthal Ion Beam

We shall now consider instabilities that arise when an azimuthal ion stream interacts with an electron-ion plasma. We shall assume that the densities of the plasma and beam are fairly low, so that we can ignore the possible development of the high-frequency instabilities considered in § 13.4.

Plasma with Cold Electrons. In the approximation $T_e \to 0$ we have the dispersion equation [cf. (17.1) and (11.33)]

$$1 - \left(\frac{\omega_{p_e} \cos\theta}{\omega}\right)^2 - \frac{(\omega_{p_i} \sin\theta)^2}{\omega^2 - \omega_{B_i}^2} - \frac{\alpha_\perp (\omega_{p_i} \sin\theta)^2}{(\omega + l\omega_{B_i})^2 - \omega_{B_i}^2} - \frac{\alpha_\| (\omega_{p_i} \cos\theta)^2}{(\omega + l\omega_{B_i})^2} = 0. \tag{17.17}$$

If we were to neglect the beam terms, we would now obtain Eq. (8.15) for the branches ω_2 and ω_3 (see also Fig. 8.2). The beam can excite only oscillations with frequencies near ω_{B_i} and the harmonics of ω_{B_i}. If $\omega \simeq \omega_{B_i}$, only the Cherenkov beam term ($\sim\cos^2\theta$) plays a role. The same general picture of branches is obtained as in the case of an anisotropic plasma (see Fig. 15.1). The growth rate of the oscillations is

$$\gamma \simeq (\alpha_{\parallel}\mu)^{1/3}\omega_{B_i}. \tag{17.18}$$

For an instability of this kind, it is necessary that the plasma density be not too high [cf. (15.6)]:

$$b_i < (\alpha_{\parallel}\mu)^{2/3}. \tag{17.19}$$

If harmonics are excited ($\omega \simeq n\omega_{B_i}$, $n \neq 1$), the condition (17.19) is not essential. In this case the Cherenkov growth rate is of the order of (17.18) (for small numbers n) and the cyclotron growth rate of order

$$\gamma \simeq \alpha_{\perp}^{1/2}\omega_{p_i}. \tag{17.20}$$

In the case of perturbations with $k_z = 0$ there is no Cherenkov excitation. Only harmonics with n = 2, 3, . . . and the growth rate (17.20) are excited.

Plasma with Hot Electrons. If the electron temperature is sufficiently high, the electrons do not play an appreciable role in the oscillations, i.e., the beam can excite purely ion oscillations. In this case the growth rates are larger than in the limit $T_e \to 0$. This follows from the dispersion equation

$$1 - \frac{(\omega_{p_i}\sin\theta)^2}{\omega^2 - \omega_{B_i}^2} - \left(\frac{\omega_{p_i}\cos\theta}{\omega}\right)^2 - \frac{\alpha_{\perp}(\omega_{p_i}\sin\theta)^2}{(\omega + l\omega_{B_i})^2 - \omega_{B_i}^2} - \frac{\alpha_{\parallel}(\omega_{p_i}\cos\theta)^2}{(\omega + l\omega_{B_i})^2} = 0. \tag{17.21}$$

Hence, we find that for $\omega \approx n\omega_{B_i}$, $n = 2, 3, \ldots$

$$\gamma \simeq \alpha^{1/3}\omega_{p_i}. \tag{17.22}$$

One can obtain more detailed information from (17.21) by proceeding as in § 11.4.

§ 17.3. Excitation of Plasma Oscillations by a Longitudinal Ion Beam with an Anisotropic Velocity Distribution

Suppose that a plasma is traversed by an ion stream moving along a magnetic field with velocity V, the stream having a strongly anisotropic velocity distribution, $T_\perp \gg T_\parallel$. It is then possible for ion-cyclotron beam-anisotropic instabilities similar to the electron-cyclotron instabilities considered in § 11.2 to develop in the plasma.

At a sufficiently high electron temperature, these instabilities are described by the dispersion equation

$$1-\left(\frac{\omega_{p_i}\cos\theta}{\omega}\right)^2-\frac{(\omega_{p_i}\sin\theta)^2}{\omega^2-\omega_{B_i}^2}-\frac{\alpha\omega_{p_i}^2\langle J_n^2\rangle\cos^2\theta}{(\omega-n\omega_{B_i}-k_zV)^2}=0. \quad (17.23)$$

Suppose $\omega_{p_i} \ll \omega_{B_i}$. In this case the plasma oscillation branches near $\omega_{p_i}\cos\theta$ and ω_{B_i}. The low-frequency oscillations are excited if

$$k_z=-n\omega_{B_i}/V \quad (17.24)$$

and have the growth rate

$$\gamma\simeq\alpha^{1/3}\omega_{p_i}. \quad (17.25)$$

In the case of perturbations with $\omega\approx\omega_{B_i}$

$$k_z=(1-n)\,\omega_{B_i}/V, \quad (17.26)$$

$$\gamma\simeq\left(\alpha\frac{\omega_{p_i}}{\omega_{B_i}}\right)^{1/3}\omega_{B_i}. \quad (17.27)$$

The presence of a velocity anisotropy renders the plasma-ion beam system more unstable than in the case $T_\perp=T_\parallel$ [cf. § 3.5].

§ 17.4. Excitation of Plasma Oscillations by an Azimuthal Ion Beam in a Magnetic Field with a Longitudinal Inhomogeneity

If the growth rate of the perturbations is small compared with the reciprocal of the free flight time of the particles, $\gamma\tau \ll 1$, the mean cyclotron frequency is a function of the particle velocity (§ 16.7). We shall take this effect into account in considering the perturbations of a plasma traversed by an azimuthal ion stream, assuming $k_{\parallel}=0$, $k_{\perp}\rho_i \ll 1$.

For simplicity we shall also assume that the anisotropy of the plasma is appreciably greater than that of the beam:

$$\left(\frac{v_{\parallel}}{v_{\perp}}\right)^2_{\text{pl}} < (v_{\parallel}/V_{\varphi})^2_{\text{beam}} \tag{17.28}$$

We can then obtain the dispersion equation

$$1-\frac{\omega_{p_0}^2}{\omega^2-\omega_{B_0}^2}-\frac{\omega_{p_1}^2}{(\omega+l\omega_{B_1})^2-\omega_{B_1}^2}+$$
$$+\varkappa_1\omega_{p_1}^2\left\{\frac{1-l}{[\omega+(l-1)\,\omega_{B_1}]^2}+\frac{1+l}{[\omega+(l+1)\,\omega_{B_1}]^2}\right\}=0. \tag{17.29}$$

Here ω_{B_0}, and ω_{B_1} are the ion-cyclotron frequencies of the plasma and the beam averaged over the period of the oscillations between the stoppers, and $\varkappa_1=-\partial\ln\omega_{B_1}/\partial\ln V_{\varphi}$. In the special case of a dependence $B_0(z)$ of the form (16.31) we have $\varkappa_1=(1/4)(v_{\parallel}/V_{\varphi})^2$, where $v_{\parallel}$ is the longitudinal beam velocity in the center of the trap.

We have neglected the plasma contribution to (17.29), which contains $\varkappa_0$, by virtue of (17.28). For the same reason we can assume $\omega_{B_0}<\omega_{B_1}$.

Suppose $\omega_{p_0}\gg\omega_{B_0}$, $(\omega_{p_1}/\omega_{p_0})^2\ll 1$. Let us consider perturbations with $\omega\approx\omega_{B_1}$. In this case the beam terms are important provided the perturbations correspond to the second azimuthal mode:

$$l=-2. \tag{17.30}$$

From (17.29) we then find that there is an instability with growth rate

$$\gamma \simeq \left(\frac{n_1}{n_0}\right)^{1/2} \varkappa_1 \omega_{B_i}. \tag{17.31}$$

The instability is due to a resonance of the type $\omega + \omega_{B_i} = -l\omega_{B_i} \equiv k_\varphi V_\varphi$. It follows from this relation that there is growth of perturbations whose phase velocity is in the direction of rotation of the ions, $\omega/(k_\varphi V_\varphi) > 0$.

Bibliography

1. P. Burt and E. G. Harris, Phys. Fluids, 4:1412 (1961).
2. V. I. Pistunovich. At. Energ., 14:72 (1963).
3. K. Jungwirth, Plasma Physics, 10:374 (1968).
4. Y. Shima and T. K. Fowler, Phys. Fluids, 8:2245 (1965). In [1-4] studies are made of the cyclotron instability of relative azimuthal motion of the electrons and ions (§17.1).
5. L. S. Hall and M. S. Grewal, Phys. Fluids, 10:1523 (1967). A study of the excitation of ion-cyclotron oscillations by a longitudinal ion beam with an anisotropic particle velocity distribution (§17.3).
6. N. L. Tsintsadze and D. G. Lominadze, Zh. Tekh. Fiz., 31:1039 (1961) [Soviet Phys. — Tech. Phys., 6:759 (1962)].
7. D. G. Lominadze and K. N. Stepanov, Zh. Tekh. Fiz., 34:1823 (1964) [Soviet Phys. — Technical Physics, 9:1408 (1965)]. In [6, 7] a study is made of the stability of an ion stream moving along a magnetic field. The temperature of the beam and plasma are taken into account.
8. D. G. Lominadze and K. N. Stepanov, Nucl. Fusion, 4:281 (1966). A study of the excitation of ion-cyclotron waves by a longitudinal stream of particles; in particular, the excitation of ion-cyclotron harmonics is considered.